VOLUME ONE HUNDRED AND FORTY EIGHT

Current Topics in DEVELOPMENTAL BIOLOGY

Mouse Models of Development and Disease

CURRENT TOPICS IN DEVELOPMENTAL BIOLOGY

"A meeting-ground for critical review and discussion of developmental processes"
A.A. Moscona and Alberto Monroy (Volume 1, 1966)

VOLUME ONE HUNDRED AND FORTY EIGHT

CURRENT TOPICS IN DEVELOPMENTAL BIOLOGY

Mouse Models of Development and Disease

Edited by

THOMAS GRIDLEY
Center for Clinical and Translational Research, Maine Medical Center Research Institute, Scarborough, ME, United States

LEIF OXBURGH
The Rogosin Institute, New York, NY, United States

ELSEVIER

ACADEMIC PRESS
An imprint of Elsevier

Academic Press is an imprint of Elsevier
50 Hampshire Street, 5th Floor, Cambridge, MA 02139, United States
525 B Street, Suite 1650, San Diego, CA 92101, United States
The Boulevard, Langford Lane, Kidlington, Oxford OX5 1GB, United Kingdom
125 London Wall, London, EC2Y 5AS, United Kingdom

First edition 2022

ISBN: 978-0-12-817131-8
ISSN: 0070-2153

For information on all Academic Press publications
visit our website at https://www.elsevier.com/books-and-journals

Publisher: Zoe Kruze
Developmental Editor: Naiza Mendoza
Production Project Manager: Vijayaraj Purushothaman
Cover Designer: Miles Hitchen

Typeset by STRAIVE, India

Contents

Contributors

Richard R. Behringer
Department of Genetics, University of Texas M.D. Anderson Cancer Center; Program in Genetics and Epigenetics, MD Anderson Cancer Center UT Health Graduate School of Biomedical Sciences; Program in Developmental Biology, Baylor College of Medicine, Houston, TX, United States

David R. Beier
Center for Developmental Biology & Regenerative Medicine, Seattle Children's Research Institute; Department of Pediatrics; Division of Genetic Medicine, Seattle Children's Hospital, University of Washington, Seattle, WA, United States

Ondine Cleaver
Department of Molecular Biology, Center for Regenerative Science and Medicine (CRSM), University of Texas Southwestern Medical Center, Dallas, TX, United States

Andrew M. Garrett
Department of Pharmacology; Department of Ophthalmology, Visual, and Anatomical Sciences, Wayne State University School of Medicine, Detroit, MI, United States

Thomas Gridley
Center for Clinical and Translational Research, Maine Medical Center Research Institute, Scarborough, ME, United States

Sarah E. Heuer
The Jackson Laboratory, Bar Harbor, ME; Graduate Biomedical Sciences, Tufts University School of Medicine, Boston, MA, United States

Samuel E. Honeycutt
Department of Cell Biology and Physiology, University of North Carolina at Chapel Hill, Chapel Hill, NC, United States

Gareth R. Howell
The Jackson Laboratory, Bar Harbor; Graduate School of Biomedical Sciences and Engineering, University of Maine, Orono, ME; Graduate Biomedical Sciences, Tufts University School of Medicine, Boston, MA, United States

Rulang Jiang
Division of Developmental Biology; Division of Plastic Surgery, Cincinnati Children's Hospital Medical Center; Department of Pediatrics, University of Cincinnati College of Medicine, Cincinnati, OH, United States

Yu Lan
Division of Developmental Biology; Division of Plastic Surgery, Cincinnati Children's Hospital Medical Center; Department of Pediatrics, University of Cincinnati College of Medicine, Cincinnati, OH, United States

Diana A. Machado
Department of Genetics, University of Texas M.D. Anderson Cancer Center; Program in Genetics and Epigenetics, MD Anderson Cancer Center UT Health Graduate School of Biomedical Sciences, Houston, TX, United States

Mark W. Majesky
Center for Developmental Biology & Regenerative Medicine, Seattle Children's Research Institute; Department of Pediatrics; Department of Laboratory Medicine & Pathology, University of Washington; Division of Pediatric Cardiology, Seattle Children's Hospital, Seattle, WA, United States

Cathy M. McLeod
Department of Pharmacology, Wayne State University School of Medicine, Detroit, MI, United States

Stephen A. Murray
The Jackson Laboratory, Bar Harbor, ME, United States

Pierre-Emmanuel Y. N'Guetta
Department of Cell Biology and Physiology, University of North Carolina at Chapel Hill, Chapel Hill, NC, United States

Elizabeth D. Nguyen
Center for Developmental Biology & Regenerative Medicine, Seattle Children's Research Institute; Department of Pediatrics, University of Washington, Seattle; Division of Pediatric Nephrology, Seattle Children's Hospital, Seattle, Washington, WA, United States

Lori L. O'Brien
Department of Cell Biology and Physiology, University of North Carolina at Chapel Hill, Chapel Hill, NC, United States

Daryl M. Okamura
Center for Developmental Biology & Regenerative Medicine, Seattle Children's Research Institute; Department of Pediatrics, University of Washington, Seattle; Division of Pediatric Nephrology, Seattle Children's Hospital, Seattle, Washington, WA, United States

Kristen D. Onos
The Jackson Laboratory, Bar Harbor, ME, United States

Alejandra E. Ontiveros
Department of Genetics, University of Texas M.D. Anderson Cancer Center; Program in Developmental Biology, Baylor College of Medicine, Houston, TX, United States

Leif Oxburgh
The Rogosin Institute, New York, NY, United States

Alaina M. Reagan
The Jackson Laboratory, Bar Harbor, ME, United States

Anne R. Ryan
Department of Molecular Biology, Center for Regenerative Science and Medicine (CRSM), University of Texas Southwestern Medical Center, Dallas, TX, United States

Michael Sasner
The Jackson Laboratory, Bar Harbor, ME, United States

Mouse models of development and disease: Preface

Scientific and technological progress has given generations of biomedical scientists insights into successively deeper levels of complexity in biological pathways. In one staggering series of scientific accomplishments spanning approximately 70 years, we have gone from understanding the biochemical basis of heredity, to sequencing the entire human genome, to defining the gene expression of every single cell in the human body. In parallel with genetics, our understanding of the biochemistry of living processes has developed on a similar trajectory. How can we critically test the meaning of these diverse mechanisms on developmental biology, physiology, and behavior and, ultimately, apply them to understanding ourselves, and how disruptions in their normal function contribute to human disease? This is the role of the model organism, and the mouse is one of the foremost tools for understanding the genetics and biochemistry of human health and disease, based on its physiological similarities to humans, its rapid breeding cycle, and the accessibility of its genome to modification through both classical breeding and genetic engineering. The utility of the mouse has not been limited to testing hypotheses generated in the test tube; large-scale gene modification and toxicological methods have also provided the means to use the mouse as a highly complex discovery tool in developmental biology, physiology, and behavior.

Origins of the field

The laboratory mouse has a long history as a model organism for the study of mammalian development, physiology, and disease. One of the earliest documented examples is the mouse candle experiment performed by Robert Boyle in the late 17th century. A mouse enclosed in a glass jar expired around the same time as the flame of a lit candle once the air was evacuated, demonstrating that elements in air are essential both for combustion and for basic life processes. Further investigations by Antoine-Laurent de Lavoisier in the 18th century using mice in conjunction with human subjects cemented the concept that combustion and respiration relied on the same substance and defined oxygen as the element of respiratory "combustion" used in the lung. The remarkable series of experiments leading to this conclusion was cut short by the unfortunate execution of the investigator in the French Revolution, but, nonetheless, it gave rise to modern respiratory physiology. Studies using mice to understand genetics started early in the 20th century when coat color mutations were used by Lucien Cuénot to confirm Gregor Mendel's observations on the genetic heritability of traits. These investigations led directly to the development of the first inbred mouse strain *Dba* (dilute, brown, non-agouti), a collaboration between geneticists William Ernest Castle and Clarence Cook Little and the mouse breeder Abbie Lathrop, who had established an enormous colony of fancy mice with specific traits around 1900. The *Dba* strain was used for foundational studies of the genetic basis of cancer and led the way to development of the many diverse inbred strains that are used in biomedical research today.

Mapping of the mouse genome started around 1915, when a group led by J.B.S. Haldane published the first analysis of linkage between genes in this species, and intense efforts in genetics and genomics led to major milestones in the 20th century. Transgenesis and homologous recombination technologies were developed, enabling insertion and deletion of specific genes, and the entire genome was sequenced, resulting in a highly resolved genetic map. The almost simultaneous sequencing of mouse and human genomes confirmed an astonishing degree of similarity at many genetic loci and also identified loci that are not orthologously conserved, providing specific insights into limitations of the mouse model for human biology. Since the 1990s, this base-pair resolution genetic map, together with technologies for targeted manipulation of genomic loci, has led to an explosion in investigation of human traits using mouse modeling, and technological refinements such as CRISPR-/Cas9-directed mutagenesis have accelerated this research and made it more accessible to investigators.

The syntenic relationship between the mouse and other mammalian genomes, including the human genome, the wide array of tools for genome modification and mutagenesis, and the availability of diverse inbred strains have all made the mouse an extremely attractive model system to study mammalian development, physiology, and disease. Comparatively low costs for housing and husbandry, combined with the ability to cryopreserve embryos, sperm, and eggs, ensure that mouse models will continue to be accessible to scientists at universities, hospitals, and research institutes, in addition to large genetic research centers and the pharmaceutical industry.

Current work in the field

This volume of *Current Topics in Developmental Biology* highlights fascinating examples of the use of the mouse as a model system by describing cutting-edge research in genetics, biochemistry, cell biology, and behavior. We present nine chapters illustrating different aspects of using mice as a model organism to provide insight into human biology and disease, as well as comparative analyses with other mammalian species.

Thomas Gridley and Stephen A. Murray describe genetic technologies and resources available to generate mouse models of human disease with specific gene mutations. The development of knockout and transgenic mouse strains was originally undertaken by individual laboratories, based on their specific scientific interests. More recently, large-scale genome modification programs with the goal of establishing repositories of both genetically engineered embryonic stem cells and mice with diverse gene mutations, which individual investigators can request based on their particular scientific interests, have transformed the field. Technology development for gene modification on a large scale is described, as are the high-throughput strategies to generate basic phenotype information on these new strains.

Yu Lan and Rulang Jiang review work on mouse models of palate development and morphogenesis. Cleft lip and cleft palate are common birth defects resulting from genetic and/or environmental perturbations of facial development in utero. Facial morphogenesis commences during early embryogenesis, with cranial neural crest cells interacting with the surface ectoderm to form initially partly separate facial primordia consisting of the medial and lateral nasal prominences and paired maxillary and mandibular processes. As these facial primordia grow around the primitive oral cavity and merge toward the ventral midline, the surface ectoderm undergoes a critical differentiation step to form an outer layer of flattened and tightly connected periderm cells with a "nonstick" apical surface that prevents epithelial adhesion. Formation of the upper lip and palate requires spatiotemporally regulated interepithelial adhesions and subsequent dissolution of the intervening epithelial seam between the maxillary and medial/lateral nasal processes and between the palatal shelves. Proper regulation of epithelial integrity plays a paramount role during human facial development, as mutations in genes encoding epithelial adhesion molecules and their regulators have been associated with syndromic and nonsyndromic orofacial clefts.

This chapter summarizes mouse genetic studies that have been instrumental in unraveling the mechanisms regulating epithelial integrity and periderm differentiation during facial and palate development. Since proper epithelial integrity also plays crucial roles in wound healing and cancer, understanding the mechanisms regulating epithelial integrity during facial development has direct implications for improvement in clinical care of craniofacial patients.

Richard R. Behringer and colleagues review the use of rodent models to study uterine morphogenesis and variations in other mammals. In eutherian and marsupial mammals, the site of embryo implantation and gestation is the uterus. Uterine morphologies vary between mammalian species. For example, laboratory mice have a bipartite uterus with two uterine horns and a single cervix, whereas humans have a simplex uterus with a single chamber and single cervix. The precursor tissue of the uterus, oviducts, and upper vagina is the Müllerian duct epithelium and its adjacent mesenchyme. Morphological variation between species is established during embryogenesis by species-specific differences in Müllerian duct fusion at the midline, growth, and differentiation. In humans, alterations in Müllerian duct development can lead to variations in uterine morphology that correlate with increased risks of miscarriage and infertility. Developmental genetic factors that regulate Müllerian duct development, uterine morphogenesis, and human uterine variation are reviewed in this chapter.

Gareth R. Howell and colleagues discuss the development and improvement of mouse models for the study of Alzheimer's disease (AD). AD is a complex neurodegenerative disease whose risk is influenced by genetic and environmental factors. Although a number of pathological hallmarks have been extensively studied over the last several decades, a complete picture of disease initiation and progression remains unclear. We now understand that numerous cell types and systems are involved in AD pathogenesis and that this cellular profile may present differently for each individual, making the creation of relevant mouse models challenging. However, with increasingly diverse data made available by genome-wide association studies, we can identify and examine new genes and pathways involved in genetic risk for AD, many of which involve vascular health and inflammation. When developing mouse models, it is critical to assess: (1) an aging timeline that represents onset and progression in humans; (2) genetic variants and context; and (3) environmental factors present in human populations that result in both neuropathological and functional changes—themes that are addressed in this chapter.

Cathy M. McLeod and Andrew M. Garrett explore mouse models for the study of clustered protocadherins (cPcdhs). Since their first description, cPcdhs have sparked interest for their potential to generate diverse cell-surface recognition cues and their widespread expression in the nervous system. Through the use of mouse models, we have learned a great deal about the functions served by cPcdhs and how their molecular diversity is regulated. cPcdhs are essential contributors to a host of processes during neural circuit formation, including neuronal survival, dendritic and axonal branching, self-avoidance and targeting, and synapse formation. Their expression is controlled by the interplay of epigenetic marks with proximal and distal elements involving high-order DNA looping, regulating transcription factor binding. Mouse models targeting the cPcdh locus and how they have been instructive in uncovering the regulation and function of cPcdhs are reviewed in this chapter.

Mark W. Majesky and colleagues discuss progress in understanding the mechanisms of regeneration in the spiny mouse (*Acomys cahirinus*). Scarless wound healing with regeneration of organ function is a long-sought goal of regenerative medicine. The work reviewed in this chapter suggests that the spiny mouse is an emerging model organism with potential to identify mechanisms of tissue and organ regenerative repair that operate in the context of mammalian physiology directed by a mammalian genome. Although these animals have distinct husbandry needs compared to laboratory mice, and their evolutionary distance from Mus presents a computational challenge for comparative transcriptomics, the remarkable wound healing properties that have evolved in the genus *Acomys* promise unique insights into potential applications for mammalian tissue regeneration.

Annie R. Ryan and Ondine Cleaver review the vascular architecture of organs and how lessons from blood vessel development might be used to inform strategies to generate laboratory-grown tissues. The formation, growth, and maintenance of our organs, such as our kidneys or pancreas, require their coordinated growth alongside the intricate vasculature that pervades them. Blood vessels course through nearly every tissue in our bodies, facilitating the essential exchange of gases, nutrition, and wastes as well as rapid circulation of hormones and other signaling molecules. Endothelial cells (ECs) that line all of our blood vessels are therefore the gatekeepers for communication between the circulation and organ-specific cell types. Important contemporary questions include: (1) how EC progenitors initially assemble to form blood vessels in the embryo and (2) how the embryonic vascular tree expands to perfuse growing organs. What has been learned from embryonic blood vessels and how this knowledge instructs our approaches to vascularize laboratory-generated tissues, such as organoids, is reviewed. General concepts in blood vessel formation and recent studies of vascularization of the growing kidney are discussed, as are challenges and limitations faced by organoid technologies, where a properly patterned vascular network is certainly essential to organ function. Understanding the ontogeny of organ-specific vasculatures will help propel regenerative therapeutic approaches.

Lori L. O'Brien and colleagues provide an overview of innervation in organogenesis. Proper innervation of peripheral organs helps to maintain physiological homeostasis and elicit responses to external stimuli. Disruptions to normal function can result in pathophysiological consequences. The establishment of connections and communication between the central nervous system and the peripheral organs is accomplished through the peripheral nervous system. Neuronal connections with target tissues arise from ganglia partitioned throughout the body. Organ innervation is initiated during development, with stimuli being conducted through several types of neurons including sympathetic, parasympathetic, and sensory. Although the physiological modulation of mature organs by these nerves is largely understood, their role in mammalian development is only beginning to be uncovered. Interactions with cells in target tissues can affect the development and eventual function of several organs, highlighting their significance. This chapter describes the origin of peripheral neurons, factors mediating organ innervation, and the composition and function of organ-specific nerves during development. This emerging field aims to identify the functional contribution of innervation to development, which will inform future investigations of normal and abnormal mammalian organogenesis, as well as contribute to regenerative and organ replacement efforts where nerve-derived signals may have significant implications for the advancement of such studies.

Leif Oxburgh reviews mechanisms that control growth of the kidney. The functional mass of kidney tissue in an adult is an important determinant of human health. Kidney formation during development is an essential determinant of the final nephron endowment of the adult organ, and no evidence has been reported that mice or humans are able to generate new nephrons after the developmental period. Mechanisms controlling organ growth after development are essential to establish the final adult organ size. The potential for organ growth is maintained in adult life, and the size of one kidney may be significantly increased by loss of the contralateral kidney. The mouse has provided a model system for investigators to critically explore genetic, cell biological, and hormonal control of developmental and juvenile kidney growth. Three basic aspects of kidney size regulation are reviewed: (1) mechanisms that control nephron formation, and how these are altered by the cessation of nephrogenesis at the end of the developmental period; (2) applicability of the general model for growth hormone–insulin-like growth factor control to kidney growth, both pre- and postnatally; (3) commonalities between mechanisms of juvenile kidney growth and the compensatory growth that is stimulated in adult life by reduction of kidney mass. Understanding the mechanisms that determine set-points for cell numbers and size in the kidney may inform ongoing efforts to generate kidney tissue from stem cells.

This exciting collection of work covering diverse fields and technologies highlights the breadth and depth of scientific understanding to which mouse models have contributed. The importance of mouse models will only increase as we develop more sophisticated technologies for manipulation of the genome, for measuring physiological functions, and for live animal imaging. Looking ahead a decade, we anticipate a thriving research field with an abundance of technologies available to answer critical questions in human health.

Thomas Gridley
Center for Clinical and Translational Research,
Maine Medical Center Research Institute,
Scarborough, ME, United States

Leif Oxburgh
The Rogosin Institute, New York, NY, United States

CHAPTER ONE

Mouse mutagenesis and phenotyping to generate models of development and disease

Thomas Gridley[a,*] **and Stephen A. Murray**[b]
[a]Center for Clinical and Translational Research, Maine Medical Center Research Institute, Scarborough, ME, United States
[b]The Jackson Laboratory, Bar Harbor, ME, United States
*Corresponding author: e-mail address: gridlt@mmc.org

Contents

Abstract

For many years, the laboratory mouse has been the favored model organism to study mammalian development, biology and disease. Among its advantages for these studies are its close concordance with human biology, the syntenic relationship between the mouse and other mammalian genomes, the existence of many inbred strains, its short gestation period, its relatively low cost for housing and husbandry, and the wide array of tools for genome modification, mutagenesis, and for cryopreserving embryos, sperm and eggs. The advent of CRISPR genome modification techniques has considerably broadened the landscape of model organisms available for study, including other mammalian species. However, the mouse remains the most popular and utilized system to model human development, biology, and disease processes. In this review, we will briefly summarize the long history of mice as a preferred mammalian genetic and model system, and review current large-scale mutagenesis efforts using genome modification to produce improved models for mammalian development and disease.

Current Topics in Developmental Biology, Volume 148
ISSN 0070-2153
https://doi.org/10.1016/bs.ctdb.2022.02.012

1. The mouse fancy and the early days of mouse genetics

Mutant mice have been collected, kept as pets, and studied for many years. In the 19th century, the collection and trading of pet mice with unusual coat colors or oddities of behavior, referred to as the "mouse fancy," provided many of the mutants analyzed by early mouse geneticists. Mice carrying these mutations were among the first mammals to be tested for conformation with Mendel's laws of heredity, after their rediscovery in 1900. For example, in 1902 Lucien Cuénot in France demonstrated Mendelian ratios for the inheritance of coat color in mice (Cuénot, 1902; Silver, 1995). In the US, William Castle at the Bussey Institute at Harvard University demonstrated segregation and independent assortment of a number of different coat color markers, including albino versus colored, spotted versus solid colored, black versus brown, and yellow versus non-yellow (Paigen, 2003a; Russell, 1985). One of Castle's students at Harvard was Clarence Cook Little, who would later go on to found the Jackson Laboratory in Maine in 1929. Little also created the first inbred mouse line, the DBA strain, in 1909 when he was still in Castle's laboratory at the Bussey Institute. Castle was a major figure in the development of mammalian genetics. He directed the Bussey Institute at Harvard until its closure in 1936. Among the geneticists who trained with Castle were L.C. Dunn, Clarence Cook Little, Sewall Wright, and George Snell (Silver, 1995).

Cuénot, in 1905, described what ultimately proved to be the first embryonic lethal mouse mutation, the A^{Y} allele of the agouti locus (Cuénot, 1905). However, Cuénot's non-Mendelian genotype segregation ratios were not fully explained until 1910, when Castle and Little demonstrated that crosses involving an embryonic lethal gene gave predictably aberrant Mendelian ratios in which the missing mice were accounted for by dead embryos (Castle & Little, 1910; Paigen, 2003a). These studies set the foundation for utilization of the mouse for the next century as a model for genetic insights into mammalian, and therefore human, development and biology.

2. Mutagenesis in mice

A variety of agents have been used to create heritable mutations in mice. These include various types of radiation, including X rays and gamma rays, and chemical mutagens such as the DNA alkylating agent N-ethyl-N-nitrosourea (ENU) (Gondo, 2008). Integration of exogenous

DNA into the mouse genome (insertional mutagenesis) can also cause heritable mutations, either by inserting directly into gene exons, or by creating concomitant insertions or deletions around the integration site. This exogenous DNA is delivered by injection of DNA constructs into the pronuclei of fertilized eggs, or by integration of retroviruses or engineered retroviral constructs (Gridley, Soriano, & Jaenisch, 1987). One type of engineered retroviral construct that has been widely utilized is the gene trap vector, which usually contains a strong splice acceptor site that disrupts normal transcription and splicing of the interrupted gene (Brown & Nolan, 1998; Joyner, 1991; Zambrowicz & Friedrich, 1998).

One of the most important advances in the field of mouse genetics was the development of embryonic stem (ES) cell technology, enabling germline transmission of cellular genomes maintained in culture, coupled with using gene targeting by homologous recombination to make mutations in these cells of a gene of interest. The 2007 Nobel Prize in Physiology or Medicine was awarded to Mario Capecchi, Oliver Smithies and Martin Evans for the development of this technology (Capecchi, 1989; Koller & Smithies, 1992). The earliest targeted mutant alleles in the Mouse Genome Informatics (MGI) database date from 1989, when eight novel mutant alleles were created using homologous recombination in ES cells, coupled with transfer of the mutant allele into the mouse germline (Bello, Perry, & Smith, 2021). Initially, these mutant alleles were constructed and analyzed by individual research laboratories. This technology spurred decades of productive work by many labs generating basic, translational and preclinical mouse models for development and disease. However, it was realized that economies of scale made it possible to use this technology to great effect in high throughput mutagenesis screens.

Initially gene targeting mostly involved generating a null allele of the gene of interest. However, the technology improved markedly over the years. It became routine to generate missense mutations, alleles tagged with reporter genes (e.g., lacZ or fluorescent proteins), and protein fusions or replacements with Cre or Flp recombinases, among other modifications. Genome modification got a further boost through the development of CRISPR technology (Clark, Dinsmore, & Soriano, 2020). CRISPR technology made it possible to no longer be dependent on ES cells (which had become routine only for mice and, to a much lesser extent, rats) for generation of modified genomes. Now modified genomes can be constructed directly in zygotes. This has both greatly opened up the range of organisms whose genomes can be experimentally modified, as well as increased the efficiency and speeded up the timeline for mutant allele generation.

3. High throughput mutagenesis and phenotyping

There is a long history of large-scale mutagenesis screens using the laboratory mouse (Brown & Nolan, 1998; Gondo, 2008; Justice, 2000). Initially, these screens used the strategy of forward genetics (i.e., phenotype-driven screens): mice are exposed to a mutagenic agent, such as X-rays or other types of radiation, or a chemical mutagen such as the DNA alkylating agent N-ethyl-N-nitrosourea (ENU), which usually generates point mutations. The mutagenized mice are then bred, using various breeding schemes to detect dominant and recessive mutations, and the progeny are phenotyped to detect deviants (the putative mutants). Heritability of the observed mutant phenotype is established through breeding tests, and the gene mutated to cause the mutant phenotype is determined through positional cloning (Paigen, 2003b). An alternative screening strategy, reverse genetics (i.e., genotype-driven screens), is utilized when using techniques such as gene traps and gene targeting (see below).

In the late 1990s through 2000s, a number of these large-scale screens using ENU as the mutagenic agent took place at research centers around the world. Two large scale ENU mutagenesis screens were launched in 1997 in Europe (Gondo, 2008; Justice, 2000). One took place at the Helmholtz Zentrum München German Research Center for Environment and Health in Germany, and a second at the Medical Research Council (MRC) in the UK. Shortly thereafter, similar screens were opened at RIKEN in Japan and at the Australian National University, followed by NIH-funded programs at several sites in the US. Other mouse ENU screens were directed at more defined phenotypic endpoints, such as developmental mutants, including embryonic lethal mutations (Anderson, 2000; Caspary & Anderson, 2006), and models of congenital cardiovascular anomalies (Li et al., 2015; Yu et al., 2004). Bruce Beutler shared the 2011 Nobel Prize in Physiology or Medicine for his discoveries concerning the role of Toll-like Receptors in the activation of innate immunity. A key element in Dr. Beutler's work was the analysis of mutant mice generated in an ENU mutagenesis screen that he carried out for many years (Beutler & Moresco, 2008). Ultimately, around twenty such screens were organized in institutions around the world (Gondo, 2008). These efforts were incredibly productive, and the Mouse Genome Informatics database has entries for 4340 alleles created using ENU mutagenesis (Bello et al., 2021).

With the development of efficient methods for creating null mutations by homologous recombination in ES cells, and transferring these mutant alleles into the mouse germline, it became clear that this technology could be utilized in a similar fashion to large scale chemical mutagenesis screens to perform high throughput mutagenesis of designed null alleles into the mouse germline. Moreover, the phenotyping pipelines developed for the ENU screens, particularly the broad-based screens deployed by MRC Harwell and the German Mouse Clinic, provided a framework for performing systematic phenotyping of targeted mutant lines. This strategy (i.e., mutating a selected gene, followed by thorough phenotypic screening) has been termed reverse genetics, and allows for prioritization of specific genes of interest, including novel human disease genes.

A Banbury Conference meeting at the Cold Spring Harbor Laboratory took place September 30—October 1, 2003, to discuss the idea of an international effort to assess phenotypes of null alleles of all protein coding genes in the mouse genome. Two white paper/commentaries were published in Nature Genetics after this conference. One white paper, entitled "The Knockout Mouse Project" (KOMP), proposed a genome-wide knockout screen to create a null mutation in every protein coding gene (Austin et al., 2004). This screen would create a resource of mutated ES cells that would be freely distributed. A subset of the mutant ES cells would be used to create mutant mice for phenotyping and distribution. The second paper, entitled "The European Dimension for the Mouse Genome Mutagenesis Program," proposed creation of a European Mouse Mutagenesis Consortium, with similar goals (Auwerx et al., 2004). However, unlike the KOMP program, the European initiative had from its beginning a strong focus on utilizing Cre/loxP technology to generate conditional null alleles, a principle that was embraced by the European Conditional Mouse Mutagenesis Program (EUCOMM). A similar initiative was started in Canada with creation of the North American Conditional Mouse Mutagenesis Project (NorCOMM). Reflecting the common goal to build and disseminate this valuable resource to the research community, individual teams coordinated their efforts under the auspices of the International Knockout Mouse Consortium (IKMC) (International Mouse Knockout Consortium et al., 2007). This coordination was critical to assuring the uniform quality of the resource, such as the use of a single C57BL/6N genetic background, and the open availability of the materials through dedicated repositories, such as the Mutant Mouse Resource and Research Centers (MMRRC) and the European Mouse Mutant Cell Repository (EuMMCR).

As the IKMC closed in on a complete genome-wide collection of targeted ES cell resources, there was increasing interest in realizing the original goals of the Banbury meeting: Tier 1 phenotyping of knockout lines to expand our understanding of the function of poorly annotated genes.

In 2011, the International Mouse Phenotyping Consortium (IMPC) was created to coordinate efforts among these and other large-scale mutagenesis programs that had been initiated around the world. The IMPC currently includes 21 research centers on five continents dedicated to the design, production, and description of the function of all human gene orthologs in the mouse genome. In addition, all data, resources and strains developed are freely distributed. Table 1 describes a number of web sites with useful information related to these efforts. In the US, the NIH-funded targeted ES cell production effort, the Knockout Mouse Project (KOMP), was expanded as the Knockout Mouse Phenotyping Program (KOMP2), with the goal of generating and phenotyping mutant mouse strains from these targeted resources. KOMP2 integrated operations with other members of the IMPC, and the introduction in 2014 of CRISPR/Cas9 technology for mutant production greatly increased the efficiency of the KOMP2 program. To date, the KOMP2 program has produced over 5500 mutant mouse strains. In November 2020, the National Institutes of Health accepted applications for a final five-year project period for the mutagenesis and phenotyping centers comprising the KOMP2 network, with the stated objectives for this final funding period to generate and phenotype at least 1200 new mutant mouse lines using CRISPR/Cas9 technology, and as with the previous phases, to make both models and data available to the scientific community.

Peterson and Murray recently summarized the progress made to date by the IMPC to generate a comprehensive catalog of null mutations for all protein coding genes in the mouse genome, what remains to be done, and how to prioritize which remaining genes to target (Peterson & Murray, 2021). To date, the IMPC has generated knockout lines for nearly 10,000 genes (Birling et al., 2021), and phenotyped knockout lines for 7824 genes, from 8457 independent mouse lines, resulting in 90,010 phenotype calls

Table 1 Web sites for mouse information and resources.

Web site	Addresses
International Mouse Phenotyping Consortium	www.mousephenotype.org
International Mouse Strain Resource	www.findmice.org
Mouse Genome Informatics	www.informatics.jax.org
Mutant Mouse Resource and Research Centers	www.mmrrc.org

(IMPC Data Release 15.1, October 11, 2021). When the IMPC knockout number is combined with ongoing efforts in the scientific community, approximately 80% of the 16,847 mouse genes with a high-confidence human ortholog have been targeted, leaving 3381 genes left to complete the mutant mouse null resource. Given the 1200 genes planned for the final phase of KOMP2 and the ongoing efforts of other IMPC members, the prospect of a complete functional annotation of the protein-coding portion of the mouse genome, a goal set forth at the Banbury conference almost two decades ago, appears to be within reach.

The members of the IMPC have made great progress with their mutagenesis and phenotyping pipeline, which has enabled the discovery of novel genes involved in specific disease or phenotypic domains (reviewed by Brown, 2021). Table 2 describes these domains, which include sensory domains such as deafness and eye development, circadian rhythms, bone biology, and metabolism, and the references for each study. Analysis of the IMPC datasets has also led to the identification genes involved in embryonic development and lethality, and in sexual dimorphism between male and female mice (Table 2). Critically, the identification of genes essential for embryonic development has proven useful for prioritization of novel human disease gene associations, complementing metrics such as constraint against loss of function mutations inferred from ongoing human sequencing efforts. Interestingly, this study showed that genes essential in mice, but not in cancer cell lines, are highly enriched for human disease genes (Cacheiro et al., 2020).

Table 2 Analyses of IMPC mutants in phenotypic and disease domains.

Disease/Phenotypic domain	References
Deafness	Bowl et al. (2017)
Eye/retina	Moore et al. (2018) Albrecht et al. (2018)
Circadian rhythms	Zhang et al. (2020)
Bone biology	Swan et al. (2020)
Metabolism	Rozman et al. (2018)
Embryonic development	Dickinson et al. (2016)
Sexual dimorphism	Karp et al. (2017)

4. Looking forward

Despite the considerable progress that has been made in generating and phenotyping null alleles of protein coding genes in the mouse genome, the functions of the majority of genes in both the mouse and human genomes remain either poorly understood, or entirely unknown. This portion of the genome has been termed the "dark genome" (Brown, 2021; Oprea, 2019). Genes encoding proteins included in the dark genome have very few scientific references, and a paucity of molecular tools, such as antibody reagents, for their analysis. High throughput mutagenesis and comprehensive phenotyping, as practiced by the members of the IMPC, are one mechanism to shed "light" on the dark genome. As described by Peterson and Murray (2021) and Brown (2021), of the genes in the dark genome that still lack null mutations, the highest priority for the IMPC should be to mutate and phenotype those genes with human orthologs, thus also providing new information on the human dark genome.

4.1 The deep genome project

With the prospect of a complete set of protein coding gene knockouts in sight, a question arises: what other genome features and/or alleles would benefit from a similar systematic approach to further our understanding of mammalian biology? In a recent commentary publication, 45 scientists, many of them IMPC investigators, proposed undertaking a Deep Genome Project, which they define as the functional biological annotation of all human orthologous genomic elements in mice (Lloyd et al., 2020). There is over 80% conserved synteny, and a high degree of gene orthology, between the mouse and human genomes. Lloyd et al. state that a comparative mouse/human functional analysis, accompanied by the generation and analysis of genetically engineered mice to understand the pathobiology of human disease, will be essential for implementation of effective precision medicine and the optimization of clinical practice. The Deep Genome Project would include analyses not only of the protein-coding orthologous elements of the mouse genome, but would also include conserved non-coding genomic elements, as well as generation and analysis in mice of Variants of Unknown Significance of human genes associated with disease. The Deep Genome Project also includes the goal of integrating functional testing of genomic variation in mouse models into the clinical decision-making process, so that they can inform diagnostic decisions and targeted therapies. Patient-derived xenograft models, also called avatars, are currently being utilized, particularly in cancer research (Durinikova, Buzo, & Arena, 2021; Tucker, George, Angelini, Bruna, & Chesler, 2021). Patient-derived avatar models also include generating patient-derived organoids, and patient-derived xenografts in the zebrafish model. However, integrating mouse avatar models into routine clinical practice will be an ambitious undertaking. Most of the goals of the Deep Genome Project are included in the current ten-year strategic plan for the IMPC, described in the next section.

4.2 IMPC strategy 2021–2030

The IMPC has recently released its strategic plan for the period 2021–2030, entitled "*The Function of Human Genetic Variation*" (https://www.mousephenotype.org/wp-content/uploads/2019/05/IMPC_Strategy_2021-30.pdf). Their key goal for this period is "to deliver broad-based genomic insights into the function of human genetic variation in both coding and non-coding sequences that underpins future developments in precision medicine, rare disease, clinical genetics, and healthy living". To accomplish this, they propose six goals for this ten-year period. These goals include: (1) completing the generation of a null mutant mouse resource for the coding genome; (2) designing and producing a genome-wide mouse strain resource of human disease-associated coding variants; (3) designing and generating mouse strains that model genetic variation in the non-coding genome; (4) phenotyping mouse strains with null mutations, human-disease coding variants, and variants in the non-coding genome to provide a comprehensive catalog of baseline information on mammalian gene function; (5) exploring genetic context to realize the wider potential for mouse functional studies; and (6) developing data integration, analysis, and visualization approaches to translate mouse functional genomics studies to the human gene and disease knowledge base, and vice versa. Through accomplishment of these ambitious goals, the IMPC proposes to further its mission during this ten-year period: "to generate a comprehensive catalogue of mammalian function and provide the foundations for the functional analysis of human genetic variation".

5. Concluding remarks

As detailed in this review, the mouse has a long history as the leading model system to provide mechanistic insight into human development, biology and disease. Beginning with the studies of Cuénot, Castle and others early in the twentieth century, and continuing through the present worldwide efforts of the IMPC to complete the generation and phenotyping of a null mutation of every protein coding gene in the mouse genome, the laboratory mouse remains the most practical, cost-effective and widely utilized model system for mammalian biology. The ambitious goals for the next 10 years of the IMPC are to catalog mammalian genome function and provide the foundations for the functional analysis of human genetic variation, ultimately impacting at multiple levels the diagnosis, management and treatment of human disease.

Acknowledgments

T.G. acknowledges support from National Institutes of Health grants U54 GM115516 and P20 GM139745. S.A.M. acknowledges support from National Institutes of Health grant UM1 OD023222.

References

Albrecht, N. E., Alevy, J., Jiang, D., Burger, C. A., Liu, B. I., Li, F., et al. (2018). Rapid and integrative discovery of retina regulatory molecules. *Cell Reports*, *24*, 2506–2519.

Anderson, K. V. (2000). Finding the genes that direct mammalian development: ENU mutagenesis in the mouse. *Trends in Genetics*, *16*, 99–201.

Austin, C. P., Battey, J. F., Bradley, A., Bucan, M., Capecchi, M., Collins, F. S., et al. (2004). The knockout mouse project. *Nature Genetics*, *36*, 921–924.

Auwerx, J., Avner, P., Baldock, R., Ballabio, A., Balling, R., Barbacid, M., et al. (2004). The European dimension for the mouse genome mutagenesis program. *Nature Genetics*, *36*, 925–927.

Bello, S. M., Perry, M. N., & Smith, C. L. (2021). Know your model: A brief history of making mutant mouse genetic models. *Laboratory Animals*, *50*, 263–266.

Beutler, B., & Moresco, E. M. Y. (2008). The forward genetic dissection of afferent innate immunity. *Current Topics in Microbiology and Immunology*, *321*, 3–26.

Birling, M. C., Yoshiki, A., Adams, D. J., Ayabe, S, Beaudet, A. L., Bottomley, J., et al. (2021). A resource of targeted mutant mouse lines for 5,061 genes. *Nature Genetics*, *53*, 416–419.

Bowl, M. R., Simon, M. M., Ingham, N. J., Greenaway, S., Santos, L., Cater, H., et al. (2017). A large scale hearing loss screen reveals an extensive unexplored genetic landscape for auditory dysfunction. *Nature Communications*, *8*, 886.

Brown, S. D. M. (2021). Advances in mouse genetics for the study of human disease. *Human Molecular Genetics*, *30*, R274–R284.

Brown, S. D. M., & Nolan, P. M. (1998). Mouse mutagenesis – Systematic studies of mammalian gene function. *Human Molecular Genetics*, *7*, 1627–1633.

Cacheiro, P., Muñoz-Fuentes, V., Murray, S. A., Dickinson, M. E., Bucan, M., Nutter, L. M. J., et al. (2020). Human and mouse essentiality screens as a resource for disease gene discovery. *Nature Communications*, *11*, 655.

Capecchi, M. (1989). The new mouse genetics: Altering the genome by gene targeting. *Trends in Genetics*, *5*, 70–76.

Caspary, T., & Anderson, K. V. (2006). Uncovering the uncharacterized and unexpected: Unbiased phenotype-driven screens in the mouse. *Developmental Dynamics*, *235*, 2412–2423.

Castle, W. E., & Little, C. C. (1910). On a modified Mendelian ratio among yellow mice. *Science*, *32*, 868–870.
Clark, J. F., Dinsmore, C. J., & Soriano, P. (2020). A most formidable arsenal: Genetic technologies for building a better mouse. *Genes & Development*, *34*, 1256–1286.
Cuénot, L. (1902). La loi de Mendel et l'hérédité de la pigmentation chez les souris. *Archives de Zoologie Expérimentale et Générale*, *3*, 27–30.
Cuénot, L. (1905). Les races pures et les combinaisons chez les souris. *Archives de Zoologie Expérimentale et Générale*, *4*, 123–132.
Dickinson, M. E., Flenniken, A. M., Ji, X., Teboul, L., Wong, M. D., White, J. K., et al. (2016). High-throughput discovery of novel developmental phenotypes. *Nature*, *537*, 508–514.
Durinikova, E., Buzo, K., & Arena, S. (2021). Preclinical models as patients' avatars for precision medicine in colorectal cancer: Past and future challenges. *Journal of Experimental & Clinical Cancer Research*, *40*, 185.
Gondo, Y. (2008). Trends in large scale mutagenesis: From genetics to functional genomics. *Nature Reviews. Genetics*, *9*, 803–810.
Gridley, T., Soriano, P., & Jaenisch, R. (1987). Insertional mutagenesis in mice. *Trends in Genetics*, *3*, 162–166.
International Mouse Knockout Consortium, Collins, F. S., Rossant, J., & Wurst, W. (2007). A mouse for all reasons. *Cell*, *128*, 9–13.
Joyner, A. L. (1991). Gene targeting and gene trap screens using embryonic stem cells: New approaches to mammalian development. *BioEssays*, *13*, 649–656.
Justice, M. (2000). Capitalizing on large-scale mouse mutagenesis screens. *Nature Reviews. Genetics*, *9*, 803–810.
Karp, N. A., Mason, J., Beaudet, A. L., Benjamini, Y., Bower, L., Braun, R. E., et al. (2017). Prevalence of sexual dimorphism in mammalian phenotypic traits. *Nature Communications*, *8*, 15475.
Koller, B. H., & Smithies, O. (1992). Altering genes in animals by gene targeting. *Annual Review of Immunology*, *10*, 705–730.
Li, Y., Klena, N. T., Gabriel, G. C., Liu, X., Kim, A. J., Lemke, K., et al. (2015). Global genetic analysis in mice unveils central role for cilia in congenital heart disease. *Nature*, *521*, 520–524.
Lloyd, K. K. C., Adams, D. J., Baynam, G., Beaudet, A. L., Bosch, F., Boycott, K. M., et al. (2020). The deep genome project. *Genome Biology*, *21*, 18.
Moore, B. A., Leonard, B. C., Sebbag, L., Edwards, S. G., Cooper, A., Imai, D. M., et al. (2018). Identification of genes required for eye development by high-throughput screening of mouse knockouts. *Communications Biology*, *1*, 236.
Oprea, T. I. (2019). Exploring the dark genome: Implications for precision medicine. *Mammalian Genome*, *30*, 192–200.
Paigen, K. (2003a). One hundred years of mouse genetics: An intellectual history. I. the classical period (1902-1980). *Genetics*, *163*, 1–7.
Paigen, K. (2003b). One hundred years of mouse genetics: An intellectual history. II. The classical period (1981-2002). *Genetics*, *163*, 1227–1235.
Peterson, K. A., & Murray, S. A. (2021). Progress towards completing the mutant mouse null resource. *Mammalian Genome*. https://doi.org/10.1007/s00335-021-09905-0.
Rozman, J., Rathkolb, B., Oestereicher, M. A., Schütt, C., Ravindranath, A. C., Leuchtenberger, S., et al. (2018). Identification of genetic elements in metabolism by high-throughput mouse phenotyping. *Nature Communications*, *9*, 288.
Russell, E. S. (1985). A history of mouse genetics. *Annual Review of Genetics*, *19*, 1–28.
Silver, L. M. (1995). *Mouse genetics*. New York: Oxford University Press.
Swan, A. L., Schütt, C., Rozman, J., del Mar Muñiz Moreno, M., Brandmaier, S., Simon, M., et al. (2020). Mouse mutant phenotyping at scale reveals novel genes controlling bone mineral density. *PLoS Genetics*, *16*, e1009190.

Tucker, E. R., George, S., Angelini, P., Bruna, A., & Chesler, L. (2021). The promise of patient-derived preclinical models to accelerate the implementation of personalised medicine for children with neuroblastoma. *Journal of Personalized Medicine, 11*, 248.
Yu, Q., Shen, Y., Chatterjee, B., Siegfried, B. H., Leatherbury, L., Rosenthal, J., et al. (2004). ENU induced mutations causing congenital cardiovascular anomalies. *Development, 131*, 6211–6223.
Zambrowicz, B. P., & Friedrich, G. A. (1998). Comprehensive mammalian genetics: History and future prospects of gene trapping in the mouse. *The International Journal of Developmental Biology, 42*, 1025–1036.
Zhang, T., Xie, P., Dong, Y., Liu, Z., Zhou, F., Pan, D., et al. (2020). High-throughput discovery of genetic determinants of circadian misalignment. *PLoS Genetics, 16*, e1008577.

CHAPTER TWO

Mouse models in palate development and orofacial cleft research: Understanding the crucial role and regulation of epithelial integrity in facial and palate morphogenesis

Yu Lan[a,b,c] and Rulang Jiang[a,b,c,*]
[a]Division of Developmental Biology, Cincinnati Children's Hospital Medical Center, Cincinnati, OH, United States
[b]Division of Plastic Surgery, Cincinnati Children's Hospital Medical Center, Cincinnati, OH, United States
[c]Department of Pediatrics, University of Cincinnati College of Medicine, Cincinnati, OH, United States
*Corresponding author: e-mail address: rulang.jiang@cchmc.org

Contents

Current Topics in Developmental Biology, Volume 148
ISSN 0070-2153
https://doi.org/10.1016/bs.ctdb.2021.12.003

Abstract

Cleft lip and cleft palate are common birth defects resulting from genetic and/or environmental perturbations of facial development *in utero*. Facial morphogenesis commences during early embryogenesis, with cranial neural crest cells interacting with the surface ectoderm to form initially partly separate facial primordia consisting of the medial and lateral nasal prominences, and paired maxillary and mandibular processes. As these facial primordia grow around the primitive oral cavity and merge toward the ventral midline, the surface ectoderm undergoes a critical differentiation step to form an outer layer of flattened and tightly connected periderm cells with a non-stick apical surface that prevents epithelial adhesion. Formation of the upper lip and palate requires spatiotemporally regulated inter-epithelial adhesions and subsequent dissolution of the intervening epithelial seam between the maxillary and medial/lateral nasal processes and between the palatal shelves. Proper regulation of epithelial integrity plays a paramount role during human facial development, as mutations in genes encoding epithelial adhesion molecules and their regulators have been associated with syndromic and non-syndromic orofacial clefts. In this chapter, we summarize mouse genetic studies that have been instrumental in unraveling the mechanisms regulating epithelial integrity and periderm differentiation during facial and palate development. Since proper epithelial integrity also plays crucial roles in wound healing and cancer, understanding the mechanisms regulating epithelial integrity during facial development have direct implications for improvement in clinical care of craniofacial patients.

1. Introduction

Craniofacial malformations account for more than one-third of all congenital birth defects in humans (Twigg & Wilkie, 2015). While many distinct classes of craniofacial anomalies exist, the most common congenital craniofacial defects involve disruption of the lip and/or palate, with cleft lip with or without cleft palate (CLP) occurring at 1:500 to 1:1000 live births, and cleft palate only (CPO) occurring at about 1:2000 live births (Dixon, Marazita, Beaty, & Murray, 2011; Nasreddine, El Hajj, & Ghassibe-Sabbagh, 2021; Seto-Salvia & Stanier, 2014). Although not a major cause of mortality in developed countries, children born with CLP or CPO require extensive surgical and dental treatments and experience a variety of health and psychosocial problems (Christensen, Juel, Herskind, & Murray, 2004; Dixon et al., 2011). The causes of orofacial clefts are heterogeneous, involving both genetic and environmental factors. Over 500 Mendelian syndromes, mostly caused by disruption of single gene loci or chromosome abnormalities, have orofacial cleft as a major clinical feature (Dixon et al., 2011; Leslie & Marazita, 2013; Nasreddine et al., 2021;

Seto-Salvia & Stanier, 2014). Several environmental factors, such as maternal smoking or exposure to teratogens including retinoic acid, phenytoin, and valproic acid, have been associated with increased risk of orofacial clefts (Dixon et al., 2011; Garland et al., 2020; Leslie & Marazita, 2013; Shi et al., 2007). Furthermore, about 70% of CLP and 50% of CPO patients show no apparent other structural anomalies, known as isolated or non-syndromic CL/P (Bishop et al., 2020; Dixon et al., 2011; Seto-Salvia & Stanier, 2014). A great deal of research efforts, including multiple genome-wide association studies and more recently large-scale whole exome or whole genome sequencing, have focused on identification of genes and genomic variants involved (Basha et al., 2018; Bishop et al., 2020; Bureau et al., 2014; Cox et al., 2018; Leslie et al., 2017; Leslie & Marazita, 2013; Nasreddine et al., 2021; Seto-Salvia & Stanier, 2014; Takahashi et al., 2018; Yu et al., 2017). Whereas these and other studies have identified mutations and variants in coding or non-coding regions in dozens of genes associated with CL/P, definitive causal mutations have accounted for less than 30% of the estimated liability of orofacial clefts (Bishop et al., 2020). For most non-syndromic orofacial cleft cases, a multifactorial model of inheritance is favored, such that variations in multiple genes are etiologically necessary, with gene–environment interactions and epigenetic modifications contributing with a threshold effect influencing the penetrance and expressivity (Dixon et al., 2011; Young, Slifer, Hecht, & Blanton, 2021).

Whereas significant progress has been made toward understanding the genetic etiology of orofacial clefts, with technological and bioinformatic advances further speeding up the identification of candidate causal mutations, these results have not significantly impacted clinical care and prevention strategies due to very limited understanding of the cellular, molecular, and developmental processes underlying cleft pathogenesis (Dixon et al., 2011; Leslie & Marazita, 2013; Nasreddine et al., 2021; Vieira, 2021). Since lip closure occurs by the end of the 6th week, whereas palatal fusion is complete by the 12th week, of human gestation (Dixon et al., 2011; Jiang, Bush, & Lidral, 2006), studies using animal models have been necessary and instrumental for understanding the cellular and molecular mechanisms of normal and abnormal craniofacial development (Bush & Jiang, 2012; Chai & Maxson, 2006; Gritli-Linde, 2007, 2008; Ji et al., 2020; Jiang et al., 2006; Juriloff & Harris, 2008; Li, Lan, & Jiang, 2017; Van Otterloo, Williams, & Artinger, 2016). The mouse is the pre-eminent model organism for unraveling the developmental mechanisms of orofacial cleft pathogenesis because it is a small mammal whose genome contains

~99% of homologous genes in human, the embryonic facial and palate development processes are strikingly similar in mouse and human, and many mouse strains carrying spontaneous or genetically engineered mutations causing cleft lip and/or cleft palate phenotypes are available (Abbott, 2010; Gritli-Linde, 2008; Jiang et al., 2006; Juriloff & Harris, 2008; Murray, 2011). In particular, the advent of mouse embryonic stem cell and gene targeting technologies in the 1980s enabled direct targeted manipulation of almost every gene in the mouse genome, which revolutionized mammalian biology and biomedical research (Capecchi, 2005). Whereas individual or combinations of null mutations in genes encoding major signaling pathway components or structural proteins often caused early embryonic lethality of homozygous mutant mice, the adaptation of site-specific DNA recombination systems, particularly the Cre/loxP system, enabled systematic genetic analysis of cell type-specific gene function at specific time or developmental stage during the life of a mouse (Jiang & Gridley, 1997; Lan, Wang, Ovitt, & Jiang, 2007; Rajewsky et al., 1996). Over 600 distinct mutant mouse lines with cleft palate phenotype have been reported (http://www.informatics.jax.org/phenotypes.shtml), of which studies have not only provided excellent candidate genes for analysis in human genetic studies, but also significantly improved the understanding of the molecular and cellular mechanisms of craniofacial and palate development (Bush & Jiang, 2012; Bush, Lan, & Jiang, 2004; Gritli-Linde, 2007, 2008; Lan et al., 2004; Lan, Xu, & Jiang, 2015; Li et al., 2017; Reynolds et al., 2020; Suzuki et al., 2018; Xu et al., 2016; Zhou, Gao, Lan, Jia, & Jiang, 2013). A number of excellent review articles have described the molecular and cellular mechanisms involving multiple signaling pathways and transcription factors controlling secondary palate development (Bush & Jiang, 2012; Gritli-Linde, 2007; Li et al., 2017). Here we summarize extensive mouse genetic studies that have revealed a paramount role of and mechanisms regulating epithelial integrity during facial and palate morphogenesis. We discuss how these studies have significantly improved the understanding of the pathogenic mechanisms underlying syndromic and non-syndromic orofacial clefting and the implications for improvement in patient care.

2. Overview of midfacial morphogenesis and periderm differentiation

Development of the face involves precisely coordinated directional growth, morphogenesis, and fusion of initially partly separate facial primordia, which first appear in the fourth week of human gestation (about embryonic day E9.5 in mice) and are composed of the frontonasal prominence (FNP) covering the ventral side of the developing forebrain, a pair of maxillary processes (Mx) and a pair of mandibular processes (Mb) forming the lateral and caudal boundaries of the primitive mouth, respectively (Hinrichsen, 1985; Jiang et al., 2006). These facial primordia initially consist of two cell types, the surface epithelium and the cranial neural crest cells that first arise at the anterior neural plate border and migrate ventrally into the space under the surface epithelium, with the FNP populated by neural crest cells from the forebrain and anterior midbrain region whereas the Mx and Mb populated by neural crest cells from the posterior midbrain and anterior hindbrain region (Fig. 1A). As development proceeds, signaling interactions involving BMP, FGF, SHH, and Wnt signaling pathways within the facial epithelium as well as between the frontonasal epithelium and the neural crest derived FNP

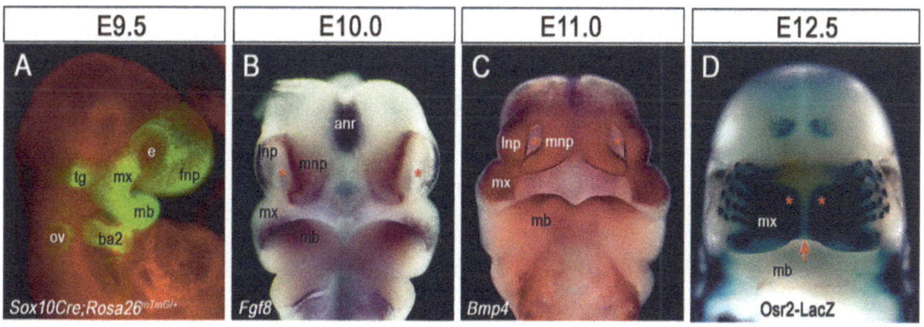

Fig. 1 Overview of facial development. (A) Lateral view of an E9.5 *Sox10Cre; Rosa26*$^{mTmG/+}$ mouse embryo showing facial primordia formed from cranial neural crest cells populating between the ventral side of the developing brain and surface ectoderm. Green color marks cranial neural crest cells while red color marks all other cells. (B) Frontal view of an E10 mouse embryonic head marked with the pattern of *Fgf8* mRNA expression (purple/blue color) in the developing facial epithelium. Red asterisks mark the nasal pits in between the medial and lateral nasal process. Note that the distal regions of the medial nasal process, lateral nasal process, and maxillary process are separate from each other at this stage. (C) Frontal view of an E11 mouse embryonic head marked with the pattern of *Bmp4* expression (purple/blue color) in the developing facial epithelium. Note that the caudal region of the medial and lateral nasal processes and the rostral region of the maxillary processes have formed the lambdoidal junction during lip fusion. (D) Frontal view of an E12.5 mouse embryonic head showing the intact upper lip formed from fusion of the medial nasal, lateral nasal, and maxillary processes. Blue staining marks tissues expressing the *Osr2* gene. Red asterisks mark the nostrils. Red arrow points to the philtrum and primary palate. anr, anterior neural ridge; ba2, second branchial arch; e, eye; fnp, frontonasal prominence; lnp, lateral nasal process; mb, mandibular arch or mandible; mnp, medial nasal process; mx, maxillary process; ov, otic vesicle; tg, trigeminal ganglion.

mesenchyme result in the outgrowth of paired medial and lateral nasal processes (MNP and LNP) and formation of the nasal pit, while a midline groove forms between the MNPs as the forebrain forms a pair of telencephalic vesicles in the 5th week of human gestation (~E10 in mice) (Fig. 1B) (Jiang et al., 2006; Marcucio, Hallgrimsson, & Young, 2015; Panaliappan et al., 2018; Reid, Yang, Melvin, Taketo, & Williams, 2011; Szabo-Rogers, Geetha-Loganathan, Nimmagadda, Fu, & Richman, 2008; Wang, Song, & Zhou, 2011; Zhu et al., 2016). Coordinated growth and morphogenesis of the facial processes driven by the signaling interactions between the epithelium and mesenchyme bring the caudal edges of the LNP and MNP into direct contact with the Mx to form the three-way fusion zone named the lambdoidal junction (Depew & Compagnucci, 2008; Li, Jones, Hooper, & Williams, 2019) (Fig. 1C). Adhesion and subsequent dissolution of the intervening epithelial seam between these processes lead to completion of lip fusion, which together with continued growth pushing the nose chambers medio-frontally and with mesenchymal proliferation backfilling the midline groove, resulting in formation of the intact upper lip by the end of sixth week of human gestation (~E12.5 in mice) (Fig. 1D) (Jiang et al., 2006). Both BMP and Wnt signaling within the epithelium play crucial roles in lip fusion, whereas interactions of FGF, SHH, and Wnt signaling regulate the directional growth while FGF signaling maintains survival of the neural crest-derived facial mesenchyme during facial primordial outgrowth (Ferretti et al., 2011; Jeong, Mao, Tenzen, Kottmann, & McMahon, 2004; Jiang et al., 2006; Jin et al., 2020; Kurosaka, Iulianella, Williams, & Trainor, 2014; Lan et al., 2006; Liu et al., 2005; Marchini et al., 2021; Ray et al., 2020; Reynolds et al., 2019, 2020).

The mammalian palate consists of the primary palate derived from the oral side of the MNP, and secondary palate derived from the oral side of the maxillary processes (Bush & Jiang, 2012). Secondary palate development commences in the 6th week of human gestation, and about E11.5 in mice, with outgrowth arising from the oral side of the maxillary processes and growing downward to form paired palatal shelves while the tongue arises from the mandible. From E12 to E13.5 in mouse embryos, the palatal shelves grow vertically flanking the developing tongue. From E13.5 to E14.5, the bilateral palatal shelves reorient to the horizontal position above the tongue, grow toward the midline, with subsequent palatal fusion involving adhesion, formation of the midline epithelial seam (MES), and MES dissolution, to form the roof of the oral cavity (Fig. 2A–C). Palatal fusion is complete by 12th week of human gestation and about E16.5 in mice.

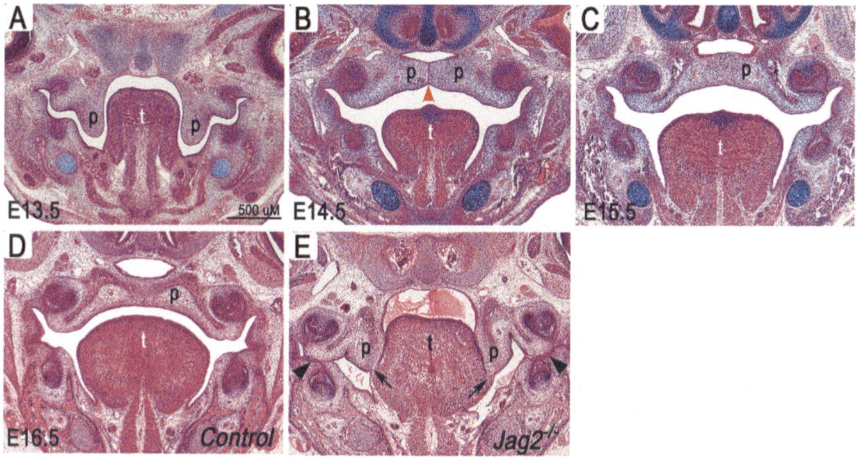

Fig. 2 Development of the secondary palate and its disruption in the *Jag2* mutant mice. (A–C) Frontal sections of embryonic mouse heads through the middle region of the developing secondary palate showing vertically oriented palatal shelves at E13.5 (A), horizontal palatal shelves that have formed the inter-shelf midline epithelial seam (red arrowhead) during palate fusion at E14.5 (B), and fused secondary palate with mesenchymal confluence through the midline in an E15.5 embryo (C). (D–E) Frontal sections through the midpalate regions of a pair of E16.5 control (D) and *Jag2*$^{-/-}$ (E) littermates showing aberrant fusion (arrow) of the medial region of the palatal shelves with the lateral sides of the tongue and aberrant adhesions (arrowheads) between the maxilla and mandible in the molar tooth germ regions of the *Jag2*$^{-/-}$ embryo. p, palatal shelf; t, tongue.

Extensive studies of the molecular mechanisms involving major signaling pathways and transcriptional regulatory networks controlling palatal shelf growth, patterning, and fusion, have been reported, and most have been summarized in several excellent review articles (Bush & Jiang, 2012; Gritli-Linde, 2007; Lan et al., 2015; Lane & Kaartinen, 2014; Li et al., 2017; Reynolds et al., 2020).

In addition to providing many of the instructive signals, including members of BMP, FGF, and Wnt families, and Shh, to direct the growth and patterning of the neural crest derived mesenchyme during facial and palatal morphogenesis (Bush & Jiang, 2012; Gritli-Linde, 2007; Jiang et al., 2006; Lan et al., 2015; Lane & Kaartinen, 2014; Li et al., 2017; Reynolds et al., 2020), the surface epithelium undergoes significant growth and differentiation at the same time (Hammond, Dixon, & Dixon, 2019). At the beginning of facial development, the surface ectoderm is a single layer of multipotent squamous epithelial cells that proliferate by symmetric cell divisions to cover the growing embryo body. At about E9.5 in mice, as the multipotent epithelial cells undergo specification and commitment to epidermal cell fate, a subset of the cells extend cellular processes over their neighbors and subsequently lose contact with the basement membrane as they migrate over the basal cells to form the outer layer of flattened epithelial cells called periderm cells (Hammond et al., 2019; M'Boneko & Merker, 1988; McGowan & Coulombe, 1998; Richardson et al., 2014). Periderm formation occurs apparently simultaneously at multiple sites and a continuous layer of periderm cells, connected by tight junctions at their apico-lateral borders, covers the oral epithelium by E12.5 and over the entire mouse embryo by E14.5 (Casey et al., 2006; Hammond et al., 2019; Morita et al., 2002; Richardson et al., 2014; Yoshida et al., 2012). Epidermal cell stratification and differentiation occur under the periderm layer to form the multilayered mature epidermis with a cornified outer layer functioning as a skin barrier, and the periderm is shed from the surface due to loss of adhesion to the underlying keratinocytes by E17.5 (Hammond et al., 2019; Saathoff, Blum, Quast, Kirfel, & Herzog, 2004). The primary function of the periderm is to provide a diffusion barrier to protect the embryo and the developing epidermis from the amnionic fluid until the barrier function of the cornified epidermis is mature (Hayward, 1983; Hu et al., 2018). In the next section, we will summarize studies that have uncovered a crucial role of the embryonic periderm in facial and palatal morphogenesis.

3. Mouse genetic studies uncover a critical role for the embryonic periderm in orofacial morphogenesis

3.1 Proper periderm differentiation is required for normal palatogenesis

Whereas formation of the upper lip and palate involves adhesion of initially separate primordial tissues to form an epithelial seam which is subsequently dissolved to allow mesenchymal confluence, the epithelial adhesion processes must be precisely regulated spatiotemporally to prevent aberrant epithelial adhesions that could disrupt lip and/or palate morphogenesis (Casey et al., 2006; Hammond et al., 2019; Jiang et al., 1998; Richardson et al., 2014). Aberrant oral adhesion as a cause of cleft palate was first discovered in studies of mice with disruption of the *Jag2* gene, which encodes a cell surface ligand for the Notch family receptors (Casey et al., 2006; Jiang et al., 1998). The *Jag2*$^{-/-}$ mice exhibit cleft palate at birth, with the palatal shelves stuck in the vertical position due to aberrant fusion of distal regions of the palatal shelves with the lateral sides of the tongue (Fig. 2D, E) (Casey et al., 2006; Jiang et al., 1998). *Jag2* is expressed throughout the developing oral epithelium and activates the Notch1 receptor in the oral periderm (Casey et al., 2006). Specific ablation of periderm cells in mouse embryos through *Krt17-Cre* mediated activation of expression of diphtheria toxin subunit-A from the *Rosa26*$^{eGFP\text{-}DTA}$ allele resulted in frequent intraoral epithelial adhesions and cleft palate in severely affected mouse embryos, indicating that the periderm layer plays a critical role in preventing aberrant epithelial adhesions (Richardson et al., 2014). In the *Jag2*$^{-/-}$ embryos, oral periderm cells formed by E12 but appeared more rounded and disorganized, with significantly reduced nuclear Notch1 activity (Casey et al., 2006). By E13.5, aberrant palate-tongue and maxillary-mandibular adhesions were detected in the *Jag2*$^{-/-}$ embryos, indicating that Jag2-Notch1 signaling plays a critical role in the regulation of oral periderm differentiation (Casey et al., 2006).

3.2 The IRF6 and GRHL3 transcription factors regulate periderm differentiation to prevent pathological epithelial adhesions that cause cleft palate and ectodermal defects in Van der Woude and popliteal pterygium syndromes

Aberrant epithelial adhesions have been associated with several CLP syndromes in humans, including Van der Woude syndrome (VWS, OMIM 119300) and popliteal pterygium syndrome (PPS, OMIM 119500) (Hammond et al., 2019; Kalay et al., 2012; Kondo et al., 2002; Kousa & Schutte, 2016). VWS is an autosomal dominant disorder characterized by CLP or CPO with lower lip pits, whereas PPS presents with similar orofacial phenotypes, plus additional features including popliteal webs, syndactyly, oral synechia, and other abnormalities (Hammond et al., 2019; Kondo et al., 2002; Kousa & Schutte, 2016). Kondo et al. (2002) identified mutations in the *Interferon Regulatory Factor-6* (*IRF6*) gene as the cause of both syndromes (Kondo et al., 2002). *IRF6* encodes a transcription factor containing a highly conserved helix-turn-helix DNA binding domain (Taniguchi, Ogasawara, Takaoka, & Tanaka, 2001). Gene expression analysis in mouse embryos showed that *Irf6* is expressed throughout the embryonic facial and oral epithelium, with highest levels in the periderm cells (Knight, Schutte, Jiang, & Dixon, 2006; Richardson et al., 2014; Richardson, Dixon, Jiang, & Dixon, 2009). Mouse embryos homozygous for either of two distinct mutations, a targeted null allele (*Irf6*$^{-}$) and the *Irf6*^{R84C} missense mutation that mimics the most common *IRF6* mutation in PPS, exhibited severe intraoral epithelial adhesions causing the vertically oriented palatal shelves to be wedged in between the lateral side of the tongue and mandible, in addition to other epidermal defects at birth (Ingraham et al., 2006; Richardson et al., 2006). Subsequent studies showed that IRF6 is required for periderm formation and maintenance (Kousa, Moussa, & Schutte, 2017; Richardson et al., 2009; Richardson et al., 2014). Transgenic expression of *Irf6* in the basal epithelial cells using the *Keratin-14* gene promoter was able to partially rescue periderm formation, although not completely rescuing cleft palate, in the *Irf6*$^{-/-}$ embryos (Kousa et al., 2017). Interestingly, whereas the palatal and tongue epithelia remained continuous in *Irf6*$^{-/-}$ and *Irf6*$^{R84C/R84C}$ embryos despite the almost complete adhesion of the palatal shelves with the tongue and mandible, the distal regions of the palatal shelves fused with the tongue, with regional dissolution of both the palatal and tongue epithelium, in the *Jag2*$^{-/-}$ and *K14-Irf6*$^{Tg/+}$;*Irf6*$^{-/-}$ embryos. This is consistent with another study identifying an essential role for IRF6 in the medial edge epithelium of palatal shelves during palatal fusion (Iwata et al., 2013) (further discussed below).

A comprehensive mutation screening of VWS and PPS patients found that mutations in *IRF6* account for about 70% of VWS and 97% of PPS cases (de Lima et al., 2009). A second VWS locus was mapped to human Chromosome 1p34, and mutations in the *Grainyhead-like-3* (*GRHL3*) gene were subsequently identified in a subset of IRF6 mutation-negative families (Koillinen et al., 2001; Peyrard-Janvid et al., 2014). *GRHL3* encodes a vertebrate homolog of the *Drosophila* Grainyhead DNA-binding transcription factor that plays an essential role in development and maintenance of the epidermal barrier (Bray & Kafatos, 1991; Sundararajan, Pang, Choolani, & Huang, 2020). Mice deficient in *Grhl3* also exhibited defective skin barrier function and deficient wound repair (Ting et al., 2005; Yu et al., 2006). While the initial report of *Grhl3* expression in mouse embryos described it as largely restricted to the surface ectoderm from E12.5 onward (Ting et al., 2005), subsequent studies found that *Grhl3* is most highly expressed in periderm cells in both zebrafish and mouse embryos, and its expression is lost in the oral epithelium in *Irf6*$^{-/-}$ embryos (de la Garza et al., 2013; Kousa et al., 2017). Studies in zebrafish embryos and human keratinocytes found that IRF6 directly regulates *GRHL3* gene expression, and that injection of zebrafish embryos with *grhl3* mRNAs partially rescued developmental defects caused by a dominant-negative *irf6* construct (de la Garza et al., 2013). Furthermore, *Grhl3*$^{-/-}$ mouse embryos exhibited defective oral periderm with reduced levels of IRF6 and Notch1 proteins in some regions and aberrant palate-tongue fusion, with some *Grhl3*$^{-/-}$ embryos displaying cleft palate (Peyrard-Janvid et al., 2014). *Grhl3*$^{-/-}$ embryos also exhibited syndactyly due to aberrant epithelial adhesions in the interdigital regions (Kashgari et al., 2020). The *Grhl3*$^{-/-}$ periderm cells displayed ectopic E-cadherin at the apical surface and mis-localization of the tight junction protein ZO-1 at the basolateral side, in contrast to the polarized apico-lateral localization of ZO-1 in between the periderm cells in the wildtype embryos (Kashgari et al., 2020). These results indicate that GRHL3 mediates part of the IRF6 function in regulating periderm integrity. Remarkably, genome wide association studies have found that mutations in *IRF6* and *GRHL3* contribute significantly to non-syndromic orofacial clefting, with mutations in *IRF6* accounting for approximately 2% of all orofacial cleft cases, whereas loss-of-function or dominant-negative mutations in *GRHL3* have been detected in both Caucasian and African patients with non-syndromic CPO (Dixon et al., 2011; Eshete et al., 2018; Leslie et al., 2016; Mangold et al., 2016).

How do IRF6 and GRHL3 regulate periderm differentiation? A recent study identified 66 high-confidence IRF6 target genes through combined RNA-seq and enhancer activities in embryonic epidermis from E15.5 *Irf6*$^{-/-}$ and wildtype embryos (Oberbeck et al., 2019). In addition to *Grhl3*, the significantly downregulated IRF6 target genes in *Irf6*$^{-/-}$ embryonic epidermis include *Occludin* (*Ocln*), encoding a component of tight junctions, and multiple genes encoding components of lipid metabolism pathways. Quantification of lipids in the skin of E16.5 *Irf6*$^{-/-}$ and wildtype embryos revealed significant decreases in six classes of lipids, including ceramides and triacylglycerols, and the *Irf6*$^{-/-}$ embryos displayed a marked absence of non-polar lipids in the outer layer of the skin (Oberbeck et al., 2019). GRHL3 is also required for lipid processing in the developing epidermis and directly regulates expression of *Tgm1*, which encodes Transglutaminase-1, involved in crosslinking cornified envelope proteins and lipid envelope (Matsuki et al., 1998; Ting et al., 2005; Yu et al., 2006). Gene expression profiling of embryonic skin from E18.5 *Grhl3*$^{-/-}$ and wildtype embryos showed that expression of several genes encoding components of tight junctions, including *Ocln*, *Cldn1*, *Cldn4*, and *Cldn23*, were significantly downregulated (Yu et al., 2006). Whether IRF6 and GRHL3 regulate periderm differentiation and integrity through directly controlling expression of tight junction proteins and/or lipid metabolism pathways remains to be elucidated.

3.3 Bartsocas-Papas syndrome is also a peridermopathy, and receptor-interacting protein kinase-4 controls epidermal and orofacial morphogenesis by posttranslational regulation of IRF6 function

Whereas IRF6 functions as a transcription factor in the regulation of *GRHL3* gene expression (de la Garza et al., 2013), immunofluorescent detection in mouse embryonic skin and human keratinocytes showed that the IRF6 protein is primarily localized in the cytoplasm (Ingraham et al., 2006; Kwa et al., 2014; Richardson et al., 2006). Kwa et al. (2014) showed that receptor-interacting protein kinase-4 (RIPK4), which is a critical regulator of keratinocyte differentiation (Holland et al., 2002), regulates IRF6 nuclear translocation and transcriptional activity through phosphorylation of its C-terminal domain at the Ser413 and Ser424 residues (Kwa et al., 2014). Loss-of-function mutations in *RIPK4* were identified in individuals affected by Bartsocas Papas Syndrome (BPS1, OMIM 263650), an autosomal recessive disorder with significant phenotypic overlap with PPS, including CL/P, popliteal webbing, and cutaneous syndactyly (Kalay et al., 2012; Kondo et al., 2002; Mitchell et al., 2012). Analysis in mouse embryos showed that *Ripk4* is expressed throughout the developing epidermis, including periderm cells (De Groote et al., 2015). *Ripk4*$^{-/-}$ mice displayed deficient skin barrier at birth and severe pathological epithelial adhesions including partial fusion of the limbs to the body and partial fusion of the mouth and nose orifices (De Groote et al., 2015; Holland et al., 2002). The epithelial lining of the oral cavity in the *Ripk4*$^{-/-}$ embryos was almost completely sealed by E13.5, with the vertical palatal shelves wedged between the mandible and tongue, similar as in *Irf6*$^{-/-}$ and *Irf6*$^{R84C/R84C}$ embryos (De Groote et al., 2015; Ingraham et al., 2006; Richardson et al., 2006). In contrast to the *Irf6* mutant embryos which exhibited lack of periderm (Kousa & Schutte, 2016; Richardson et al., 2014), the *Ripk4*$^{-/-}$ embryos had a complete periderm layer (De Groote et al., 2015). Whereas the periderm cells are shed from the surface of the epidermis by E17.5 in wildtype mouse embryos due to loss of adhesion to the underlying keratinocytes as the cornified layer forms a competent protective barrier (Hammond et al., 2019; Richardson et al., 2014), the *Ripk4*$^{-/-}$ embryos displayed persistent Krt17-positive periderm cells over the skin by birth (De Groote et al., 2015). The periderm cells displayed ectopic E-cadherin at the apical surface in *Ripk4*$^{-/-}$ embryos from E13.5 onward, whereas E-cadherin was absent from the apical surface of the flat periderm cells in wildtype embryos (De Groote et al., 2015).

Remarkably, epithelial-specific inactivation of the *Cdh1* gene, which codes for E-cadherin, partly rescued the *Ripk4*$^{-/-}$ embryos, such that aberrant epithelial adhesions only occurred in areas of incomplete *Cdh1* deletion (Urwyler-Rosselet et al., 2018). Mice with conditional inactivation of *Ripk4* using a *K14-Cre* driver (*Ripk4*^{EKO}) displayed an open mouth with aberrant epithelial adhesions observed between the mandible and tongue only in the anterior region of the oral cavity, and two-thirds of the E18.5 *Ripk4*^{EKO} embryos displayed fused palate (Urwyler-Rosselet et al., 2018). Mice with conditional inactivation of *Ripk4* or *Irf6* using another *K14-Cre* driver were described as displaying no soft tissue fusions (Oberbeck et al., 2019). It has been reported that another commonly used *K14-Cre* transgenic mouse line exhibited strong Cre activity in the oral epithelium except in the palatal periderm (Lane et al., 2014). The differences in epithelial adhesion phenotypes in the *Ripk4*$^{-/-}$ and *Ripk4*^{EKO} mouse embryos suggest that RIPK4 plays a critical cell-autonomous role in the oral periderm in preventing aberrant intraoral epithelial adhesions.

Oberbeck et al. (2019) recently generated and analyzed *Ripk4*$^{D161N/D161N}$ mice that express only catalytically inactive RIPK4 and showed that they phenocopied *Ripk4*$^{-/-}$ mice, indicating that the RIPK4 kinase activity is crucial for periderm and epidermal development (Oberbeck et al., 2019). They compared RNA-seq gene expression profiling of IRF6-dependent genes with that of RIPK4-dependent genes in E15.5 embryonic epidermis, and found that all the significant changes in gene expression seen in the *Ripk4*$^{D161N/D161N}$ epidermis were contained within the changes seen in *Irf6*$^{-/-}$ embryonic epidermis, suggesting that RIPK4 primarily acts through regulating IRF6 function during epidermal development (Oberbeck et al., 2019). Furthermore, generation and analysis of *Irf6*$^{S413A,S424A/S413A,S424A}$ mice carrying mutations in both the RIPK4 phosphorylation sites in the IRF6 protein confirmed that phosphorylation at these sites is essential for IRF6 function *in vivo*. Moreover, Oberbeck et al. (2019) generated and analyzed *Irf6*$^{S413E,S424E/S413E,S424E}$ mice, analysis of which led to identification of an additional RIPK4-dependent phosphorylation site at Ser90 in the IRF6 protein essential for its *in vivo* function in epidermal development (Oberbeck et al., 2019). A Ser90Gly mutation in *IRF6* has been reported in VWS patients whereas a Ser424Leu mutation has been reported in a PPS patient (Kondo et al., 2002; Matsuzawa et al., 2010), indicating that RIPK4 directly regulates IRF6 function in human orofacial and epidermal development.

3.4 IKKα is required for periderm differentiation, and loss of IKKα function causes ectodermal dysplasia with orofacial clefts

Three independent gene-knockout mutations in *Chuk* (conserved helix–loop–helix ubiquitous kinase), encoding the Inhibitor of nuclear factor kappa-B kinase subunit alpha (IKKα) (DiDonato, Hayakawa, Rothwarf, Zandi, & Karin, 1997), caused similarly severe epidermal developmental defects in homozygous mutants, with rudimentary limbs encased inside a shiny body skin, absent or severely truncated tail, and short rounded face (Hu et al., 1999; Li et al., 1999; Takeda et al., 1999). Richardson et al. (2014) performed direct comparative analysis of $Irf6^{R84C/R84C}$ and $Chuk^{-/-}$ mouse embryos at multiple developmental stages and found that they displayed similar defects in periderm formation and aberrant epithelial adhesions (Richardson et al., 2014). A recent report on studies of tongue papillae development showed similar intraoral adhesions, with vertically oriented palatal shelves wedged in between the lateral sides of the tongue and mandible, in $Chuk^{-/-}$ embryos (Kawasaki et al., 2016). Interestingly, although IKKa was initially identified as a kinase that phosphorylates Inhibitor of kB in response to cytokine signaling (DiDonato et al., 1997), IKKa function in epidermal development does not depend on the kinase activity nor acting through NFkB (Cao et al., 2001; Hu et al., 2001). Sil, Maeda, Sano, Roop, and Karin (2004) observed that induction of keratinocyte differentiation was associated with nuclear localization of IKKa and identified a nuclear localization signal sequence necessary for IKKa to induce keratinocyte differentiation (Sil et al., 2004). Expression of *Irf6* and *Jag2* mRNAs in the oral epithelium was unaltered in $Chuk^{-/-}$ embryos in comparison with wildtype embryos, and $Chuk^{-/-};Irf6^{-/-}$ double mutants showed similar intraoral epithelial adhesions as the $Chuk^{-/-}$ and $Irf6^{-/-}$ single mutants (Kawasaki et al., 2016). How IKKa and IRF6 converge on regulating periderm differentiation is not known. Xin, Lu, and Li (2011) showed that primary keratinocytes from E18.5 $Chuk^{-/-}$ embryos displayed a significantly reduced amount of cleaved Notch1 intracellular domain (NICD), as well as reduced expression of Notch pathway target genes. Furthermore, they showed that lentiviral mediated overexpression of NICD enhanced spinous cell differentiation and tight junction formation in $Chuk^{-/-}$ keratinocytes (Xin et al., 2011). Whether IKKa regulates periderm differentiation through modulating Notch signaling remains to be determined.

Lahtela et al. (2010) studied aborted fetuses with a lethal popliteal pterygia condition, termed cocoon syndrome, that showed similar phenotypes as the *Chuk*$^{-/-}$ mice, and discovered a homozygous nonsense mutation (p.Gln422*) in *CHUK* exon-12 (Lahtela et al., 2010). Richardson et al. (2014) showed that the *CHUK*$^{G422*/G422*}$ human fetal epidermal tissues lacked periderm formation and displayed pathological epithelial adhesions similar to *Chuk*$^{-/-}$ mouse embryos, confirming that loss of IKKa function causes peridermopathy in human as well (Richardson et al., 2014). Furthermore, Leslie et al. (2015) performed whole exome sequencing of a patient diagnosed with BPS1 with bilateral CLP but did not carry a mutation in *IRF6* or *RIPK4*; they identified a homozygous mutation in the splice acceptor site of *CHUK* exon-10 (Leslie et al., 2015). Khandelwal et al. (2017) reported a *de novo* missense mutation in *CHUK* (c.425A > G, p. His142Arg) in a 10-year-old patient who showed complex ectodermal dysplasia with cleft palate and oral synechia (Khandelwal et al., 2017). More recently, Cadieux-Dion et al. (2018) reported trans-heterozygous pathogenic mutations in *CHUK* in a patient initially diagnosed as having Ankyloblepharon-Ectodermal defects-Cleft lip/palate (AEC) syndrome (Cadieux-Dion et al., 2018). Thus, mutations in *CHUK* have been associated with several related ectodermal dysplasia syndromes with orofacial clefting.

3.5 P63 is a key regulator of periderm differentiation, and *TP63* mutations have been associated with syndromic and non-syndromic orofacial clefts

The mammalian *Trp63* gene (the human homolog is named *TP63*), encoding transformation related protein 63 (p63), was first identified in 1997 from a unexpected PCR product in a study amplifying the rat homolog of the mouse *Glut4* gene, that showed high sequence homology to *Trp53* (Schmale & Bamberger, 1997). In contrast to the ubiquitous expression of *Trp53*, *Trp63* expression was found highly restricted in epithelial tissues, particularly basal epithelial cells, and the gene was initially named keratinocyte transcription factor (*KET*) (Schmale & Bamberger, 1997). Yang et al. (1998) reported that *Trp63* produces at least six different transcripts due to the use of two distinct transcription start sites (TA and ΔN) and three alternative splicing routes (α, β, γ) in mouse and human tissues (Soares & Zhou, 2018; Yang et al., 1998). RT-PCR and immunohistochemical analyses revealed that ΔNp63α is the predominant isoform expressed in epithelial tissues (Yang et al., 1998). Mice deficient in p63 exhibited a thin aplastic epidermis and lack of epidermal derivatives (Mills et al., 1999; Yang et al., 1999). Mice with specific deletion of the ΔNp63 isoforms largely phenocopied the epithelial defects in *Trp63*$^{-/-}$ mice, including lack of oral epithelial stratification, indicating that ΔNp63 is responsible for most p63 function in epithelial lineages (Romano et al., 2012). Subsequent studies have shown that p63 plays critical roles in multiple stages of epidermal development, including promoting proliferation, cell adhesion, and stratification while suppressing terminal differentiation (Carroll et al., 2006; Koster et al., 2007; Nguyen et al., 2006; Pellegrini et al., 2001; Senoo, Pinto, Crum, & McKeon, 2007; Soares & Zhou, 2018; Truong, Kretz, Ridky, Kimmel, & Khavari, 2006).

The *Trp63*$^{-/-}$ mice displayed complete penetrance of bilateral cleft lip and palate (Thomason, Dixon, & Dixon, 2008). At E11.5, *Trp63*$^{-/-}$ embryos exhibited disorganized facial epithelium with increased apoptosis. While the outer surface of the facial prominences in wildtype embryos appeared smooth and covered by the newly formed flat periderm cells, *Trp63*$^{-/-}$ embryos exhibited rounded cells that were sloughing off from the lateral nasal and maxillary processes (Thomason et al., 2008). At E12.5, when wildtype embryos formed a layer of Krt17+ periderm cells with characteristic flattened morphology at the surface of the palatal shelves, *Trp63*$^{-/-}$ palatal shelves displayed patchy rounded cells above a disorganized basal epithelium with no detectable Krt17 expression, indicating that p63 is required for periderm formation (Richardson et al., 2017). Thomason et al. showed that p63 is involved in activation of *Irf6* gene expression through binding to an upstream enhancer, and that *Irf6* mRNA expression was significantly reduced in the oral epithelium in *Trp63*$^{-/-}$ embryos (Thomason et al., 2010). In the *Irf6*$^{R84C/R84C}$, *Ripk4*$^{-/-}$, and *Chuk*$^{-/-}$ embryos, high levels of p63 protein persisted in the suprabasal cells in the epidermis, in contrast to the down-regulation of p63 in differentiating suprabasal epidermal cells in wildtype embryos (De Groote et al., 2015; Richardson et al., 2006, 2014). Moretti et al. (2010) reported that IRF6 promoted proteosome-mediated p63 protein degradation in human keratinocytes (Moretti et al., 2010). In addition, Notch signaling has been shown to repress p63 expression in the surface ectoderm at the time of periderm specification (Tadeu & Horsley, 2013). Notch signaling could also activate *Irf6* expression and IRF6 mediated p63 down-regulation downstream of Notch in cultured keratinocytes (Restivo et al., 2011). Together, these data suggest that IRF6 converges with Notch signaling to drive periderm and epidermal cell differentiation through feedback downregulation of p63.

Mice homozygous for a classic radiation-induced mutation, *repeated epilation* (*Er*), which was later found to be a 1-bp insertion in the coding sequence of the Stratifin (*Sfn*) gene, displayed very similar intraoral epithelial adhesions and epidermal defects as in the *Irf6*$^{R84C/R84C}$ and *Ripk4*$^{-/-}$ mutants (De Groote et al., 2015; Herron et al., 2005; Li, Lu, Estepa, & Verma, 2005; Richardson et al., 2006; Tassin, Salzgeber, & Guenet, 1983). *Sfn* encodes a cell cycle regulator, and is highly expressed throughout the developing epidermis (Richardson et al., 2006). However, two independent gene-knockout alleles, including one deleting the entire *Sfn* gene, resulted in surviving homozygous mice with no apparent epidermal development defect except disorganized hair (Su et al., 2011; Winter, Lodygin, Verdoodt, & Hermeking, 2016), indicating that disruption of periderm and epidermal development in *Sfn*$^{Er/Er}$ mice resulted from a neomorphic, unnatural function of the mutant gene product. One study suggested that Sfn could interact with ΔNp63 and mediate its nuclear export for proteosome-mediated degradation upon DNA damage in cancer cells (Fomenkov et al., 2004). The persistence of high p63 levels in suprabasal epidermal cells in the *Sfn*$^{Er/Er}$ embryos (Richardson et al., 2014) suggests that a truncated Sfn product caused aberrant ΔNp63 stabilization during periderm and epidermal development, and may serve as a tool for dissecting the biochemical mechanism of IRF6-mediated p63 regulation.

Through a forward genetic screen, Lee, Kong, and Weatherbee (2013) discovered that mice deficient in *Kdf1* (*keratinocyte differentiation factor-1*) displayed pathological adhesions and epidermal development defects similar to the *Sfn*$^{Er/Er}$ mice (Lee et al., 2013). *Kdf1* encodes a novel cytoplasmic protein expressed throughout the developing epidermis. *Kdf1*$^{-/-}$ embryos displayed strong p63 protein expression throughout the epidermal layers and failed to develop a functional epidermal barrier (Lee et al., 2013). Epidermal development was largely rescued in *Kdf1*$^{-/-}$;*Trp63*$^{+/-}$ embryos (Lee et al., 2013), suggesting that Kdf1 also regulates periderm and epidermal development through modulating p63 expression or stabilization.

Germline mutations in *TP63* have been associated with five autosomal dominant ectodermal dysplasia syndromes with orofacial cleft as a hallmark feature in four of them, including AEC (also called Hay-Wells syndrome, OMIM 106260), Ectrodactyly, Ectodermal dysplasia, and Cleft lip/palate syndrome (EEC, OMIM 604292), Limb-Mammary Syndrome (LMS, OMIM 603543), and Rapp-Hodgkin Syndrome (RHS, OMIM 129400) (Osterburg, Osterburg, Zhou, Missero, & Dotsch, 2021; Soares & Zhou, 2018). The *TP63* mutations, mostly missense, that cause these diseases are believed to act *via* a dominant-negative or gain-of-function mechanism, and analyses of mouse models with knock-in missense mutations found in AEC or EEC patients resembled the human phenotypes (Celli et al., 1999; Ferone et al., 2012; McGrath et al., 2001; Rinne, Bolat, Meijer, Scheffer, & van Bokhoven, 2009; Vernersson Lindahl, Garcia, & Mills, 2013). However, several heterozygous deletions and frame-shift or nonsense mutations have recently been identified in multiple families with orofacial clefting and mild ectodermal defects (Khandelwal et al., 2019). Interestingly, Khandelwal et al. showed that those *TP63* loss-of-function variants were also present in a unaffected parent, indicating that *TP63* loss-of-function is a predisposing risk factor for orofacial clefting (Khandelwal et al., 2019). Another study identified a heterozygous 2-bp insertion that caused a frameshift in the p63 protein in a father and son with non-syndromic CLP (Basha et al., 2018). A genome-wide meta-analysis of CLP reported strong association of a SNP in an enhancer sequence in the last intron of the *TP63* gene with non-syndromic CLP (Leslie et al., 2017). Thus, mutations in *TP63* are a significant contributor to syndromic and non-syndromic CLP.

4. Connections between epithelial integrity and epithelial–mesenchymal interactions regulating orofacial morphogenesis

Studies of mice deficient in *Irf6*, *Chuk*, *Jag2*, and *Ripk4* have clearly demonstrated that the embryonic periderm forms a protective barrier to prevent pathological epithelial adhesions that may disrupt normal orofacial morphogenesis (Casey et al., 2006; Hammond et al., 2019; Richardson et al., 2014). The CLP phenotype of the *Trp63*$^{-/-}$ mutant mice, on the other hand, indicates that disruption of facial epithelial integrity prior to formation of the periderm is also detrimental to facial and palate morphogenesis (Thomason et al., 2008). A recent single-cell RNA-seq analysis of the dorsal skin from E13.5 *Trp63*$^{-/-}$ and wildtype embryos showed that a portion of the embryonic epithelial cells displayed strong enrichment of expression of mesenchymal genes, reminiscent of epithelial-to-mesenchymal transition signatures observed in squamous cell carcinoma during tumorigenic epithelial-mesenchymal transition (EMT) (Fan et al., 2018). By studying epidermal development from specification of multipotent simple epithelium to basal stratified epithelia to mature epidermis using human induced pluripotent stem cells (iPSCs), Soares et al. found that EEC iPSCs displayed defects during the specification switch from simple epithelium to stratified epithelia and aberrant activation of mesodermal gene expression (Soares et al., 2019), consistent with the data from mutant mouse studies. These results indicate that p63 plays a critical role in maintaining epithelial identity and stability of the embryonic surface epithelium, consistent with recent finding that p63 directly regulates expression of a number of epithelial cell adhesion molecules during palate development (Richardson et al., 2017). An important role of p63-mediated regulation of epithelial cell adhesion in facial morphogenesis is also supported by the identification of homozygous nonsense or frameshift mutations in *NECTIN1* (also known as *PVRL1, poliovirus receptor like-1*), a p63 direct target gene that encodes an immunoglobulin-like adhesion molecule important in desmosome assembly, as the cause of cleft lip/palate-ectodermal dysplasia syndrome (CLPED1, OMIM 225060) (Barron et al., 2008; Mollo et al., 2015; Suzuki et al., 2000). *NECTIN1* heterozygosity is also a significant risk factor for non-syndromic CLP (Sozen et al., 2001).

The p63 transcription factor also plays essential roles in regulating epithelial-mesenchymal interactions controlling facial primordial growth. *Trp63*$^{-/-}$ embryos exhibited marked changes in expression of several key signaling molecules, including Bmp4 and Fgf8, in the facial epithelium, accompanied by increased apoptosis of the epithelium and significantly decreased proliferation of the mesenchyme in the lateral nasal and maxillary processes (Thomason et al., 2008). By ChIP-seq and RNA-seq analysis of *Trp63*$^{-/-}$ and wildtype mouse embryonic epidermal cells, Fan et al. showed that p63 directly regulates expression of several components of the BMP, FGF, Notch, and Wnt pathways at the onset of epidermal fate specification, including *Bmp7*, *Fgfr2*, *Fgfr3*, *Jag2*, *Notch1*, *Wnt3a*, *Wnt10b*, and *Fzd6* (Fan et al., 2018). These signaling pathways play essential roles in early facial primordial growth (Jiang et al., 2006; Lan et al., 2006; Reynolds et al., 2020). In addition, Ferreti et al. showed that expression of *Trp63* in the early facial ectoderm is regulated by Wnt signaling, and suggested a Pbx-Wnt-p63-Irf6 regulatory module controlling facial morphogenesis (Ferretti et al., 2011).

Ferone et al. showed that the *Trp63*$^{+/L514F}$ mice, heterozygous for the same missense mutation as found in AEC patients, were born with cleft palate and several other ectodermal tissue defects including thinner epidermis, and blistering skin lesions in the basal and suprabasal layers of the epidermis (Ferone et al., 2012). At E13, the *Trp63*$^{+/L514F}$ embryos displayed significantly smaller palatal shelves with disorganized palatal epithelium containing patchy Krt17+ periderm cells intermingled with Krt14+ basal epithelial cells, and the periderm cells lacked the characteristic flattened morphology observed in wildtype embryos (Ferone et al., 2012). The *Trp63*$^{+/L514F}$ embryos exhibited significantly reduced proliferation of both the palatal epithelium and mesenchyme, resembling that in *Fgfr2b*$^{-/-}$ and *Fgf10*$^{-/-}$ mouse embryos, which also displayed defective palatal periderm and reduced palatal shelf growth (Alappat et al., 2005; Rice et al., 2004). Ferone et al. (2012) further demonstrated that p63 directly regulated *Fgfr2* and *Fgfr3* expression through binding to their cis-regulatory elements (Ferone et al., 2012). Fgfr2b signaling, likely through regulating epithelial cell proliferation and integrity, maintains *Shh* expression in the palatal epithelium, and Shh signaling regulates expression of *Fgf10* and *Fgf18* in the palatal mesenchyme to coordinate palatal shelf growth (Lan & Jiang, 2009; Rice et al., 2004; Xu et al., 2016). In addition, Fgf10-Fgfr2 signaling also regulates palatal epithelial integrity, as *Fgf10*$^{-/-}$ embryos exhibited ectopic expression of *Tgfb3* in the lateral side of the palatal epithelium with subsequent fusion of the palatal shelves to the mandible (Alappat et al., 2005). Interestingly, mouse embryos lacking the BMP antagonist Noggin displayed a similar palate-mandible fusion defect without affecting *Fgf10* expression, suggesting that Fgf10-Fgfr2 signaling may regulate palatal epithelial integrity through modulating BMP signaling (He et al., 2010). Thus, in addition to directly regulating expression of cell adhesion molecules in the palatal epithelium (Richardson et al., 2017), p63 also regulates both palatal epithelial integrity and palatal shelf growth through Fgfr2b signaling.

Another example that maintaining epithelial integrity plays a key role in facial primordial growth and morphogenesis has been provided by recent studies of *Grhl2* mutant mice. *Grhl2* is expressed in the surface ectoderm starting at E8.5 and directly regulates expression of a number of genes essential for epithelial fate, including epithelial adherens and tight junction molecules E-cadherin and Cldn4, and the epithelial-specific splicing factors Esrp1 and Esrp2 (Carpinelli et al., 2020; Ray & Niswander, 2016; Tanimizu & Mitaka, 2013; Werth et al., 2010). *Grhl2*$^{-/-}$ mouse embryos exhibited a neural tube closure defect and a cleft face, with significantly smaller maxillary processes by E10.5 (Carpinelli et al., 2020; Pyrgaki, Liu, & Niswander, 2011). Through RNA-seq analysis, Carpinelli et al. found that the *Grhl2*$^{-/-}$ embryos displayed significantly increased expression of *Zeb1* and many mesenchymal-specific transcripts, in addition to significant downregulation of epithelial genes (Carpinelli et al., 2020). Zeb1 is a major regulator of EMT and is a repressor of *Cdh1* expression (Nieto, Huang, Jackson, & Thiery, 2016; Vannier, Mock, Brabletz, & Driever, 2013). Remarkably, in contrast to the mid-gestation lethality of and severe craniofacial defects in *Grhl2*$^{-/-}$ embryos, the *Grhl2*$^{-/-}$;*Zeb1*$^{-/-}$ embryos displayed a nearly normal nasal septum and fused secondary palate (Carpinelli et al., 2020), indicating that a key role for GRHL2 in facial and palate development is to maintain epithelial integrity through suppression of EMT. Together with a previous study identifying a single nucleotide mutation in the first intron of the *Zeb1* gene as the cause of cleft lip and palate in the *Twirler* mutant mice (Kurima et al., 2011), these results indicate that gain of function of *Zeb1* is a risk factor for orofacial clefting through disruption of epithelial integrity during facial morphogenesis.

Mice lacking Esrp1 exhibited complete penetrance of cleft lip and palate (Bebee et al., 2015). Lee et al. showed that the *Esrp1*$^{-/-}$ embryos displayed significantly reduced cell proliferation in both the epithelium and mesenchyme of the medial and lateral nasal processes, accompanied by alterations in expression of genes in several signaling pathways, particularly multiple Wnt family ligands and Shh, at E10.5 (Lee et al., 2020). They performed RNA-seq analysis of E12.5 facial epithelium and mesenchyme, and identified 1467 alternative splicing changes in *Esrp1*$^{-/-}$ embryos, with many RNAs switched from epithelial to mesenchymal isoforms. The validated epithelial-to-mesenchymal isoform switching in the *Esrp1*$^{-/-}$ embryos disrupted multiple molecules crucial for adherens and tight junction assembly or maintenance, including Ctnnd1 (p120 catenin), Arhgap17, Magi1, and Scrib (Lee et al., 2020; Warzecha & Carstens, 2012). Of these, Ctnnd1 directly binds to E-cadherin and is a major component of the adherens junction complex (Pieters, van Roy, & van Hengel, 2012). A crucial role for maintaining epithelial adherens junction stability in facial morphogenesis is supported by human genetic studies identifying mutations in *CDH1* and *CTNND1* in several syndromes with cleft lip or cleft palate as a variable feature (Alharatani et al., 2020; Frebourg et al., 2006; Ghoumid et al., 2017; Selvanathan et al., 2020). Loss of function variants in *CDH1* and *CTNND1*, respectively, have also been associated with non-syndromic CLP (Cox et al., 2018; Dixon et al., 2011; Song & Zhang, 2011). In addition, conditional inactivation of *Ctnnd1* in the embryonic facial ectoderm caused cleft lip or palate in mice (Cox et al., 2018). Thus, Esrp1 mediated epithelial-specific splicing plays crucial roles in facial morphogenesis at least in part though maintaining epithelial integrity.

5. The roles of the periderm and the IRF6-p63 pathway in palatal fusion

Whereas the periderm acts as a protective barrier to prevent epithelial adhesion during palatal shelf growth, the bilateral palatal shelves must adhere to each other at the midline, and the resultant inter-shelf midline epithelial seam (MES) has to be dissolved, during palatal fusion to form the intact roof of the oral cavity (Lan et al., 2015; Li et al., 2017). Palatal fusion is driven by TGFβ3, whose expression is specifically activated in the palatal shelf medial edge epithelium (MEE) prior to palatal shelf adhesion (Fitzpatrick, Denhez, Kondaiah, & Akhurst, 1990; Kaartinen et al., 1995; Pelton, Hogan, Miller, & Moses, 1990; Proetzel et al., 1995). In *Tgfb3*$^{-/-}$ embryos, the palatal shelf growth and reorientation to the horizontal position occurred normally, but the bilateral palatal shelves adhered poorly and exhibited intact periderm layer over their surface (Hu et al., 2015; Kaartinen, Cui, Heisterkamp, Groffen, & Shuler, 1997; Richardson et al., 2017). Interestingly, the *Snai1*$^{+/-}$;*Snai2*$^{-/-}$ mouse embryos exhibited cleft palate with an intact periderm layer over the palatal shelves, similar to the *Tgfb3*$^{-/-}$ embryos, but they displayed apparently normal patterns of expression of *Tgfb3* mRNAs in the MEE (Murray, Oram, & Gridley, 2007). Snai1 and Snai2 have been implicated in driving EMT by downregulating expression of cell adhesion molecules including E-cadherin (Carver, Jiang, Lan, Oram, & Gridley, 2001; Liu et al., 2012; Peinado, Olmeda, & Cano, 2007). Although the role of EMT in palatal fusion has long been controversial since no genetically labeled palatal epithelial cells contributed to the post-fusion palatal mesenchyme structures (Bush & Jiang, 2012; Gritli-Linde, 2007; Jiang et al., 2006; Nawshad, 2008; Xu et al., 2006), the palatal defect in the *Snai1*$^{+/-}$*Snai2*$^{-/-}$ embryos suggest that Snai1 and Snai2 act partly redundantly downstream of TGFβ3 signaling to drive palatal fusion.

How do the periderm cells respond to TGFβ3-Snai1/2 signaling, and do they play an active role in palatal fusion? Whereas some studies have suggested that the medial edge periderm cells undergo programmed cell death and are sloughed off immediately prior to palatal shelf adhesion (Fitchett & Hay, 1989; Hu et al., 2015), other studies showed that the medial edge periderm cells change shape and extend filapodial or lamellipodial processes at the beginning of palatal shelf adhesion, and migrate away from the MES to the oral and nasal surfaces during palatal fusion (Richardson et al., 2017; Taya, O'Kane, & Ferguson, 1999; Waterman, Ross, & Meller, 1973). Li et al. recently reported that periderm cells were also present and changed shape in the fusion zone between the medial and lateral nasal processes during lip fusion in mouse embryos (Li et al., 2019). Using live imaging of mouse palatal explants, Kim et al. (2015) demonstrated that MEE cells extend protrusions to contact the opposite shelf to establish cellular bridges that give rise to a transient multilayered epithelial seam where periderm cells were still present (Kim et al., 2015). Subsequently, palatal fusion proceeded through concurrent cellular intercalation and displacement in the oronasal axis by convergent extension driven by an actomyosin pathway involving non-muscle myosin-IIA activation (Kim et al., 2015). While it was an elegant study that has provided unprecedented insight into the cellular mechanisms of palatal fusion, the study used *K14-Cre* mediated lineage labeling, which does not label the medial edge periderm cells (Lane et al., 2014) and, hence, did not address specifically periderm cell behavior during palatal fusion. Richardson et al. examined periderm cell fate during palatal fusion by using confocal imaging of palatal shelf explants from mouse embryos carrying a *Krt17-GFP* transgenic reporter and observed migration of periderm cells out of the MES into the oral and nasal epithelia of the wildtype palatal shelves but not of the *Tgfb3*$^{-/-}$ palatal shelf explants, despite forced contact of the palatal shelves (Richardson et al., 2017). Furthermore, Kim et al. observed that *Tgfb3*$^{Cre/+}$;*Myh9*$^{lox/lox}$ embryos displayed failure of cell-intercalation at the MES by E15.5 when the MES had almost completely dissolved in wildtype embryos, whereas the *K14-Cre; Myh9*$^{lox/lox}$ embryos also exhibited significant MES dissolution (Kim et al., 2015). Since the *K14-Cre* mice exhibit robust Cre activity in basal epithelial cells but lack Cre-mediated recombination in palatal periderm cells, whereas *Tgfb3*^{Cre} induces recombination in both the basal and periderm layers of the MEE (Lane et al., 2014), could the difference in MES dissolution observed in between the *Tgfb3*$^{Cre/+}$;*Myh9*$^{lox/lox}$ and *K14-Cre; Myh9*$^{lox/lox}$ embryos be due to a contribution of the periderm cells in facilitating MES formation? In this regard, it is worth noting that eyelid closure in the mouse fetus also requires myosin-IIA driven cell intercalation and a *Krt17-Cre* lentivirus-mediated selective inactivation of myosin-IIA in the eyelid front cells in *Myh9*$^{lox/lox}$ embryos caused failure of eyelid closure (Heller, Kumar, Grill, & Fuchs, 2014).

An active role for the medial edge periderm cells in MES formation in response to TGFβ3 signaling could also be deduced from previous studies of the *K14-Cre;Tgfbr1*$^{f/f}$ and *K14-Cre;Tgfbr2*$^{f/f}$ mice. In contrast to the lack of proper adhesion of the bilateral palatal shelves in the *Tgfb3*$^{-/-}$ embryos resulting in complete cleft palate at birth, the palatal shelves in the *K14-Cre;Tgfbr1*$^{f/f}$ and *K14-Cre;Tgfbr2*$^{f/f}$ embryos adhered and formed the MES spanning about 90% of the anteroposterior length of the secondary palate by E14.5, but the MES persisted in these conditional mutant embryos resulting in submucous cleft in the anterior two-thirds and complete cleft of the posterior part of the palate (Dudas et al., 2006; Xu et al., 2006). Whereas the milder palatal phenotype in the *K14-Cre;Tgfbr1*$^{f/f}$ and *K14-Cre;Tgfbr2*$^{f/f}$ mice than in the *Tgfb3*$^{-/-}$ mice was likely due to remaining TGFβ signaling in the periderm in the *K14-Cre;Tgfbr1*$^{f/f}$ and *K14-Cre; Tgfbr2*$^{f/f}$ embryos (Lane et al., 2014), the *K14-Cre;Tgfbr1*$^{f/f}$ and *K14-Cre; Tgfbr2*$^{f/f}$ embryos have not been analyzed to specifically address the role of the periderm in palatal fusion. Interestingly, while the lack of contribution of K14-Cre lineage cells to the post-fusion palatal mesenchyme and the lack of apoptosis in the MES in the *K14-Cre;Tgfbr1*$^{f/f}$ and *K14-Cre;Tgfbr2*$^{f/f}$ embryos were presented as strong evidence supporting TGFβ signaling induced apoptosis, rather than EMT, as the predominant cellular mechanism of palatal fusion (Gritli-Linde, 2007; Vaziri Sani et al., 2005), no apoptosis was detected in the periderm in either *K14-Cre;Tgfbr1*$^{f/f}$ or the *K14-Cre; Tgfbr2*$^{f/f}$ embryos despite the belief that formation of the MES in these embryos, as opposed to the failure of MES formation in the *Tgfb3*$^{-/-}$ embryos, was probably due to TGFβ-induced shedding of the medial edge periderm cells (Dudas et al., 2006; Lane et al., 2014; Xu et al., 2006). Furthermore, both *K14-Cre;Tgfbr1*$^{f/f}$ and *K14-Cre;Tgfbr2*$^{f/f}$ embryos displayed an epithelial bridge in between the palatal shelves that was progressively thinner from the midpalate toward the posterior palate until the gap in the extreme posterior region, but there was no observation of rupture of the MES or discontinuity of the periderm across the midline bridge (Dudas et al., 2006; Xu et al., 2006). Altogether, the existing data do not support the hypothesis that periderm undergoes apoptosis and slough off prior to palatal shelf adhesion, but rather are consistent with the periderm playing an active role in initiating palatal shelf adhesion in response to TGFβ3 signaling and in maintaining palatal epithelial integrity through the palatal fusion processes.

While the exact role of and the mechanism involving the medial edge periderm cells in palatal fusion remain to be resolved, it is remarkable that the mammalian embryo re-deploys the IRF6-p63 pathway used for periderm differentiation to facilitate dissolution of the MES during palatal fusion. Elegant mouse genetic studies have shown that activation of *Irf6* expression in the basal cells of the MEE is essential for TGFβ3-induced palatal fusion, and that IRF6 acts downstream of TGFβ-Smad4 signaling to shut down the p63-mediated proliferation and cell-adhesion regulatory network in the MES cells to facilitate MES breakdown (Iwata et al., 2013; Richardson et al., 2017). Interestingly, the failure of MES breakdown in the *Tgfb3*$^{Cre/+}$*;Myh9*$^{lox/lox}$ embryos was accompanied by persistent high levels of p63 protein in the MES cell nuclei (Kim et al., 2015), suggesting that biomechanical forces generated by myosin-IIA driven cell intercalation movement play an important role in regulating IRF6-mediated down-regulation of p63 in the MES cells. How these biomechanical, biochemical, and cellular mechanisms are integrated downstream of TGFβ3 signaling to achieve palatal fusion while maintaining epithelial tissue integrity remains to be resolved.

6. Concluding remarks

Formation of the intact upper lip and palate involves not only coordinated directional growth and patterning of the facial primordia but also timely adhesion and fusion of initially separate primordia, which make facial development particularly susceptible to the influence of any deficiency in epithelial integrity. A critical role of epithelial integrity in orofacial development is demonstrated by findings that loss-of-function mutations in *CDH1* and *CTNND1* are associated with syndromic and non-syndromic CLP in human patients, and that conditional inactivation of *Ctnnd1* in the embryonic facial ectoderm resulted in cleft lip and palate in mice (Alharatani et al., 2020; Cox et al., 2018; Frebourg et al., 2006; Ghoumid et al., 2017; Selvanathan et al., 2020). In addition to providing the structural foundation for the spatiotemporal regulation of many signaling molecules for directing the growth and patterning of the facial primordia, epithelial integrity is critical for generating the biomechanical forces in the surface epithelium to push the maxillary and lateral nasal processes medio-frontally toward meeting the medial nasal process to initiate contact and fusion at the lambdoidal junction as well as to fill the midline groove between the bilateral medial nasal processes. A recent study showed that Grhl2-mediated regulation of surface ectoderm integrity is also required for the cranial neural folds to move toward the midline for neural tube closure in mouse embryos (Ray & Niswander, 2016). Previous epidemiological studies suggested that there is a significant etiological overlap between neural tube closure defects and CLP, with the presence of one of these two traits in a family seeming to increase risk of occurrence of the other trait in other individuals (Juriloff & Harris, 2008). GHRL2 regulates the integrity of the surface epithelium during both neural tube closure and facial morphogenesis through activating expression of genes encoding epithelial cell–cell adhesion molecules, as well as suppressing expression of genes driving EMT (Carpinelli et al., 2020; Ray & Niswander, 2016). The rescue of facial and palate development in the *Ghrl2*$^{-/-}$;*Zeb1*$^{-/-}$ embryos indicate that multiple molecular pathways, including other transcription factors and epithelial-specific splicing factors, converge on regulating epithelial cell adhesion to ensure epithelial integrity during facial morphogenesis (Carpinelli et al., 2020). Both GHRL2 and p63 directly regulate expression of genes encoding E-cadherin and other epithelial cell junction molecules as well as *Esrp1*, whereas Esrp1 function is important in generating many components and regulators of epithelial cell junction assembly (Carpinelli et al., 2020; Lin-Shiao et al., 2019; Richardson et al., 2017; Warzecha & Carstens, 2012; Werth et al., 2010). The expression and function of these regulators of epithelial cell–cell adhesion are further modulated by complex regulatory processes including epigenetic modifications such as histone modifications, DNA methylation, and miRNAs, as well as additional post-transcriptional and post-translational modifications (Nieto et al., 2016). While current orofacial cleft research is largely focused on identifying the genetic variants that cause the cleft phenotype, elucidating the regulatory networks that control epithelial integrity is necessary in order to understand the complex etiology and pathogenic mechanisms of orofacial clefting.

A key event for maintaining integrity of the surface epithelium during facial development is the formation of the periderm. The periderm provides a diffusion barrier and non-stick apical surface covering the entire embryo. As studies of the *Irf6*$^{-/-}$, *Jag2*$^{-/-}$, *Ripk4*$^{-/-}$ mutant mice and of the mouse embryos with specific genetic ablation of periderm have demonstrated, defects in periderm formation and/or maintenance cause pathological epithelial adhesions that disrupt orofacial morphogenesis. Whereas these studies have identified several key regulators of periderm formation or maintenance, our understanding of periderm differentiation and maintenance is still quite primitive. How is the periderm cell fate specified, and what regulates the stratification process of periderm layer formation? How do the IRF6-GRHL3 and Jag2-Notch pathways regulate periderm differentiation and integrity? Does the periderm layer expand as the embryo grows by proliferation within the periderm layer or by addition of new periderm cells differentiated from the basal epithelial cells? Whereas several studies have implicated TFAP2A as another crucial transcription factor in the transcriptional regulation of *Irf6* expression, and other studies have shown that mice deficient in several cytoplasmic proteins, such as ARHGAP29 and SPECC1L, exhibited aberrant oral epithelial adhesions (Hall et al., 2020; McDade et al., 2012; Paul et al., 2017; Rahimov et al., 2008), more research efforts and innovative technologies directed at elucidating the molecular network and cellular dynamics regulating periderm differentiation and maintenance are needed to understand pathogenic mechanisms involving periderm dysfunction. Furthermore, the role and regulation of periderm behavior in palatal fusion remain to be resolved.

Lastly, understanding the significant contribution of epithelial instability to orofacial cleft pathogenesis has strong implications in patient care. Defects in epithelial differentiation and/or integrity significantly affect wound healing and cancer metastasis (Frisch, Farris, & Pifer, 2017; Nieto et al., 2016; Sundararajan et al., 2020). It has been reported that children with VWS have an increased risk of wound complications after cleft repair (Jones et al., 2010). Both Grhl2 and Grhl3 have been shown to play critical roles in embryonic wound repair in mice (Boglev et al., 2011). Variants in many regulators of epithelial cell differentiation, as well as epithelial cell adhesion molecules such as E-cadherin, have been associated with cancer (Nieto et al., 2016). Epidemiological studies have shown that orofacial cleft patients have increased cancer risk (Leslie & Marazita, 2013; Seto-Salvia & Stanier, 2014). Thus, translating the understanding of the contributions of epithelial instability to orofacial cleft pathogenesis will help improve clinical management and family counseling in craniofacial patient care.

Acknowledgments

We thank Dr. Tom Gridley for encouragement and discussions. We thank Zhaoming Wu and Jingyue Xu for providing the mouse embryo images in Fig. 1A and B, respectively. The research work on craniofacial development in the Jiang and Lan Laboratories was supported by National Institutes of Health/National Institute of Dental and Craniofacial Research (NIDCR grants DE027046, DE029417, and DE030193) and Eunice Kennedy Shriver National Institute of Child Health and Human Development (NICHD grant HD091263).

References

Abbott, B. D. (2010). The etiology of cleft palate: A 50-year search for mechanistic and molecular understanding. *Birth Defects Research. Part B, Developmental and Reproductive Toxicology*, *89*, 266–274.

Alappat, S. R., Zhang, Z., Suzuki, K., Zhang, X., Liu, H., Jiang, R., et al. (2005). The cellular and molecular etiology of the cleft secondary palate in Fgf10 mutant mice. *Developmental Biology*, *277*, 102–113.

Alharatani, R., Ververi, A., Beleza-Meireles, A., Ji, W., Mis, E., Patterson, Q. T., et al. (2020). Novel truncating mutations in CTNND1 cause a dominant craniofacial and cardiac syndrome. *Human Molecular Genetics*, *29*, 1900–1921.

Barron, M. J., Brookes, S. J., Draper, C. E., Garrod, D., Kirkham, J., Shore, R. C., et al. (2008). The cell adhesion molecule nectin-1 is critical for normal enamel formation in mice. *Human Molecular Genetics*, *17*, 3509–3520.

Basha, M., Demeer, B., Revencu, N., Helaers, R., Theys, S., Bou Saba, S., et al. (2018). Whole exome sequencing identifies mutations in 10% of patients with familial non-syndromic cleft lip and/or palate in genes mutated in well-known syndromes. *Journal of Medical Genetics*, *55*, 449–458.

Bebee, T. W., Park, J. W., Sheridan, K. I., Warzecha, C. C., Cieply, B. W., Rohacek, A. M., et al. (2015). The splicing regulators Esrp1 and Esrp2 direct an epithelial splicing program essential for mammalian development. *eLife*, *4*.

Bishop, M. R., Diaz Perez, K. K., Sun, M., Ho, S., Chopra, P., Mukhopadhyay, N., et al. (2020). Genome-wide enrichment of de novo coding mutations in orofacial cleft trios. *American Journal of Human Genetics*, *107*, 124–136.

Boglev, Y., Wilanowski, T., Caddy, J., Parekh, V., Auden, A., Darido, C., et al. (2011). The unique and cooperative roles of the Grainy head-like transcription factors in epidermal development reflect unexpected target gene specificity. *Developmental Biology*, *349*, 512–522.

Bray, S. J., & Kafatos, F. C. (1991). Developmental function of Elf-1: An essential transcription factor during embryogenesis in Drosophila. *Genes & Development*, *5*, 1672–1683.

Bureau, A., Parker, M. M., Ruczinski, I., Taub, M. A., Marazita, M. L., Murray, J. C., et al. (2014). Whole exome sequencing of distant relatives in multiplex families implicates rare variants in candidate genes for oral clefts. *Genetics*, *197*, 1039–1044.

Bush, J. O., & Jiang, R. (2012). Palatogenesis: Morphogenetic and molecular mechanisms of secondary palate development. *Development*, *139*, 231–243.

Bush, J. O., Lan, Y., & Jiang, R. (2004). The cleft lip and palate defects in Dancer mutant mice result from gain of function of the Tbx10 gene. *Proceedings of the National Academy of Sciences of the United States of America*, *101*, 7022–7027.

Cadieux-Dion, M., Safina, N. P., Engleman, K., Saunders, C., Repnikova, E., Raje, N., et al. (2018). Novel heterozygous pathogenic variants in CHUK in a patient with AEC-like phenotype, immune deficiencies and 1q21.1 microdeletion syndrome: A case report. *BMC Medical Genetics*, *19*, 41.

Cao, Y., Bonizzi, G., Seagroves, T. N., Greten, F. R., Johnson, R., Schmidt, E. V., et al. (2001). IKKalpha provides an essential link between RANK signaling and cyclin D1 expression during mammary gland development. *Cell*, *107*, 763–775.

Capecchi, M. R. (2005). Gene targeting in mice: Functional analysis of the mammalian genome for the twenty-first century. *Nature Reviews. Genetics*, *6*, 507–512.

Carpinelli, M. R., de Vries, M. E., Auden, A., Butt, T., Deng, Z., Partridge, D. D., et al. (2020). Inactivation of Zeb1 in GRHL2-deficient mouse embryos rescues mid-gestation viability and secondary palate closure. *Disease Models & Mechanisms*, *13*.

Carroll, D. K., Carroll, J. S., Leong, C. O., Cheng, F., Brown, M., Mills, A. A., et al. (2006). p63 regulates an adhesion programme and cell survival in epithelial cells. *Nature Cell Biology*, *8*, 551–561.

Carver, E. A., Jiang, R., Lan, Y., Oram, K. F., & Gridley, T. (2001). The mouse snail gene encodes a key regulator of the epithelial-mesenchymal transition. *Molecular and Cellular Biology*, *21*, 8184–8188.

Casey, L. M., Lan, Y., Cho, E. S., Maltby, K. M., Gridley, T., & Jiang, R. (2006). Jag2-Notch1 signaling regulates oral epithelial differentiation and palate development. *Developmental Dynamics*, *235*, 1830–1844.

Celli, J., Duijf, P., Hamel, B. C., Bamshad, M., Kramer, B., Smits, A. P., et al. (1999). Heterozygous germline mutations in the p53 homolog p63 are the cause of EEC syndrome. *Cell*, *99*, 143–153.

Chai, Y., & Maxson, R. E. (2006). Recent advances in craniofacial morphogenesis. *Developmental Dynamics*, *235*, 2353–2375.

Christensen, K., Juel, K., Herskind, A. M., & Murray, J. C. (2004). Long term follow up study of survival associated with cleft lip and palate at birth. *BMJ*, *328*, 1405.

Cox, L. L., Cox, T. C., Moreno Uribe, L. M., Zhu, Y., Richter, C. T., Nidey, N., et al. (2018). Mutations in the epithelial cadherin-p120-catenin complex cause mendelian

non-syndromic cleft lip with or without cleft palate. *American Journal of Human Genetics, 102*, 1143–1157.
De Groote, P., Tran, H. T., Fransen, M., Tanghe, G., Urwyler, C., De Craene, B., et al. (2015). A novel RIPK4-IRF6 connection is required to prevent epithelial fusions characteristic for popliteal pterygium syndromes. *Cell Death and Differentiation, 22*, 1012–1024.
de la Garza, G., Schleiffarth, J. R., Dunnwald, M., Mankad, A., Weirather, J. L., Bonde, G., et al. (2013). Interferon regulatory factor 6 promotes differentiation of the periderm by activating expression of Grainyhead-like 3. *The Journal of Investigative Dermatology, 133*, 68–77.
de Lima, R. L., Hoper, S. A., Ghassibe, M., Cooper, M. E., Rorick, N. K., Kondo, S., et al. (2009). Prevalence and nonrandom distribution of exonic mutations in interferon regulatory factor 6 in 307 families with Van der Woude syndrome and 37 families with popliteal pterygium syndrome. *Genetics in Medicine, 11*, 241–247.
Depew, M. J., & Compagnucci, C. (2008). Tweaking the hinge and caps: Testing a model of the organization of jaws. *Journal of Experimental Zoology. Part B, Molecular and Developmental Evolution, 310*, 315–335.
DiDonato, J. A., Hayakawa, M., Rothwarf, D. M., Zandi, E., & Karin, M. (1997). A cytokine-responsive IkappaB kinase that activates the transcription factor NF-kappaB. *Nature, 388*, 548–554.
Dixon, M. J., Marazita, M. L., Beaty, T. H., & Murray, J. C. (2011). Cleft lip and palate: Understanding genetic and environmental influences. *Nature Reviews. Genetics, 12*, 167–178.
Dudas, M., Kim, J., Li, W. Y., Nagy, A., Larsson, J., Karlsson, S., et al. (2006). Epithelial and ectomesenchymal role of the type I TGF-beta receptor ALK5 during facial morphogenesis and palatal fusion. *Developmental Biology, 296*, 298–314.
Eshete, M. A., Liu, H., Li, M., Adeyemo, W. L., Gowans, L. J. J., Mossey, P. A., et al. (2018). Loss-of-function GRHL3 variants detected in African patients with isolated cleft palate. *Journal of Dental Research, 97*, 41–48.
Fan, X., Wang, D., Burgmaier, J. E., Teng, Y., Romano, R. A., Sinha, S., et al. (2018). Single cell and open chromatin analysis reveals molecular origin of epidermal cells of the skin. *Developmental Cell, 47*, 133.
Ferone, G., Thomason, H. A., Antonini, D., De Rosa, L., Hu, B., Gemei, M., et al. (2012). Mutant p63 causes defective expansion of ectodermal progenitor cells and impaired FGF signalling in AEC syndrome. *EMBO Molecular Medicine, 4*, 192–205.
Ferretti, E., Li, B., Zewdu, R., Wells, V., Hebert, J. M., Karner, C., et al. (2011). A conserved Pbx-Wnt-p63-Irf6 regulatory module controls face morphogenesis by promoting epithelial apoptosis. *Developmental Cell, 21*, 627–641.
Fitchett, J. E., & Hay, E. D. (1989). Medial edge epithelium transforms to mesenchyme after embryonic palatal shelves fuse. *Developmental Biology, 131*, 455–474.
Fitzpatrick, D. R., Denhez, F., Kondaiah, P., & Akhurst, R. J. (1990). Differential expression of TGF beta isoforms in murine palatogenesis. *Development, 109*, 585–595.
Fomenkov, A., Zangen, R., Huang, Y. P., Osada, M., Guo, Z., Fomenkov, T., et al. (2004). RACK1 and stratifin target DeltaNp63alpha for a proteasome degradation in head and neck squamous cell carcinoma cells upon DNA damage. *Cell Cycle, 3*, 1285–1295.
Frebourg, T., Oliveira, C., Hochain, P., Karam, R., Manouvrier, S., Graziadio, C., et al. (2006). Cleft lip/palate and CDH1/E-cadherin mutations in families with hereditary diffuse gastric cancer. *Journal of Medical Genetics, 43*, 138–142.
Frisch, S. M., Farris, J. C., & Pifer, P. M. (2017). Roles of Grainyhead-like transcription factors in cancer. *Oncogene, 36*, 6067–6073.
Garland, M. A., Sun, B., Zhang, S., Reynolds, K., Ji, Y., & Zhou, C. J. (2020). Role of epigenetics and miRNAs in orofacial clefts. *Birth Defects Research, 112*, 1635–1659.

Ghoumid, J., Stichelbout, M., Jourdain, A. S., Frenois, F., Lejeune-Dumoulin, S., Alex-Cordier, M. P., et al. (2017). Blepharocheilodontic syndrome is a CDH1 pathway-related disorder due to mutations in CDH1 and CTNND1. *Genetics in Medicine*, *19*, 1013–1021.

Gritli-Linde, A. (2007). Molecular control of secondary palate development. *Developmental Biology*, *301*, 309–326.

Gritli-Linde, A. (2008). The etiopathogenesis of cleft lip and cleft palate: Usefulness and caveats of mouse models. *Current Topics in Developmental Biology*, *84*, 37–138.

Hall, E. G., Wenger, L. W., Wilson, N. R., Undurty-Akella, S. S., Standley, J., Augustine-Akpan, E. A., et al. (2020). SPECC1L regulates palate development downstream of IRF6. *Human Molecular Genetics*, *29*, 845–858.

Hammond, N. L., Dixon, J., & Dixon, M. J. (2019). Periderm: Life-cycle and function during orofacial and epidermal development. *Seminars in Cell & Developmental Biology*, *91*, 75–83.

Hayward, A. F. (1983). The permeability of the epithelium of the skin of fetal rats demonstrated with a lanthanum-containing solution. *Journal of Anatomy*, *136*, 379–388.

He, F., Xiong, W., Wang, Y., Matsui, M., Yu, X., Chai, Y., et al. (2010). Modulation of BMP signaling by Noggin is required for the maintenance of palatal epithelial integrity during palatogenesis. *Developmental Biology*, *347*, 109–121.

Heller, E., Kumar, K. V., Grill, S. W., & Fuchs, E. (2014). Forces generated by cell intercalation tow epidermal sheets in mammalian tissue morphogenesis. *Developmental Cell*, *28*, 617–632.

Herron, B. J., Liddell, R. A., Parker, A., Grant, S., Kinne, J., Fisher, J. K., et al. (2005). A mutation in stratifin is responsible for the repeated epilation (Er) phenotype in mice. *Nature Genetics*, *37*, 1210–1212.

Hinrichsen, K. (1985). The early development of morphology and patterns of the face in the human embryo. *Advances in Anatomy, Embryology, and Cell Biology*, *98*, 1–79.

Holland, P., Willis, C., Kanaly, S., Glaccum, M., Warren, A., Charrier, K., et al. (2002). RIP4 is an ankyrin repeat-containing kinase essential for keratinocyte differentiation. *Current Biology*, *12*, 1424–1428.

Hu, Y., Baud, V., Delhase, M., Zhang, P., Deerinck, T., Ellisman, M., et al. (1999). Abnormal morphogenesis but intact IKK activation in mice lacking the IKKalpha subunit of IkappaB kinase. *Science*, *284*, 316–320.

Hu, Y., Baud, V., Oga, T., Kim, K. I., Yoshida, K., & Karin, M. (2001). IKKalpha controls formation of the epidermis independently of NF-kappaB. *Nature*, *410*, 710–714.

Hu, M. S., Borrelli, M. R., Hong, W. X., Malhotra, S., Cheung, A. T. M., Ransom, R. C., et al. (2018). Embryonic skin development and repair. *Organogenesis*, *14*, 46–63.

Hu, L., Liu, J., Li, Z., Ozturk, F., Gurumurthy, C., Romano, R. A., et al. (2015). TGFbeta3 regulates periderm removal through DeltaNp63 in the developing palate. *Journal of Cellular Physiology*, *230*, 1212–1225.

Ingraham, C. R., Kinoshita, A., Kondo, S., Yang, B., Sajan, S., Trout, K. J., et al. (2006). Abnormal skin, limb and craniofacial morphogenesis in mice deficient for interferon regulatory factor 6 (Irf6). *Nature Genetics*, *38*, 1335–1340.

Iwata, J., Suzuki, A., Pelikan, R. C., Ho, T. V., Sanchez-Lara, P. A., Urata, M., et al. (2013). Smad4-Irf6 genetic interaction and TGFβ-mediated IRF6 signaling cascade are crucial for palatal fusion in mice. *Development*, *140*, 1220–1230.

Jeong, J., Mao, J., Tenzen, T., Kottmann, A. H., & McMahon, A. P. (2004). Hedgehog signaling in the neural crest cells regulates the patterning and growth of facial primordia. *Genes & Development*, *18*, 937–951.

Ji, Y., Garland, M. A., Sun, B., Zhang, S., Reynolds, K., McMahon, M., et al. (2020). Cellular and developmental basis of orofacial clefts. *Birth Defects Research*, *112*, 1558–1587.

Jiang, R., Bush, J. O., & Lidral, A. C. (2006). Development of the upper lip: Morphogenetic and molecular mechanisms. *Developmental Dynamics*, *235*, 1152–1166.
Jiang, R., & Gridley, T. (1997). Gene targeting: Things go better with Cre. *Current Biology*, *7*, R321–R323.
Jiang, R., Lan, Y., Chapman, H. D., Shawber, C., Norton, C. R., Serreze, D. V., et al. (1998). Defects in limb, craniofacial, and thymic development in Jagged2 mutant mice. *Genes & Development*, *12*, 1046–1057.
Jin, Y. R., Han, X. H., Nishimori, K., Ben-Avraham, D., Oh, Y. J., Shim, J. W., et al. (2020). Canonical WNT/beta-catenin signaling activated by WNT9b and RSPO2 cooperation regulates facial morphogenesis in mice. *Frontiers in Cell and Development Biology*, *8*, 264.
Jones, J. L., Canady, J. W., Brookes, J. T., Wehby, G. L., L'Heureux, J., Schutte, B. C., et al. (2010). Wound complications after cleft repair in children with Van der Woude syndrome. *The Journal of Craniofacial Surgery*, *21*, 1350–1353.
Juriloff, D. M., & Harris, M. J. (2008). Mouse genetic models of cleft lip with or without cleft palate. *Birth Defects Research. Part A, Clinical and Molecular Teratology*, *82*, 63–77.
Kaartinen, V., Cui, X. M., Heisterkamp, N., Groffen, J., & Shuler, C. F. (1997). Transforming growth factor-beta3 regulates transdifferentiation of medial edge epithelium during palatal fusion and associated degradation of the basement membrane. *Developmental Dynamics*, *209*, 255–260.
Kaartinen, V., Voncken, J. W., Shuler, C., Warburton, D., Bu, D., Heisterkamp, N., et al. (1995). Abnormal lung development and cleft palate in mice lacking TGF-beta 3 indicates defects of epithelial-mesenchymal interaction. *Nature Genetics*, *11*, 415–421.
Kalay, E., Sezgin, O., Chellappa, V., Mutlu, M., Morsy, H., Kayserili, H., et al. (2012). Mutations in RIPK4 cause the autosomal-recessive form of popliteal pterygium syndrome. *American Journal of Human Genetics*, *90*, 76–85.
Kashgari, G., Meinecke, L., Gordon, W., Ruiz, B., Yang, J., Ma, A. L., et al. (2020). Epithelial migration and non-adhesive periderm are required for digit separation during mammalian development. *Developmental Cell*, *52*, 764–778. e764.
Kawasaki, M., Kawasaki, K., Oommen, S., Blackburn, J., Watanabe, M., Nagai, T., et al. (2016). Regional regulation of Filiform tongue papillae development by Ikkalpha/ Irf6. *Developmental Dynamics*, *245*, 937–946.
Khandelwal, K. D., Ockeloen, C. W., Venselaar, H., Boulanger, C., Brichard, B., Sokal, E., et al. (2017). Identification of a de novo variant in CHUK in a patient with an EEC/AEC syndrome-like phenotype and hypogammaglobulinemia. *American Journal of Medical Genetics. Part A*, *173*, 1813–1820.
Khandelwal, K. D., van den Boogaard, M. H., Mehrem, S. L., Gebel, J., Fagerberg, C., van Beusekom, E., et al. (2019). Deletions and loss-of-function variants in TP63 associated with orofacial clefting. *European Journal of Human Genetics*, *27*, 1101–1112.
Kim, S., Lewis, A. E., Singh, V., Ma, X., Adelstein, R., & Bush, J. O. (2015). Convergence and extrusion are required for normal fusion of the mammalian secondary palate. *PLoS Biology*, *13*, e1002122.
Knight, A. S., Schutte, B. C., Jiang, R., & Dixon, M. J. (2006). Developmental expression analysis of the mouse and chick orthologues of IRF6: The gene mutated in Van der Woude syndrome. *Developmental Dynamics*, *235*, 1441–1447.
Koillinen, H., Wong, F. K., Rautio, J., Ollikainen, V., Karsten, A., Larson, O., et al. (2001). Mapping of the second locus for the Van der Woude syndrome to chromosome 1p34. *European Journal of Human Genetics*, *9*, 747–752.
Kondo, S., Schutte, B. C., Richardson, R. J., Bjork, B. C., Knight, A. S., Watanabe, Y., et al. (2002). Mutations in IRF6 cause Van der Woude and popliteal pterygium syndromes. *Nature Genetics*, *32*, 285–289.

Koster, M. I., Dai, D., Marinari, B., Sano, Y., Costanzo, A., Karin, M., et al. (2007). p63 induces key target genes required for epidermal morphogenesis. *Proceedings of the National Academy of Sciences of the United States of America*, *104*, 3255–3260.

Kousa, Y. A., Moussa, D., & Schutte, B. C. (2017). IRF6 expression in basal epithelium partially rescues Irf6 knockout mice. *Developmental Dynamics*, *246*, 670–681.

Kousa, Y. A., & Schutte, B. C. (2016). Toward an orofacial gene regulatory network. *Developmental Dynamics*, *245*, 220–232.

Kurima, K., Hertzano, R., Gavrilova, O., Monahan, K., Shpargel, K. B., Nadaraja, G., et al. (2011). A noncoding point mutation of Zeb1 causes multiple developmental malformations and obesity in Twirler mice. *PLoS Genetics*, *7*, e1002307.

Kurosaka, H., Iulianella, A., Williams, T., & Trainor, P. A. (2014). Disrupting hedgehog and WNT signaling interactions promotes cleft lip pathogenesis. *The Journal of Clinical Investigation*, *124*, 1660–1671.

Kwa, M. Q., Huynh, J., Aw, J., Zhang, L., Nguyen, T., Reynolds, E. C., et al. (2014). Receptor-interacting protein kinase 4 and interferon regulatory factor 6 function as a signaling axis to regulate keratinocyte differentiation. *The Journal of Biological Chemistry*, *289*, 31077–31087.

Lahtela, J., Nousiainen, H. O., Stefanovic, V., Tallila, J., Viskari, H., Karikoski, R., et al. (2010). Mutant CHUK and severe fetal encasement malformation. *The New England Journal of Medicine*, *363*, 1631–1637.

Lan, Y., & Jiang, R. (2009). Sonic hedgehog signaling regulates reciprocal epithelial-mesenchymal interactions controlling palatal outgrowth. *Development*, *136*, 1387–1396.

Lan, Y., Ovitt, C. E., Cho, E. S., Maltby, K. M., Wang, Q., & Jiang, R. (2004). Odd-skipped related 2 (Osr2) encodes a key intrinsic regulator of secondary palate growth and morphogenesis. *Development*, *131*, 3207–3216.

Lan, Y., Ryan, R. C., Zhang, Z., Bullard, S. A., Bush, J. O., Maltby, K. M., et al. (2006). Expression of Wnt9b and activation of canonical Wnt signaling during midfacial morphogenesis in mice. *Developmental Dynamics*, *235*, 1448–1454.

Lan, Y., Wang, Q., Ovitt, C. E., & Jiang, R. (2007). A unique mouse strain expressing Cre recombinase for tissue-specific analysis of gene function in palate and kidney development. *Genesis*, *45*, 618–624.

Lan, Y., Xu, J., & Jiang, R. (2015). Cellular and molecular mechanisms of palatogenesis. *Current Topics in Developmental Biology*, *115*, 59–84.

Lane, J., & Kaartinen, V. (2014). Signaling networks in palate development. *Wiley Interdisciplinary Reviews. Systems Biology and Medicine*, *6*, 271–278.

Lane, J., Yumoto, K., Pisano, J., Azhar, M., Thomas, P. S., & Kaartinen, V. (2014). Control elements targeting Tgfb3 expression to the palatal epithelium are located intergenically and in introns of the upstream Ift43 gene. *Frontiers in Physiology*, *5*, 258.

Lee, S., Kong, Y., & Weatherbee, S. D. (2013). Forward genetics identifies Kdf1/1810019J16Rik as an essential regulator of the proliferation-differentiation decision in epidermal progenitor cells. *Developmental Biology*, *383*, 201–213.

Lee, S., Sears, M. J., Zhang, Z., Li, H., Salhab, I., Krebs, P., et al. (2020). Cleft lip and cleft palate in Esrp1 knockout mice is associated with alterations in epithelial-mesenchymal crosstalk. *Development*, *147*.

Leslie, E. J., Carlson, J. C., Shaffer, J. R., Butali, A., Buxo, C. J., Castilla, E. E., et al. (2017). Genome-wide meta-analyses of nonsyndromic orofacial clefts identify novel associations between FOXE1 and all orofacial clefts, and TP63 and cleft lip with or without cleft palate. *Human Genetics*, *136*, 275–286.

Leslie, E. J., Liu, H., Carlson, J. C., Shaffer, J. R., Feingold, E., Wehby, G., et al. (2016). A genome-wide association study of nonsyndromic cleft palate identifies an etiologic missense variant in GRHL3. *American Journal of Human Genetics*, *98*, 744–754.

Leslie, E. J., & Marazita, M. L. (2013). Genetics of cleft lip and cleft palate. *American Journal of Medical Genetics. Part C, Seminars in Medical Genetics*, *163C*, 246–258.

Leslie, E. J., O'Sullivan, J., Cunningham, M. L., Singh, A., Goudy, S. L., Ababneh, F., et al. (2015). Expanding the genetic and phenotypic spectrum of popliteal pterygium disorders. *American Journal of Medical Genetics. Part A*, *167A*, 545–552.

Li, H., Jones, K. L., Hooper, J. E., & Williams, T. (2019). The molecular anatomy of mammalian upper lip and primary palate fusion at single cell resolution. *Development*, *146*.

Li, C., Lan, Y., & Jiang, R. (2017). Molecular and cellular mechanisms of palate development. *Journal of Dental Research*, *96*, 1184–1191.

Li, Q., Lu, Q., Estepa, G., & Verma, I. M. (2005). Identification of 14-3-3sigma mutation causing cutaneous abnormality in repeated-epilation mutant mouse. *Proceedings of the National Academy of Sciences of the United States of America*, *102*, 15977–15982.

Li, Q., Lu, Q., Hwang, J. Y., Buscher, D., Lee, K. F., Izpisua-Belmonte, J. C., et al. (1999). IKK1-deficient mice exhibit abnormal development of skin and skeleton. *Genes & Development*, *13*, 1322–1328.

Lin-Shiao, E., Lan, Y., Welzenbach, J., Alexander, K. A., Zhang, Z., Knapp, M., et al. (2019). p63 establishes epithelial enhancers at critical craniofacial development genes. *Science Advances*, *5*, eaaw0946.

Liu, Y. N., Abou-Kheir, W., Yin, J. J., Fang, L., Hynes, P., Casey, O., et al. (2012). Critical and reciprocal regulation of KLF4 and SLUG in transforming growth factor beta-initiated prostate cancer epithelial-mesenchymal transition. *Molecular and Cellular Biology*, *32*, 941–953.

Liu, W., Sun, X., Braut, A., Mishina, Y., Behringer, R. R., Mina, M., et al. (2005). Distinct functions for Bmp signaling in lip and palate fusion in mice. *Development*, *132*, 1453–1461.

Mangold, E., Bohmer, A. C., Ishorst, N., Hoebel, A. K., Gultepe, P., Schuenke, H., et al. (2016). Sequencing the GRHL3 coding region reveals rare truncating mutations and a common susceptibility variant for nonsyndromic cleft palate. *American Journal of Human Genetics*, *98*, 755–762.

Marchini, M., Hu, D., Lo Vercio, L., Young, N. M., Forkert, N. D., Hallgrimsson, B., et al. (2021). Wnt signaling drives correlated changes in facial morphology and brain shape. *Frontiers in Cell and Development Biology*, *9*, 644099.

Marcucio, R., Hallgrimsson, B., & Young, N. M. (2015). Facial morphogenesis: Physical and molecular interactions between the brain and the face. *Current Topics in Developmental Biology*, *115*, 299–320.

Matsuki, M., Yamashita, F., Ishida-Yamamoto, A., Yamada, K., Kinoshita, C., Fushiki, S., et al. (1998). Defective stratum corneum and early neonatal death in mice lacking the gene for transglutaminase 1 (keratinocyte transglutaminase). *Proceedings of the National Academy of Sciences of the United States of America*, *95*, 1044–1049.

Matsuzawa, N., Kondo, S., Shimozato, K., Nagao, T., Nakano, M., Tsuda, M., et al. (2010). Two missense mutations of the IRF6 gene in two Japanese families with popliteal pterygium syndrome. *American Journal of Medical Genetics. Part A*, *152A*, 2262–2267.

M'Boneko, V., & Merker, H. J. (1988). Development and morphology of the periderm of mouse embryos (days 9-12 of gestation). *Acta Anatomica (Basel)*, *133*, 325–336.

McDade, S. S., Henry, A. E., Pivato, G. P., Kozarewa, I., Mitsopoulos, C., Fenwick, K., et al. (2012). Genome-wide analysis of p63 binding sites identifies AP-2 factors as co-regulators of epidermal differentiation. *Nucleic Acids Research*, *40*, 7190–7206.

McGowan, K. M., & Coulombe, P. A. (1998). Onset of keratin 17 expression coincides with the definition of major epithelial lineages during skin development. *The Journal of Cell Biology*, *143*, 469–486.

McGrath, J. A., Duijf, P. H., Doetsch, V., Irvine, A. D., de Waal, R., Vanmolkot, K. R., et al. (2001). Hay-Wells syndrome is caused by heterozygous missense mutations in the SAM domain of p63. *Human Molecular Genetics*, *10*, 221–229.

Mills, A. A., Zheng, B., Wang, X. J., Vogel, H., Roop, D. R., & Bradley, A. (1999). p63 is a p53 homologue required for limb and epidermal morphogenesis. *Nature*, *398*, 708–713.

Mitchell, K., O'Sullivan, J., Missero, C., Blair, E., Richardson, R., Anderson, B., et al. (2012). Exome sequence identifies RIPK4 as the Bartsocas-Papas syndrome locus. *American Journal of Human Genetics*, *90*, 69–75.

Mollo, M. R., Antonini, D., Mitchell, K., Fortugno, P., Costanzo, A., Dixon, J., et al. (2015). p63-dependent and independent mechanisms of nectin-1 and nectin-4 regulation in the epidermis. *Experimental Dermatology*, *24*, 114–119.

Moretti, F., Marinari, B., Lo Iacono, N., Botti, E., Giunta, A., Spallone, G., et al. (2010). A regulatory feedback loop involving p63 and IRF6 links the pathogenesis of 2 genetically different human ectodermal dysplasias. *The Journal of Clinical Investigation*, *120*, 1570–1577.

Morita, K., Furuse, M., Yoshida, Y., Itoh, M., Sasaki, H., Tsukita, S., et al. (2002). Molecular architecture of tight junctions of periderm differs from that of the maculae occludentes of epidermis. *The Journal of Investigative Dermatology*, *118*, 1073–1079.

Murray, S. A. (2011). Mouse resources for craniofacial research. *Genesis*, *49*, 190–199.

Murray, S. A., Oram, K. F., & Gridley, T. (2007). Multiple functions of Snail family genes during palate development in mice. *Development*, *134*, 1789–1797.

Nasreddine, G., El Hajj, J., & Ghassibe-Sabbagh, M. (2021). Orofacial clefts embryology, classification, epidemiology, and genetics. *Mutation Research, Reviews in Mutation Research*, *787*, 108373.

Nawshad, A. (2008). Palatal seam disintegration: To die or not to die? That is no longer the question. *Developmental Dynamics*, *237*, 2643–2656.

Nguyen, B. C., Lefort, K., Mandinova, A., Antonini, D., Devgan, V., Della Gatta, G., et al. (2006). Cross-regulation between Notch and p63 in keratinocyte commitment to differentiation. *Genes & Development*, *20*, 1028–1042.

Nieto, M. A., Huang, R. Y., Jackson, R. A., & Thiery, J. P. (2016). Emt: 2016. *Cell*, *166*, 21–45.

Oberbeck, N., Pham, V. C., Webster, J. D., Reja, R., Huang, C. S., Zhang, Y., et al. (2019). The RIPK4-IRF6 signalling axis safeguards epidermal differentiation and barrier function. *Nature*, *574*, 249–253.

Osterburg, C., Osterburg, S., Zhou, H., Missero, C., & Dotsch, V. (2021). Isoform-specific roles of mutant p63 in human diseases. *Cancers (Basel)*, *13*.

Panaliappan, T. K., Wittmann, W., Jidigam, V. K., Mercurio, S., Bertolini, J. A., Sghari, S., et al. (2018). Sox2 is required for olfactory pit formation and olfactory neurogenesis through BMP restriction and Hes5 upregulation. *Development*, *145*.

Paul, B. J., Palmer, K., Sharp, J. C., Pratt, C. H., Murray, S. A., & Dunnwald, M. (2017). ARHGAP29 mutation is associated with abnormal oral epithelial adhesions. *Journal of Dental Research*, *96*, 1298–1305.

Peinado, H., Olmeda, D., & Cano, A. (2007). Snail, Zeb and bHLH factors in tumour progression: An alliance against the epithelial phenotype? *Nature Reviews. Cancer*, *7*, 415–428.

Pellegrini, G., Dellambra, E., Golisano, O., Martinelli, E., Fantozzi, I., Bondanza, S., et al. (2001). p63 Identifies keratinocyte stem cells. *Proceedings of the National Academy of Sciences of the United States of America*, *98*, 3156–3161.

Pelton, R. W., Hogan, B. L., Miller, D. A., & Moses, H. L. (1990). Differential expression of genes encoding TGFs beta 1, beta 2, and beta 3 during murine palate formation. *Developmental Biology*, *141*, 456–460.

Peyrard-Janvid, M., Leslie, E. J., Kousa, Y. A., Smith, T. L., Dunnwald, M., Magnusson, M., et al. (2014). Dominant mutations in GRHL3 cause Van der Woude Syndrome and disrupt oral periderm development. *American Journal of Human Genetics*, *94*, 23–32.
Pieters, T., van Roy, F., & van Hengel, J. (2012). Functions of p120ctn isoforms in cell-cell adhesion and intracellular signaling. *Frontiers in Bioscience*, *17*, 1669–1694.
Proetzel, G., Pawlowski, S. A., Wiles, M. V., Yin, M., Boivin, G. P., Howles, P. N., et al. (1995). Transforming growth factor-beta 3 is required for secondary palate fusion. *Nature Genetics*, *11*, 409–414.
Pyrgaki, C., Liu, A., & Niswander, L. (2011). Grainyhead-like 2 regulates neural tube closure and adhesion molecule expression during neural fold fusion. *Developmental Biology*, *353*, 38–49.
Rahimov, F., Marazita, M. L., Visel, A., Cooper, M. E., Hitchler, M. J., Rubini, M., et al. (2008). Disruption of an AP-2alpha binding site in an IRF6 enhancer is associated with cleft lip. *Nature Genetics*, *40*, 1341–1347.
Rajewsky, K., Gu, H., Kuhn, R., Betz, U. A., Muller, W., Roes, J., et al. (1996). Conditional gene targeting. *The Journal of Clinical Investigation*, *98*, 600–603.
Ray, A. T., Mazot, P., Brewer, J. R., Catela, C., Dinsmore, C. J., & Soriano, P. (2020). FGF signaling regulates development by processes beyond canonical pathways. *Genes & Development*, *34*, 1735–1752.
Ray, H. J., & Niswander, L. A. (2016). Grainyhead-like 2 downstream targets act to suppress epithelial-to-mesenchymal transition during neural tube closure. *Development*, *143*, 1192–1204.
Reid, B. S., Yang, H., Melvin, V. S., Taketo, M. M., & Williams, T. (2011). Ectodermal Wnt/beta-catenin signaling shapes the mouse face. *Developmental Biology*, *349*, 261–269.
Restivo, G., Nguyen, B. C., Dziunycz, P., Ristorcelli, E., Ryan, R. J., Ozuysal, O. Y., et al. (2011). IRF6 is a mediator of Notch pro-differentiation and tumour suppressive function in keratinocytes. *The EMBO Journal*, *30*, 4571–4585.
Reynolds, K., Kumari, P., Sepulveda Rincon, L., Gu, R., Ji, Y., Kumar, S., et al. (2019). Wnt signaling in orofacial clefts: Crosstalk, pathogenesis and models. *Disease Models & Mechanisms*, *12*.
Reynolds, K., Zhang, S., Sun, B., Garland, M. A., Ji, Y., & Zhou, C. J. (2020). Genetics and signaling mechanisms of orofacial clefts. *Birth Defects Research*, *112*, 1588–1634.
Rice, R., Spencer-Dene, B., Connor, E. C., Gritli-Linde, A., McMahon, A. P., Dickson, C., et al. (2004). Disruption of Fgf10/Fgfr2b-coordinated epithelial-mesenchymal interactions causes cleft palate. *The Journal of Clinical Investigation*, *113*, 1692–1700.
Richardson, R. J., Dixon, J., Jiang, R., & Dixon, M. J. (2009). Integration of IRF6 and Jagged2 signalling is essential for controlling palatal adhesion and fusion competence. *Human Molecular Genetics*, *18*, 2632–2642.
Richardson, R. J., Dixon, J., Malhotra, S., Hardman, M. J., Knowles, L., Boot-Handford, R. P., et al. (2006). Irf6 is a key determinant of the keratinocyte proliferation-differentiation switch. *Nature Genetics*, *38*, 1329–1334.
Richardson, R. J., Hammond, N. L., Coulombe, P. A., Saloranta, C., Nousiainen, H. O., Salonen, R., et al. (2014). Periderm prevents pathological epithelial adhesions during embryogenesis. *The Journal of Clinical Investigation*, *124*, 3891–3900.
Richardson, R., Mitchell, K., Hammond, N. L., Mollo, M. R., Kouwenhoven, E. N., Wyatt, N. D., et al. (2017). p63 exerts spatio-temporal control of palatal epithelial cell fate to prevent cleft palate. *PLoS Genetics*, *13*, e1006828.
Rinne, T., Bolat, E., Meijer, R., Scheffer, H., & van Bokhoven, H. (2009). Spectrum of p63 mutations in a selected patient cohort affected with ankyloblepharon-ectodermal defects-cleft lip/palate syndrome (AEC). *American Journal of Medical Genetics. Part A*, *149A*, 1948–1951.

Romano, R. A., Smalley, K., Magraw, C., Serna, V. A., Kurita, T., Raghavan, S., et al. (2012). DeltaNp63 knockout mice reveal its indispensable role as a master regulator of epithelial development and differentiation. *Development, 139*, 772–782.

Saathoff, M., Blum, B., Quast, T., Kirfel, G., & Herzog, V. (2004). Simultaneous cell death and desquamation of the embryonic diffusion barrier during epidermal development. *Experimental Cell Research, 299*, 415–426.

Schmale, H., & Bamberger, C. (1997). A novel protein with strong homology to the tumor suppressor p53. *Oncogene, 15*, 1363–1367.

Selvanathan, A., Nixon, C. Y., Zhu, Y., Scietti, L., Forneris, F., Uribe, L. M. M., et al. (2020). CDH1 mutation distribution and type suggests genetic differences between the etiology of orofacial clefting and gastric cancer. *Genes (Basel), 11*.

Senoo, M., Pinto, F., Crum, C. P., & McKeon, F. (2007). p63 Is essential for the proliferative potential of stem cells in stratified epithelia. *Cell, 129*, 523–536.

Seto-Salvia, N., & Stanier, P. (2014). Genetics of cleft lip and/or cleft palate: Association with other common anomalies. *European Journal of Medical Genetics, 57*, 381–393.

Shi, M., Christensen, K., Weinberg, C. R., Romitti, P., Bathum, L., Lozada, A., et al. (2007). Orofacial cleft risk is increased with maternal smoking and specific detoxification-gene variants. *American Journal of Human Genetics, 80*, 76–90.

Sil, A. K., Maeda, S., Sano, Y., Roop, D. R., & Karin, M. (2004). IkappaB kinase-alpha acts in the epidermis to control skeletal and craniofacial morphogenesis. *Nature, 428*, 660–664.

Soares, E., Xu, Q., Li, Q., Qu, J., Zheng, Y., Raeven, H. H. M., et al. (2019). Single-cell RNA-seq identifies a reversible mesodermal activation in abnormally specified epithelia of p63 EEC syndrome. *Proceedings of the National Academy of Sciences of the United States of America, 116*, 17361–17370.

Soares, E., & Zhou, H. (2018). Master regulatory role of p63 in epidermal development and disease. *Cellular and Molecular Life Sciences, 75*, 1179–1190.

Song, Y., & Zhang, S. (2011). Association of CDH1 promoter polymorphism and the risk of non-syndromic orofacial clefts in a Chinese Han population. *Archives of Oral Biology, 56*, 68–72.

Sozen, M. A., Suzuki, K., Tolarova, M. M., Bustos, T., Fernandez Iglesias, J. E., & Spritz, R. A. (2001). Mutation of PVRL1 is associated with sporadic, non-syndromic cleft lip/palate in northern Venezuela. *Nature Genetics, 29*, 141–142.

Su, Y. W., Hao, Z., Hirao, A., Yamamoto, K., Lin, W. J., Young, A., et al. (2011). 14-3-3σ Regulates B-cell homeostasis through stabilization of FOXO1. *Proceedings of the National Academy of Sciences of the United States of America, 108*, 1555–1560.

Sundararajan, V., Pang, Q. Y., Choolani, M., & Huang, R. Y. (2020). Spotlight on the granules (Grainyhead-like proteins)—From an evolutionary conserved controller of epithelial trait to pioneering the chromatin landscape. *Frontiers in Molecular Biosciences*, 7, 213.

Suzuki, A., Abdallah, N., Gajera, M., Jun, G., Jia, P., Zhao, Z., et al. (2018). Genes and microRNAs associated with mouse cleft palate: A systematic review and bioinformatics analysis. *Mechanisms of Development, 150*, 21–27.

Suzuki, K., Hu, D., Bustos, T., Zlotogora, J., Richieri-Costa, A., Helms, J. A., et al. (2000). Mutations of PVRL1, encoding a cell-cell adhesion molecule/herpesvirus receptor, in cleft lip/palate-ectodermal dysplasia. *Nature Genetics, 25*, 427–430.

Szabo-Rogers, H. L., Geetha-Loganathan, P., Nimmagadda, S., Fu, K. K., & Richman, J. M. (2008). FGF signals from the nasal pit are necessary for normal facial morphogenesis. *Developmental Biology, 318*, 289–302.

Tadeu, A. M., & Horsley, V. (2013). Notch signaling represses p63 expression in the developing surface ectoderm. *Development, 140*, 3777–3786.

Takahashi, M., Hosomichi, K., Yamaguchi, T., Nagahama, R., Yoshida, H., Maki, K., et al. (2018). Whole-genome sequencing in a pair of monozygotic twins with discordant cleft lip and palate subtypes. *Oral Diseases, 24*, 1303–1309.

Takeda, K., Takeuchi, O., Tsujimura, T., Itami, S., Adachi, O., Kawai, T., et al. (1999). Limb and skin abnormalities in mice lacking IKKalpha. *Science*, *284*, 313–316.
Taniguchi, T., Ogasawara, K., Takaoka, A., & Tanaka, N. (2001). IRF family of transcription factors as regulators of host defense. *Annual Review of Immunology*, *19*, 623–655.
Tanimizu, N., & Mitaka, T. (2013). Role of grainyhead-like 2 in the formation of functional tight junctions. *Tissue Barriers*, *1*, e23495.
Tassin, M. T., Salzgeber, B., & Guenet, J. L. (1983). Studies on "repeated epilation" mouse mutant embryos: I. Development of facial malformations. *Journal of Craniofacial Genetics and Developmental Biology*, *3*, 289–307.
Taya, Y., O'Kane, S., & Ferguson, M. W. (1999). Pathogenesis of cleft palate in TGF-beta3 knockout mice. *Development*, *126*, 3869–3879.
Thomason, H. A., Dixon, M. J., & Dixon, J. (2008). Facial clefting in Tp63 deficient mice results from altered Bmp4, Fgf8 and Shh signaling. *Developmental Biology*, *321*, 273–282.
Thomason, H. A., Zhou, H., Kouwenhoven, E. N., Dotto, G. P., Restivo, G., Nguyen, B. C., et al. (2010). Cooperation between the transcription factors p63 and IRF6 is essential to prevent cleft palate in mice. *The Journal of Clinical Investigation*, *120*, 1561–1569.
Ting, S. B., Caddy, J., Wilanowski, T., Auden, A., Cunningham, J. M., Elias, P. M., et al. (2005). The epidermis of grhl3-null mice displays altered lipid processing and cellular hyperproliferation. *Organogenesis*, *2*, 33–35.
Truong, A. B., Kretz, M., Ridky, T. W., Kimmel, R., & Khavari, P. A. (2006). p63 regulates proliferation and differentiation of developmentally mature keratinocytes. *Genes & Development*, *20*, 3185–3197.
Twigg, S. R., & Wilkie, A. O. (2015). New insights into craniofacial malformations. *Human Molecular Genetics*, *24*, R50–R59.
Urwyler-Rosselet, C., Tanghe, G., Leurs, K., Gilbert, B., De Rycke, R., De Bruyne, M., et al. (2018). Keratinocyte-specific ablation of RIPK4 allows epidermal cornification but impairs skin barrier formation. *The Journal of Investigative Dermatology*, *138*, 1268–1278.
Van Otterloo, E., Williams, T., & Artinger, K. B. (2016). The old and new face of craniofacial research: How animal models inform human craniofacial genetic and clinical data. *Developmental Biology*, *415*, 171–187.
Vannier, C., Mock, K., Brabletz, T., & Driever, W. (2013). Zeb1 regulates E-cadherin and Epcam (epithelial cell adhesion molecule) expression to control cell behavior in early zebrafish development. *The Journal of Biological Chemistry*, *288*, 18643–18659.
Vaziri Sani, F., Hallberg, K., Harfe, B. D., McMahon, A. P., Linde, A., & Gritli-Linde, A. (2005). Fate-mapping of the epithelial seam during palatal fusion rules out epithelial-mesenchymal transformation. *Developmental Biology*, *285*, 490–495.
Vernersson Lindahl, E., Garcia, E. L., & Mills, A. A. (2013). An allelic series of Trp63 mutations defines TAp63 as a modifier of EEC syndrome. *American Journal of Medical Genetics. Part A*, *161A*, 1961–1971.
Vieira, A. R. (2021). A need for updating the research agenda for cleft lip and palate and extending the cleft team model to dental medicine. *Frontiers in Dental Medicine*, *2*, 678477.
Wang, Y., Song, L., & Zhou, C. J. (2011). The canonical Wnt/beta-catenin signaling pathway regulates Fgf signaling for early facial development. *Developmental Biology*, *349*, 250–260.
Warzecha, C. C., & Carstens, R. P. (2012). Complex changes in alternative pre-mRNA splicing play a central role in the epithelial-to-mesenchymal transition (EMT). *Seminars in Cancer Biology*, *22*, 417–427.
Waterman, R. E., Ross, L. M., & Meller, S. M. (1973). Alterations in the epithelial surface of A-Jax mouse palatal shelves prior to and during palatal fusion: A scanning electron microscopic study. *The Anatomical Record*, *176*, 361–375.

Werth, M., Walentin, K., Aue, A., Schonheit, J., Wuebken, A., Pode-Shakked, N., et al. (2010). The transcription factor Grainyhead-like 2 regulates the molecular composition of the epithelial apical junctional complex. *Development*, *137*, 3835–3845.
Winter, M., Lodygin, D., Verdoodt, B., & Hermeking, H. (2016). Deletion of 14-3-3sigma sensitizes mice to DMBA/TPA-induced papillomatosis. *Oncotarget*, 7, 46862–46870.
Xin, Y., Lu, Q., & Li, Q. (2011). IKK1 control of epidermal differentiation is modulated by notch signaling. *The American Journal of Pathology*, *178*, 1568–1577.
Xu, X., Han, J., Ito, Y., Bringas, P., Jr., Urata, M. M., & Chai, Y. (2006). Cell autonomous requirement for Tgfbr2 in the disappearance of medial edge epithelium during palatal fusion. *Developmental Biology*, *297*, 238–248.
Xu, J., Liu, H., Lan, Y., Aronow, B. J., Kalinichenko, V. V., & Jiang, R. (2016). A Shh-Foxf-Fgf18-Shh molecular circuit regulating palate development. *PLoS Genetics*, *12*, e1005769.
Yang, A., Kaghad, M., Wang, Y., Gillett, E., Fleming, M. D., Dotsch, V., et al. (1998). p63, a p53 homolog at 3q27-29, encodes multiple products with transactivating, death-inducing, and dominant-negative activities. *Molecular Cell*, *2*, 305–316.
Yang, A., Schweitzer, R., Sun, D., Kaghad, M., Walker, N., Bronson, R. T., et al. (1999). p63 is essential for regenerative proliferation in limb, craniofacial and epithelial development. *Nature*, *398*, 714–718.
Yoshida, M., Shimono, Y., Togashi, H., Matsuzaki, K., Miyoshi, J., Mizoguchi, A., et al. (2012). Periderm cells covering palatal shelves have tight junctions and their desquamation reduces the polarity of palatal shelf epithelial cells in palatogenesis. *Genes to Cells*, *17*, 455–472.
Young, J. I., Slifer, S., Hecht, J. T., & Blanton, S. H. (2021). DNA methylation variation is identified in monozygotic twins discordant for non-syndromic cleft lip and palate. *Frontiers in Cell and Development Biology*, *9*, 656865.
Yu, Z., Lin, K. K., Bhandari, A., Spencer, J. A., Xu, X., Wang, N., et al. (2006). The Grainyhead-like epithelial transactivator Get-1/Grhl3 regulates epidermal terminal differentiation and interacts functionally with LMO4. *Developmental Biology*, *299*, 122–136.
Yu, Y., Zuo, X., He, M., Gao, J., Fu, Y., Qin, C., et al. (2017). Genome-wide analyses of non-syndromic cleft lip with palate identify 14 novel loci and genetic heterogeneity. *Nature Communications*, *8*, 14364.
Zhou, J., Gao, Y., Lan, Y., Jia, S., & Jiang, R. (2013). Pax9 regulates a molecular network involving Bmp4, Fgf10, Shh signaling and the Osr2 transcription factor to control palate morphogenesis. *Development*, *140*, 4709–4718.
Zhu, X. J., Liu, Y., Yuan, X., Wang, M., Zhao, W., Yang, X., et al. (2016). Ectodermal Wnt controls nasal pit morphogenesis through modulation of the BMP/FGF/JNK signaling axis. *Developmental Dynamics*, *245*, 414–426.

CHAPTER THREE

Mammalian uterine morphogenesis and variations

Diana A. Machado[a,b], Alejandra E. Ontiveros[a,c], and Richard R. Behringer[a,b,c,*]

[a]Department of Genetics, University of Texas M.D. Anderson Cancer Center, Houston, TX, United States
[b]Program in Genetics and Epigenetics, MD Anderson Cancer Center UT Health Graduate School of Biomedical Sciences, Houston, TX, United States
[c]Program in Developmental Biology, Baylor College of Medicine, Houston, TX, United States
[*]Corresponding author: e-mail address: rrb@mdanderson.org

Contents

Current Topics in Developmental Biology, Volume 148
ISSN 0070-2153
https://doi.org/10.1016/bs.ctdb.2021.12.004

Abstract

In eutherian and marsupial mammals, the site of embryo implantation and gestation is the uterus. Uterine morphologies vary between mammalian species. For example, laboratory mice have a bipartite uterus with two uterine horns and a single cervix, whereas humans have a simplex uterus with a single chamber and single cervix. The precursor tissue of the uterus, oviducts, and upper vagina is the Müllerian duct epithelium and its adjacent mesenchyme. Morphological variation between species is established during embryogenesis by species-specific differences in Müllerian duct fusion at the midline, growth, and differentiation. In humans, alterations in Müllerian duct development can lead to variations in uterine morphology that correlate with increased risks of miscarriage and infertility. Here we review the developmental genetic factors that regulate Müllerian duct development, uterine morphogenesis, and human uterine variation.

1. Introduction

1.1 The female reproductive tract organs

The reproductive tract organs of female eutherian and marsupial mammals include the Fallopian tubes/oviducts, uterus, cervix, and vagina (Fig. 1). The oviduct is the organ that receives ovulated oocytes, serves as a conduit for sperm, is the site of fertilization, and is a tube for preimplantation embryo transport to the uterus. The uterus is the site for embryo implantation, placentation, and most of gestation. At term, the fetus is delivered through the

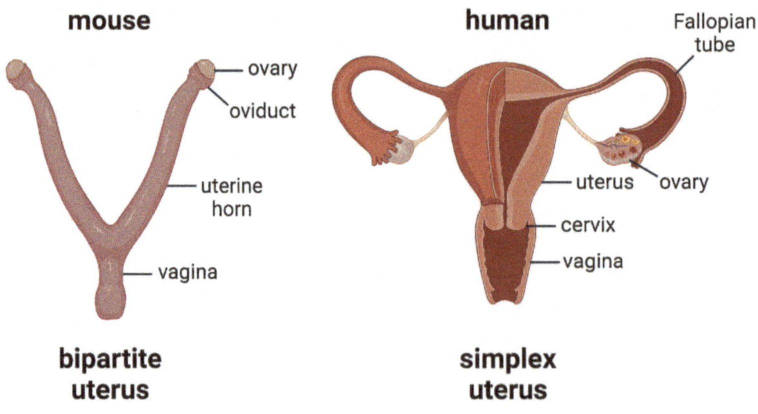

Fig. 1 Bipartite and simplex uterus morphologies. The mouse has been used extensively as a research organism to study female reproductive tract organ development. The mouse has a bipartite uterine morphology with two uterine horns and a single cervix. The human has a simplex uterine morphology with a single uterine body and a single cervix.

vaginal canal. Variation from the typical female reproductive tract organs for a given species can limit or block reproduction.

1.2 Wolffian and Müllerian ducts give rise to reproductive tract organs

During the development of the mammalian embryo, both XX and XY individuals develop two pairs of genital ducts, the Wolffian (mesonephric) and Müllerian (paramesonephric) ducts, within the mesonephroi adjacent to the differentiating gonads (Belle et al., 2017; Mullen & Behringer, 2014) (Fig. 2). The Wolffian ducts with their associated mesenchyme can

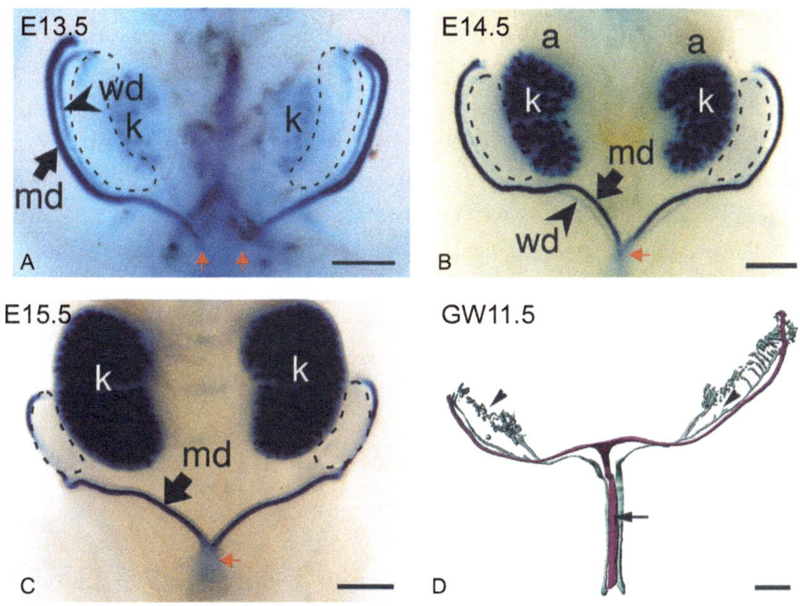

Fig. 2 Müllerian duct development. (A–C) X-gal stained *Lhx1-lacZ* female urogenital region from E13.5 to E15.5. At E13.5, the Müllerian ducts have elongated and fused to the urogenital sinus (red arrows). At E14.5 and 15.5, the two Müllerian ducts have fused to each other adjacent to the urogenital sinus (red arrows). Dotted outline, ovary; a, adrenals; k, metanephric kidneys; md, Müllerian ducts; wd, Wolffian duct. Scale bar, 500 μm. (D) Three-dimensional reconstruction of female human embryo at gestation week (GW) 11.5, showing fluorescently-labelled Müllerian (pink) and Wolffian ducts (blue). Arrow, fused Müllerian ducts; arrowheads, degenerating Wolffian ducts. Scale bar, 500 μm. *With permission from Kobayashi, A., Shawlot, W., Kania, A., & Behringer, R. R. (2004). Requirement of Lim1 for female reproductive tract development.* Development, 131, *539–549; Belle, M., Godefroy, D., Couly, G., Malone, S.A., Collier, F., Giacobini, P., & Chedotal, A. (2017). Tridimensional visualization and analysis of early human development.* Cell, 169.

differentiate into male-specific organs, the epididymides, vas deferentia, and seminal vesicles. The Müllerian ducts with their associated mesenchyme can differentiate into female-specific organs, the oviducts/fallopian tubes, uterus, cervix, and upper portion of the vagina. Because these structures occur together in both sexes, each individual has the potential to develop both male and female reproductive tract organs independent of their genetic sex during early embryonic development. However, only one genital duct pair will differentiate while the other genital duct pair will degenerate as developmental time progresses. In males, the Wolffian ducts will differentiate, whereas the Müllerian ducts will be eliminated by the action of anti-Müllerian hormone (AMH) secreted by Sertoli cells of the fetal testes. In females, the Müllerian ducts will differentiate into the oviducts, uterus, and upper vagina and the Wolffian ducts will be eliminated by the lack of testosterone.

Female reproductive tract formation can be divided into multiple steps: (1) Müllerian duct cell specification of the anterior mesonephric surface epithelium, (2) Müllerian duct invagination, (3) Müllerian duct elongation along the Wolffian duct, (4) Müllerian duct fusion to the urogenital sinus, (5) Müllerian duct fusion to each other at the midline, (6) regression of the uterine septum (Orvis & Behringer, 2007) (Fig. 2A–C). The early stages of Müllerian duct development are conserved across mammalian species; however, the last two steps vary across mammalian species (Kobayashi & Behringer, 2003).

1.3 Wolffian ducts are required for Müllerian duct formation

Although the Wolffian ducts are ultimately lost in female embryos, there is an intimate and functional role for the Wolffian ducts in Müllerian duct development. The Wolffian ducts form within the intermediate mesoderm. In the mouse, the Wolffian ducts begin forming around embryonic day (E) 9.0 and complete formation occurs at E11.0. Soon thereafter at E11.5, Müllerian duct development begins with the differentiation of surface epithelial cells at the anterior end of the mesonephros (Orvis & Behringer, 2007). There is an initial invagination of the surface epithelium of the anterior mesonephros at E11.75. Müllerian duct formation proceeds in an anterior to posterior direction, elongating an epithelial cord/tube within the mesonephric mesenchyme to fuse with the urogenital sinus by E13.5 (Guioli, Sekido, & Lovell-Badge, 2007; Huang, Orvis, Kwan, & Behringer, 2014; Orvis & Behringer, 2007) (Fig. 2A). During Müllerian duct elongation, the posterior tip of the Müllerian duct is located adjacent to the Wolffian duct (Orvis & Behringer, 2007) and the Müllerian duct uses the Wolffian duct as a physical guide towards the urogenital sinus. These observations indicate a very close physical relationship between the forming Müllerian duct and the Wolffian duct. Experiments in chick embryos demonstrated that alteration of the Wolffian duct led to defects in Müllerian duct development, i.e. Müllerian duct formation requires the Wolffian duct, as alteration of the Wolffian duct led to defects in Müllerian duct development (Grünwald, 1937). Further evidence that the Wolffian duct is required for Müllerian duct development comes from studies showing Müllerian duct loss secondary to Wolffian duct degeneration observed in mice with a Wolffian duct-specific knockout of *Lhx1* (Kobayashi et al., 2005).

Along with physical interactions, molecular signals from the Wolffian duct are also known to be crucial for Müllerian duct development. One factor expressed by the Wolffian duct epithelium that is essential for the elongation of the Müllerian duct is WNT9B. *Wnt9b* is expressed in the Wolffian duct epithelium but not in the adjacent Müllerian ducts (Carroll, Park, Hayashi, Majumdar, & McMahon, 2005). *Wnt9b*-null mice form Wolffian ducts but the Müllerian ducts do not elongate, suggesting that WNT9B secreted by the Wolffian duct is required for Müllerian duct elongation.

1.4 Müllerian duct fusion to the urogenital sinus and subsequent morphogenesis

The tip of the Müllerian duct elongates caudally in response to WNT9B signaling from the Wolffian duct while maintaining physical contact with Wolffian duct until it fuses with the urogenital sinus at E13.5 (Carroll et al., 2005; Kobayashi, Shawlot, Kania, & Behringer, 2004). The Müllerian duct and urogenital sinus form a transitional plate called the sinovaginal bulb which will become the upper two thirds of the vagina (Cunha et al., 2018). When this transitional plate forms, it is speculated that fusion between the two Müllerian ducts initiates. The two ducts fuse to form a single cavity that will become the caudal portion of the uterus (uterine corpus) in mice (Cunha et al., 2018). Another important step during Müllerian duct fusion is the regression of the uterine, cervical, and vaginal septum (Cunha et al., 2018). Failure of Müllerian duct fusion at the midline or regression of the septum results in variations in uterine morphology, for example, a double cervix and/or double vagina.

1.5 Genes that regulate Müllerian duct formation

1.5.1 Transcription factors

Emx2, *Lhx1*, and *Pax2* encode transcription factors that are expressed in the Müllerian duct and Wolffian duct epithelium (Dressler, Deutsch, Chowdhury, Nornes, & Gruss, 1990; Kobayashi et al., 2004; Miyamoto, Yoshida, Kuratani, Matsuo, & Aizawa, 1997). Individual whole-body knockouts in these genes result in an absence of kidneys and the loss of male and female reproductive tract organs (Kobayashi et al., 2004; Miyamoto et al., 1997; Torres, Gomez-Pardo, Dressler, & Gruss, 1995). The loss of both Wolffian duct- and Müllerian duct-derived tissues suggests that the Müllerian defects may be secondary to Wolffian duct loss. These three transcription factor genes are co-expressed and their loss-of-function results in very similar reproductive tract phenotypes, suggesting that they likely function in the same genetic pathway.

1.5.2 Lhx1

Lhx1 (also known as *Lim1*) encodes a LIM homeodomain transcription factor. *Lhx1*-null mice lack anterior head structures, kidneys, and male and female reproductive tract organs (Kobayashi et al., 2005, 2004; Shawlot & Behringer, 1995). During mouse urogenital development, *Lhx1* is expressed in the intermediate mesoderm at E7.5 and expression is subsequently restricted to the Wolffian duct (Barnes, Crosby, Jones, Wright, & Hogan, 1994). *Lhx1* is also expressed in the epithelium of the Müllerian duct (Kobayashi et al., 2004), the developing mesonephros and metanephros (Barnes et al., 1994; Fujii et al., 1994; Karavanov, Karavanova, Perantoni, & Dawid, 1998). These expression data indicate that *Lhx1* is involved in multiple aspects of urogenital system development, including the reproductive tract organs.

Conditional deletion of *Lhx1* in the Wolffian duct after the Wolffian duct has initially formed resulted in the loss of the Wolffian duct, demonstrating that *Lhx1* is required intrinsically for maintaining the Wolffian duct (Kobayashi et al., 2005). The loss of the Wolffian duct also resulted in abnormalities in the Müllerian ducts, supporting the notion that the Wolffian duct is required for Müllerian duct formation.

The function of *Lhx1* specifically in the Müllerian duct epithelium was tested using *Wnt7a-Cre* transgenic mice that express Cre recombinase in the Müllerian duct epithelium (Huang et al., 2014; Winuthayanon, Hewitt, Orvis, Behringer, & Korach, 2010). Müllerian duct epithelium-specific deletion of *Lhx1* resulted in viable mutant females. However, the mutant females were infertile because they lacked a uterus. They also had short oviducts that appeared grossly normal in structure and histology. Developmental studies revealed that the most anterior Müllerian duct formed but was much shorter than controls, suggesting a block in Müllerian duct elongation (Huang et al., 2014). This block in elongation resulted in the loss of the endometrium and inner circular myometrium of the uterus, but the outer longitudinal myometrium still formed. Time-lapse analysis of Müllerian duct elongation in controls and mutants and molecular studies suggested that *Lhx1* acts autonomously in epithelial progenitor cells for Müllerian duct elongation. These experiments identify LHX1 as a transcription factor that is specifically required in Müllerian duct epithelial cells for uterine and oviduct development. These studies also indicate that epithelial-mesenchymal interactions are essential for the differentiation of tissue compartments in the uterus (Cunha, Cooke, & Kurita, 2004).

1.6 Secreted proteins

1.6.1 Wnt4

Wnt4 is expressed in the developing urogenital system (Kispert, Vainio, & McMahon, 1998; Stark, Vainio, Vassileva, & McMahon, 1994). *Wnt4*-null male and female mice do not form Müllerian ducts (Vainio, Heikkila, Kispert, Chin, & McMahon, 1999). Studies of *Lhx1* expression in *Wnt4*-null mice suggest that the initial specification of Müllerian duct precursor cells occurs but there is a block in elongation (Kobayashi et al., 2004). These studies suggest that *Wnt4* is a mesenchyme-secreted factor that is required for Müllerian duct development, highlighting the importance of mesenchymal-epithelial interactions.

1.7 Membrane-associated proteins

1.7.1 Greb1l

GREB1L (growth regulation by estrogen in breast cancer 1-like) encodes a predicted transmembrane protein that is believed to be involved in retinoic acid signaling (Brophy et al., 2017). Autosomal dominant mutations in *GREBL1* are found in individuals with Deafness Autosomal Dominant 80 and Renal Hypodysplasia/Aplasia 3 (De Tomasi et al., 2017; Sanna-Cherchi et al., 2017; Schrauwen et al., 2018, 2020). *GREBL1* mutations are also associated with uterine aplasia (Barffour & Kwarkoh, 2021; Herlin et al., 2019; Jacquinet et al., 2020). *Greb1l*-null male and female mice lack kidneys and reproductive tract organs. At E13.5, Wolffian and Müllerian ducts were not present (De Tomasi et al., 2017). The lack of Müllerian ducts may be secondary due to the absence of the Wolffian ducts. This mouse model suggests that humans with *GREBL1* mutations develop uterine aplasia due to a primary defect in Wolffian duct formation alterations (Table 1).

1.7.2 Lhfpl2

Vgim (*vaginal imperforation*) is a spontaneous recessive mutation isolated at the Jackson Laboratory that causes eye and distal reproductive tract abnormalities in male and female mice. Whole exome sequencing of *vgim* mice identified a G102E missense mutation in *Lhfpl2* (Fairfield et al., 2011). *Lhfpl2* (lipoma HMGIC fusion partner-like 2) encodes a predicted transmembrane protein whose function is currently unknown. *Lhfpl2 G102E* homozygous female mice had an imperforate vagina and two separate uterine horns filled with fluid (Zhao, Zhou, Li, Dudley, & Ye, 2016). The Müllerian ducts of the mutants did not elongate sufficiently to reach and fuse with the urogenital sinus. In wild-type embryos, the Müllerian ducts will fuse to each other at the midline to form the uterine corpus (Fig. 2B and C). In the *Lhfpl2 G102E* homozygous female embryos, the Müllerian ducts did not fuse to each other. This consequently led to an "unmerged upper vagina" and lower vaginal agenesis (Zhao et al., 2016). *Lhfpl2* transcripts were found in every adult tissue screened by RT-PCR, suggesting that it is expressed ubiquitously. *In situ* hybridization of 4-week old wild-type females showed that *Lhfpl2* mRNA was detected in epithelial cells of the oviduct, luminal and glandular epithelial cells of the uterus, and epithelial cells of the vagina. In *Lhfpl2* mutant male mice, the Wolffian ducts did not fuse to the urogenital sinus, resulting in defects in the connection between the vas deferens and urethra. These findings suggest that fusion of the genital ducts to the urogenital sinus is essential for vagina formation in females. In addition, the blockage of Müllerian duct elongation precludes ductal fusion to each other to form the uterine corpus. The *Lhfpl2* mutant mice provide a unique model for male and female distal reproductive tract development.

1.7.3 Dlgh1

Dlgh1 (*discs large homolog 1*) is the mammalian ortholog of the *Drosophila* tumor suppressor gene *discs large 1* (Burgess, Rafael, Meisler, & Chamberlain, 1996; Woods, Hough, Peel, Callaini, & Bryant, 1996).

Table 1 Mouse models of female reproductive tract development.

Gene	Molecule encoded	Expression (mouse)	Reproductive tract phenotype (references)
Dlgh1	Scaffolding protein	WE	two separate uterine horns, cervical and vaginal aplasia (Iizuka-Kogo et al., 2007)
Emx2	Homeodomain transcription factor	ME, WE	No FRT (Miyamoto et al., 1997)
Greb1l	Transmembrane protein	unknown	No FRT (De Tomasi et al., 2017)
Itga3; Itga6	Transmembrane proteins	Not determined	Caudal regions of Wolffian and Müllerian ducts absent (De Arcangelis, Mark, Kreidberg, Sorokin, & Georges-Labouesse, 1999)
Lhfpl2	Transmembrane protein	Postnatal epithelial cells of the oviducts, uterus and vagina	vaginal agenesis and two separate uterine horns filled with fluid (Zhao et al., 2016)
Pax2	Homeodomain transcription factor	ME, WE	No FRT (Torres et al., 1995)
Rara; Rarb	Nuclear hormone receptor	Not determined	No FRT (Ghyselinck et al., 1997)
Wnt4	Wnt secreted protein	MM	No FRT (Vainio et al., 1999)
Wnt7a	Wnt secreted protein	ME	Partial homeotic transformation of oviduct to uterus, uterus to vagina, no uterine glands, abnormal mesenchyme differentiation Miller and Sassoon (1998); Parr & McMahon, 1998
Wnt9b	Wnt secreted protein	WE	Absence of uterus and upper vagina (Carroll et al., 2005)

FRT, female reproductive tract; ME, Mullerian duct epithelium; MM, Mullerian duct mesenchyme; WE, Wolffian duct epithelium.

Dlgh1 encodes a membrane-associated guanylate kinase (Roberts, Delury, & Marsh, 2012). In *Drosophila*, *dlg1* regulates cell polarity, cell-cell adhesion, and cell proliferation (Laprise, Viel, & Rivard, 2004; Woods & Bryant, 1991; Woods et al., 1996). *Dlgh1* mutant mice die soon after birth with cleft palate, hypoplastic kidneys, shortened ureters, and in females there was vaginal aplasia (Caruana & Bernstein, 2001; Iizuka-Kogo, Ishidao, Akiyama, & Senda, 2007). Female *Dlgh1*-null mice showed a truncation of the Müllerian duct without fusion to the urogenital sinus by E13.5. Additionally, the Müllerian ducts did not fuse laterally to each other at the midline by E15.5 (Iizuka-Kogo et al., 2007). At E18.5 female *Dlgh1*-null mice had two separate uterine horns and cervical and vaginal aplasia. *Dlgh1*-null male and female mice appeared to have proper Wolffian duct formation and elongation. However, the Wolffian ducts do not separate from the ureter or fuse to the urogenital sinus and connect with the bladder. These findings demonstrate that *Dlgh1* is essential for Müllerian duct elongation for cervical and vaginal development. The block in Müllerian duct elongation precluded fusion of the two Müllerian ducts to each other to form the uterine corpus. These studies suggest that *Dlgh1* is part of a genetic pathway for uterine development and morphogenesis.

1.8 Genes that regulate Müllerian duct differentiation

Wnt genes encode secreted glycoproteins that regulate stem cell fate, cell differentiation, and tissue growth through the canonical beta-catenin signaling pathway and the noncanonical Ca+2/PCP signaling pathway. WNT4, WNT5A, WNT7A, and WNT9B are important signaling molecules for Müllerian duct development (Deutscher & Hung-Chang Yao, 2007).

1.8.1 Wnt5a

Wnt5a is expressed in the mesenchyme associated with the uterus, cervix and vagina (Miller, Pavlova, & Sassoon, 1998). *Wnt5a-null* mice die soon after birth with shortened body and limb axes (Yamaguchi, Bradley, McMahon, & Jones, 1999). *Wnt5a*-null female mice were found to have shortened coiled uterine horns and also lack a cervix and vagina (Mericskay, Kitajewski, & Sassoon, 2004). Transplant experiments demonstrated that *Wnt5a* is required for uterine gland formation (Mericskay et al., 2004). Thus, mesenchyme-expressed *Wnt5a* influences the development of the female reproductive tract organs.

1.8.2 *Wnt7a*

Wnt7a is expressed in the Müllerian duct epithelium and subsequently the epithelium of the postnatal uterus (Hayashi et al., 2011; Parr & McMahon, 1998). *Wnt7a*-null mice are viable with dorsal-ventral limb alterations (Parr & McMahon, 1995). *Wnt7a* is also required during urogenital development in both males and females for distinct reasons (Parr & McMahon, 1998). *Wnt7a-null* adult female mice have shortened uterine horns that are thinner in diameter compared to controls. In addition, there is a posterior shift in the identity of the segments of the reproductive tract organs. The histoarchitecture of the oviduct appears like the uterus. The mutant uterus has a stratified epithelium like the vagina instead of a simple columnar epithelium. These findings suggest that *Wnt7a* regulates the regional identity of the female reproductive tract organs. Interestingly, *Hoxa10* and *Hoxa11* expression is down-regulated in the postnatal uterus of *Wnt7a-null* females.

Wnt7a-null male mice have fully differentiated male external and internal features but they also have female reproductive organs (uterus and oviducts) because the Müllerian ducts do not regress (Parr & McMahon, 1998). *Wnt7a*-null male embryos do not express anti-Müllerian hormone receptor type 2 (*Amhr2*) in the Müllerian duct mesenchyme. Thus, loss of *Wnt7a* makes the Müllerian duct mesenchyme incompetent to respond to AMH for Müllerian duct regression (Mishina et al., 1996). These findings indicate that WNT7A secreted by the Müllerian duct epithelium induces the transcription of *Amhr2* in the adjacent mesenchyme, inducing competency to respond to AMH.

1.9 Human Müllerian duct development

Müller (1830) published his studies describing the genital ducts in diverse species of vertebrates, including embryos (Müller, 1830). Since this initial description, relatively little has been discovered about Müllerian duct development in human embryos because of the scarcity of specimens and the challenges of determining the gestational age of human embryos from abortus specimens (Cunha et al., 2018; Morgan, 2004). O'Rahilly and Müller (1987) detailed the development of the reproductive organs in the Carnegie staging system to provide a detailed morphological account of its development in the human embryo. Between stages 16 and 17, the Müllerian ducts appear as invaginations of thickened coelomic epithelium over the mesonephros (O'Rahilly, Müller, & Streeter, 1987). At stage 18, the Müllerian ducts have elongated in a caudal direction on the lateral side of the urogenital ridges (O'Rahilly et al., 1987). The infoldings of the Müllerian ducts become the ostium of the "uterine tube" (O'Rahilly et al., 1987). By stage 23, the Müllerian ducts have reached the urogenital sinus and fused with each other along the medial plane (Koff, 1933; O'Rahilly et al., 1987; Pillet, 1967).

Recently, Belle et al. (2017) used fluorescent light sheet microscopy to generate three-dimensional reconstructions of developing organs in human embryos. These studies included the developing urogenital system in male and female embryos. By gestation week 7 (GW7), the urogenital ridges are undifferentiated and both Wolffian and Müllerian ducts are present. By GW8, the caudal tip of the Müllerian duct is in contact with the Wolffian duct. At this stage, the Müllerian ducts have elongated ~75% of the distance but have not yet reached the urogenital sinus (Belle et al., 2017). By GW9.5, the caudal tips of the Müllerian ducts further elongate caudally but still have not fused with the urogenital sinus. Right before GW10, the Müllerian ducts have reached the urogenital sinus and subsequently fused to each other, creating a single structure caudal to the point of fusion. At GW10, this structure extends medially in between the Wolffian ducts and a uterine septum separates the lumina of the two conjoined Müllerian ducts. Concurrently, the fused Müllerian ducts form the uterovaginal canal to the urogenital sinus. Similar Müllerian and Wolffian duct morphologies were observed by E-cadherin whole mount immunofluorescent imaging in a GW9.5 female embryo (Cunha et al., 2018). At GW11.5, there is significant growth where the Müllerian ducts have fused. It is speculated that Müllerian duct fusion initiates near the caudal tips and fuses cranially (Campbell, 1952; Davies & Walpole, 1949). At GW13, the cranial region of the Müllerian ducts will differentiate into the Fallopian tubes (Belle et al., 2017). Despite what we know about Müllerian duct development in the human embryo, many steps after Müllerian duct formation are poorly understood, including: what drives Müllerian ducts to fuse to the urogenital sinus? What mechanisms induce a vaginal septum? When and how do the uterine, cervical, and vaginal septum regress? The improper fusion of the Müllerian ducts or regression of the uterine, cervical, or vaginal septum is the cause of multiple congenital reproductive tract variations.

1.10 Other mammalian uterine morphologies

Mammals are distinguished by having mammary glands, a neocortex, hair or fur, and three middle ear bones. The class of mammals is extensive and highly variable. Perhaps it is not surprising then that there are numerous uterine types among mammalian species (Kobayashi & Behringer, 2003). Humans have a simplex uterus, a single uterus and single cervix, and the laboratory mouse has a bipartite uterus, two uterine horns with a single cervix (Fig. 1). Müllerian duct formation is similar across species, however differences in uterine types result from differences in the extent of Müllerian duct midline fusion (Mossman, 1987). It is thought that the anterior extent of Müllerian duct fusion determines the gross morphological uterine type. The extent of Müllerian duct fusion may be absent, minimal, partial, or extensive, and this will define the gross morphological characteristics of the adult female reproductive tract: duplex, bipartite, bicornuate, or simplex (Fig. 3). Morphological variations of the female reproductive tract organs can also be found within a single order of mammals (Mossman, 1987). For example, in bats, *Carollia perspicillata* has a simplex uterus, while *Myotis nigricans* has a bicornuate uterus with long uterine horns, and *Mormopterus planiceps* also has a bicornuate uterus but with short uterine horns (Beguelini, Santiago, Morielle-Versute, & Taboga, 2020; Krutzsch & Crichton, 1987; Rasweiler, Oliveira, & Badwaik, 2002).

1.10.1 Duplex uterus

A duplex uterus has two separate uterine horns each with a cervix. Among eutherian mammals, lagomorphs (rabbits, hares, and pikas) have a duplex uterus (Hafez, 1970). The two uterine horns descend into the cloaca, a

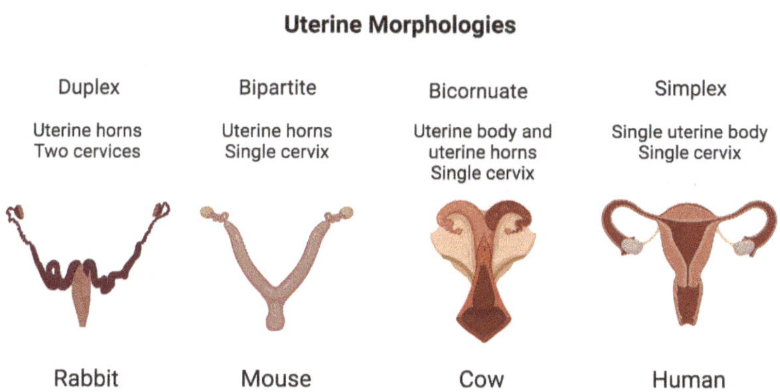

Fig. 3 Types of uterine morphologies among eutherian mammals. There are four types of uterine morphologies among eutherian mammals: duplex, bipartite, bicornuate, and simplex. These uterine morphology types are defined by cervix number and the presence or absence of uterine horns and uterine body.

common cavity, where the digestive, urinary, and reproductive tracts meet. The cloaca is the common posterior orifice in these animals.

Marsupial mammals, including kangaroos, opossums, koalas, and wombats (Rodger, 1991; Schimming, Cesario, & Matheus, 2018), have an interesting female reproductive tract compared to eutherian mammals. Marsupials also have a duplex uterus with two separate uteri each with a cervix, though unlike eutherian mammals, they also have a vaginal cul-de-sac with two lateral vaginas and one median vagina (the future birth canal), and a common urogenital canal (Renfree, 1993; Wick & Kress, 2002).

1.10.2 Bipartite uterus

A bipartite uterus consists of two uterine horns with a single cervix connected to the vagina. Rodents such as mice, rats, guinea pigs, squirrels, hamsters, and capybaras have a bipartite uterus morphology (Miglino, dos Santos, Kanashiro, & dos Santos Ferraz, 2013; Spencer, Hayashi, Hu, & Carpenter, 2005). Müllerian duct fusion is very limited which results in two separate uterine horns. The laboratory mouse is a primary research organ for biomedical studies and has a bipartite uterine morphology.

1.10.3 Bicornuate uterus

A bicornuate uterus consists of two uterine horns, a common uterine corpus, a single cervix, and a single vagina. Mammals with a bicornuate uterus morphology include porcines, canines, felines, ruminants, equines, and prosimian primates (Patten, 1948). Müllerian duct fusion extends more anteriorly to varying degrees. Shorter fusion results in long uterine horns (porcine), while greater fusion results in relatively short uterine horns (ruminants) (Spencer et al., 2005). Congenital uterine variations are also common across different mammalian species. Uterus didelphys in ruminants is associated with infertility (Raggio, Lefebvre, Poitras, Vaillancourt, & Champagne, 2006; Timurkaan & Ozer, 2002). Ishiyama et al. (2019) examined 1054 Holstein cattle and found 2.83% of individuals had evidence of incomplete Müllerian duct fusion. The rate of incomplete Müllerian duct fusion varied across breeds: Meuse Rhine Issel (3.77%), Swedish Highland (2.47%), and Simmental (0.008%) (Ishiyama et al., 2019).

1.10.4 Simplex uterus

A simplex uterus consists of a single uterine body, a single cervix, and a single vagina. In humans and non-human primates, Müllerian duct fusion and septal resorption are extensive and result in a single uterine body.

Other mammals with a simplex uterus include some bat species, including *Carollia perspicillata*, the short-tailed fruit bat that serves as an important research organism for evolutionary and developmental biology studies (Cretekos et al., 2005). This unique and accessible bat species may be useful to model human simplex uterus development.

2. Human uterine variation

2.1 Prevalence of human uterine variation and correlation with miscarriage and infertility

Human uterine variations are those morphologies that diverge from the typical simplex uterus. The American Society of Reproductive Medicine has defined human uterine variation into 6 morphologically-based classes: Class I, Müllerian agenesis; Class II, Unicornuate uterus; Class III, Uterus didelphys; Class IV, Bicornuate uterus; Class V, Septated uterus; Class VI, DES uterus. These variations can occur if there are alterations in Müllerian duct formation, lack or partial fusion between the Müllerian ducts, and/or regression of the uterine, cervical, and vaginal septum. Divergence in any of the steps from typical development can give rise to uterine variation, including uterine aplasia, unicornuate uterus, septate uterus, bicornuate uterus, or didelphys uterus (Saravelos, Cocksedge, & Li, 2008) (Fig. 4).

Uterine variations may increase the rate of miscarriage, preterm delivery, and other complications of pregnancy. A systematic review of the literature encompassing nearly 90,000 women found that 5.5% of the general population had a uterine morphology variation (Chan et al., 2011). For those women with a history of miscarriage, the incidence of uterine variation was 13.3% but in women with miscarriage in association with infertility was 24.5% (Chan et al., 2011). These data indicate that women with a medical history of miscarriage and infertility have significantly higher chance of having a uterine variation.

2.2 Müllerian duct formation

Failure of Müllerian duct specification, invagination, elongation or persistence can result in uterine aplasia. However, if only one Müllerian duct forms, then this variation is called unicornuate uterus (Fig. 4). Only 0.1% of the general population have a unicornuate uterus but those who have a history of miscarriage combined with infertility have a higher prevalence of this uterine variation at 3.1% (Chan et al., 2011).

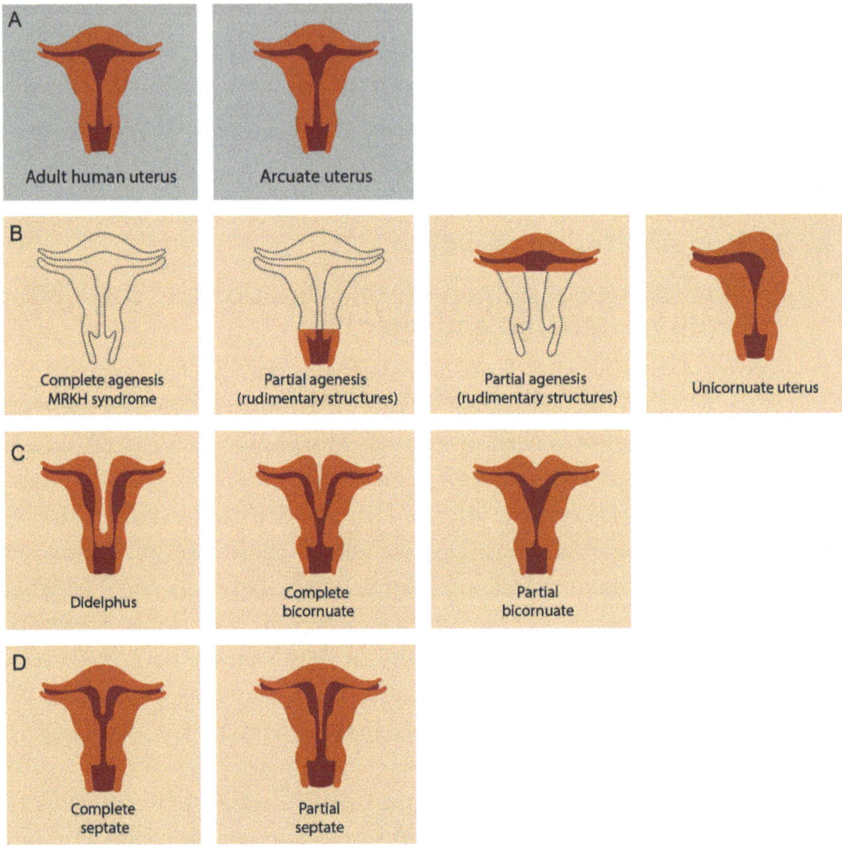

Fig. 4 Human uterine variation. Diagrams of different types of uterine variation grouped into (A), typical and mild variant (arcuate), (B), formation variants, (C), fusion variants, and (D), septal variants.

2.3 Müllerian duct fusion

If the Müllerian ducts fail to fuse to each other, then patients may have a didelphys uterus (Fig. 4). Didelphys uterus consists of two separate vaginas, cervices, and uterine horns. Didelphys uterus is present in 0.3% of the general population but is significantly higher 2.1% in women with a combined history of miscarriage and infertility (Chan et al., 2011). Furthermore, if the Müllerian ducts initiate fusion but do not fuse to completion, a common uterine corpus with two uterine horns results in a bicornuate uterus (Fig. 4). Although it is uncommon in the general population with 0.4% of prevalence, it is found in 4.7% of women with a combined history of miscarriage and infertility (Chan et al., 2011).

2.4 Uterine septal regression

When Müllerian ducts fuse to each other, there is an epithelial wall in the middle of what would become the lumen of the uterus. This uterine septum must be reabsorbed to generate a single uterine cavity. However, the uterine septum is not always reabsorbed and the resulting uterus is called a septate uterus (Fig. 4). A septate uterus may be partial, the lower region fused having a uterine corpus with uterine horns, or complete, where there are two cervixes and two uterine horns. A septate uterus is found in 5.3% of women with a history of miscarriage but in 15.4% of women with miscarriage combined with infertility, as compared to 2.3% in the general population (Chan et al., 2011).

3. Human uterine variation syndromes

3.1 Mayer-Rokitansky-Küster-Hauser syndrome

Mayer-Rokitansky-Küster-Hauser (MRKH) Syndrome is found in women with normal external sex differentiation and ovarian function. However, there is an absence of the uterus and upper portion of the vagina (Fontana, Gentilin, Fedele, Gervasini, & Miozzo, 2017; Liszewska-Kaplon, Strozik, Kotarski, Baglaj, & Hirnle, 2020). Thus, women with MRKH Syndrome are infertile. However, the presence of normal ovaries provides the opportunity for assisted reproductive technologies for these women to have a biological child. MRKH Syndrome is classified into two types. In MRKH Type I individuals the uterus and vagina are affected but not other organs, while MRKH Type II individuals also have other organs affected. These individuals are typically diagnosed in adolescence because of a failure to menstruate. MRKH is a rare condition found in 1:4000–5000 girls (Ledig & Wieacker, 2018). MRKH cases are sporadic; however, familial occurrence has been reported suggesting an autosomal-dominant inheritance pattern with low penetrance.

Microarray-based comparative genomic hybridization and multiplex ligation-dependent probe amplification analysis on MRKH patients have found consistent chromosomal duplications and deletions in regions 1q21.1, 16p11.2, 17q12, 22q11.21, and Xp22 (Fontana et al., 2017). Candidate genes for this syndrome found in these regions include *LHX1*, *TBX6*, and *RBM8A*. *LHX1* is a prime candidate gene for contributing to MRKH syndrome because Müllerian duct-specific knockout of *Lhx1* in mice results in uterine aplasia with a blind vaginal pouch

(Huang et al., 2014; Ledig & Wieacker, 2018). Mouse studies are yet to be performed to determine whether *Tbx6* or *Rbm8a* and other candidate genes are required for Mullerian duct development. More recently, exome sequencing identified variants in genes important for Müllerian and Wolffian duct development, including *BMP4*, *BMP7*, *EMX2*, *HOXA10*, *PAX8*, *TBX6* and *WNT9B* (Chen et al., 2021) (Table 2).

Table 2 Human syndromes with uterine variation.

Syndrome	Reproductive tract phenotype	Gene	References
Al-Awadi-Raas-Rothschild and Fuhrmann syndromes	Uterine aplasia and external genitalia variation	*WNT7A*	Woods et al. (2006), Kantaputra, Mundlos and Sripathomsawat (2010), Eyaid, Al-Qattan, Al Abdulkareem, Fetaini and Al Balwi (2011), Garavelli et al. (2011), Al-Qattan (2013)
Hand-Foot-Genital Syndrome	Varying degrees of incomplete Mullerian duct fusion, including longitudinal vaginal septum	*HOXA13*	Stern, Gall, Perry and Stimson (1969), Mortlock and Innis (1997), Goodman et al. (2000), Innis (2006)
Herlyn-Werner-Wünderlich syndrome	Uterovaginal duplication with an obstructed hemivagina and ipsilateral renal agenesis	Not determined	Purslow (1922), Orazi et al. (2007), Mittal (2016), Liang, Fu, Yin and Chang (2020)
Mayer-Rokitansky-Küster-Hauser Syndrome	Lack of uterus and upper vagina	*LHX1* and not determined	Bernardini et al. (2009), Cheroki et al. (2008), Ledig et al. (2012)
Renal cysts and diabetes syndrome	Uterine variation, including bicornuate uterus, uterine aplasia, unicornuate uterus, didelphic uterus, and hypoplastic uterus	*HNF1B*	Bellanne-Chantelot et al. (2004), Lindner et al. (1999), Nakayama et al. (2010), Oram et al. (2010)

3.2 Hand-foot-genital syndrome

Hand-foot-genital syndrome (HFGS) is an autosomal dominantly inherited condition that consists of distal limb variations and urogenital alterations. The urogenital alterations show varying degrees of incomplete Müllerian duct fusion to each other, including a longitudinal vaginal septum (Innis, 1993). Individuals are diagnosed by examination of hands and feet with additional imaging of the urogenital system. Removal of the longitudinal vaginal or uterine septum or the joining of a bicornuate uterus is often not recommended as the risks outweigh the benefits. HFGS individuals are at a significantly higher risk of premature labor, fetal loss, or stillbirth because of structural variation of the uterus. HFGS syndrome is caused by heterozygous mutations in *HOXA13*, encoding a homeodomain transcription factor (Mortlock & Innis, 1997). Often, a *HOXA13* pathogenic variant encoding an N-terminal polyalanine expansion is used to establish diagnosis because 60% of HFGS patients carry this variant (Innis, 1993) (Table 2).

3.3 Al-Awadi-Raas-Rothschild and Fuhrmann syndromes

Al-Awadi-Raas-Rothschild (limb/pelvis/uterus-hypoplasia/aplasia) syndrome (AARRS) and Fuhrmann syndrome (FS) are autosomal recessive syndromes that consist of limb and urogenital system variation (Al-Qattan, 2013). Fuhrmann Syndrome individuals show milder versions of these variations. AARRS and FS individuals have normal ovarian function with uterine aplasia in addition to altered external genitalia. Complete loss of *WNT7A* function has been associated with AARRS, whereas partial loss of *WNT7A* function appears to result in a milder presentation of the organ variations for Fuhrmann syndrome (Woods et al., 2006) (Table 2).

3.4 Herlyn-Werner-Wünderlich syndrome

Herlyn-Werner-Wünderlich syndrome (HWWS) was first described by Purslow (1922) as a uterovaginal duplication with an obstructed hemivagina accompanied by ipsilateral renal agenesis (Purslow, 1922). HWWS individuals first present with recurrent pelvic pain and dysmenorrhea at the onset of puberty shortly after menarche. Upon medical examination, these individuals are found to have didelphys or bicornuate uterus, absent right kidney, and a vaginal obstructing septum (Orazi, Lucchetti, Schingo, Marchetti, & Ferro, 2007). These conditions appear to be consistent with variable fusion of the Müllerian ducts at the midline. HWWS individuals undergo surgery to remove the vaginal septum and allow the outflow of menstruation to provide relief of painful complications. To date there are no gene mutations associated with this syndrome (Table 2).

3.5 HNF1 homeobox B

HNF1B (hepatic nuclear factor 1 homeobox B) encodes a homeodomain transcription factor (Bach, Mattei, Cereghini, & Yaniv, 1991). In one study, autosomal dominant mutations in *HNFB1* were found in women with uterine variation, including bicornuate uterus, uterine aplasia, unicornuate uterus, didelphys uterus, and hypoplastic uterus (Oram et al., 2010). The variety of phenotypes found in individuals with *HNF1B* mutations suggests diverse mechanisms that alter Müllerian duct development. No mutations were found in a cohort of 45 women with MRKH syndrome (Oram et al., 2010). These findings suggest that heterozygous mutations in *HNF1B* can alter different steps in uterine morphogenesis. *Hnf1b-null* mice die soon after implantation (Barbacci et al., 1999). Genital duct-specific knockouts could be used to examine the role of *Hnf1b* in uterine development.

4. Conclusion and final remarks

Uterine variations are found at a higher frequency among women who have infertility or miscarriage. The morphology of the uterus can be determined using non-invasive imaging modalities, but the developmental genetic cause of the variation cannot be determined because the event(s) has already occurred. However, the variation in uterine morphology can suggest potential developmental genetic mechanisms of causation. Although the laboratory mouse has a bipartite uterus and humans have a simplex uterus, mouse models have been very useful in identifying the various genetic and developmental mechanisms that cause female reproductive tract organ.

The Wolffian ducts are essential for guiding the Müllerian ducts and stimulating elongation to the urogenital sinus. Thus, a subset of human uterine variation may be due to primary alterations in the Wolffian duct that lead to secondary consequences in Müllerian duct elongation and fusion to the urogenital sinus. Once the Müllerian ducts have fused to the urogenital sinus the Müllerian ducts must then fuse at the midline to initiate the formation of the uterine corpus. Failure of Müllerian duct fusion to each other leads to a didelphys uterus. Finally, the extent of Müllerian duct fusion and septal resorption can lead to women with different forms of septate uterus.

Null alleles in mice have proven very useful to identify urogenital phenotypes, including those of the female reproductive tract organs (Carroll et al., 2005; Dressler et al., 1990; Kobayashi et al., 2004; Miyamoto et al., 1997). Likewise, conditional null alleles have been useful to bypass embryonic lethal phenotypes before reproductive tract formation (Huang et al., 2014; Kobayashi et al., 2005). The field has relied primarily on two Cre-expressing transgenic mouse lines for Wolffian duct- and Müllerian duct-specific knockouts. These include *Hoxb7-Cre* (Wolffian duct) and *Wnt7a-Cre* (Müllerian duct) (Winuthayanon et al., 2010; Yu, Carroll, & McMahon, 2002). *Hoxb7-Cre* is active in the Wolffian duct once it is formed but not earlier. In addition, Cre is also expressed in other tissues and organs. *Wnt7a-Cre* is active in the Müllerian duct as soon as it is formed but is also expressed in the limb, neural and other tissues. Cre expression in these other tissues may lead to lethality after birth, complicating postnatal analyses of the adult female reproductive tract organs. Thus, more transgenic/knock-in mouse lines that express Cre in a genital duct-restricted manner are needed.

Finally, exome sequencing of women with uterine variation should result in a catalog of gene alterations (Chen et al., 2021; Mikhael et al., 2021). These mutations could then be tested in mouse models to understand the developmental mechanisms that lead to the formation of these female human uterine variations.

References

Al-Qattan, M. M. (2013). Molecular basis of the clinical features of Al-Awadi-Raas-Rothschild (limb/pelvis/uterus-hypoplasia/aplasia) syndrome (AARRS) and Fuhrmann syndrome. *American Journal of Medical Genetics. Part A, 161A*, 2274–2280.

Bach, I., Mattei, M. G., Cereghini, S., & Yaniv, M. (1991). Two members of an HNF1 homeoprotein family are expressed in human liver. *Nucleic Acids Research, 19*, 3553–3559.

Barbacci, E., Reber, M., Ott, M. O., Breillat, C., Huetz, F., & Cereghini, S. (1999). Variant hepatocyte nuclear factor 1 is required for visceral endoderm specification. *Development, 126*, 4795–4805.

Barffour, I. K., & Kwarkoh, R. K. B. (2021). GREB1L as a candidate gene of Mayer-Rokitansky-Kuster-Hauser syndrome. *European Journal of Medical Genetics, 64*, 104158.

Barnes, J. D., Crosby, J. L., Jones, C. M., Wright, C. V. E., & Hogan, B. L. M. (1994). Embryonic expression of Lim-1, the mouse homolog of Xenopus Xlim-1, suggests a role in lateral mesoderm differentiation and neurogenesis. *Developmental Biology, 161*, 168–178.

Beguelini, M. R., Santiago, C. S., Morielle-Versute, E., & Taboga, S. R. (2020). Morphophysiological variations of the female reproductive organs of the vespertilionid bat Myotis nigricans during its different reproductive phases. *Theriogenology, 158*, 121–137.

Bellanne-Chantelot, C., Chauveau, D., Gautier, J.-F., Dubois-Laforgue, D., Clauin, S., Beaufils, S., et al. (2004). Clinical spectrum associated with hepatocyte nuclear factor-1B mutations. *Annals of Internal Medicine, 140*, 510–517.

Belle, M., Godefroy, D., Couly, G., Malone, S. A., Collier, F., Giacobini, P., et al. (2017). Tridimensional visualization and analysis of early human development. *Cell, 169*.

Bernardini, L., Gimelli, S., Gervasini, C., Carella, M., Baban, A., Frontino, G., et al. (2009). Recurrent microdeletion at 17q12 as a cause of Mayer-Rokitansky-Kuster-Hauser (MRKH) syndrome: Two case reports. *Orphanet Journal of Rare Diseases, 4*, 25. https://doi.org/10.1186/1750-1172-4-25. PMID: 19889212.

Brophy, P. D., Rasmussen, M., Parida, M., Bonde, G., Darbro, B. W., Hong, X. J., et al. (2017). A gene implicated in activation of retinoic acid receptor targets is a novel renal agenesis gene in humans. *Genetics, 207*, 215–228.

Burgess, D. L., Rafael, J. A., Meisler, M. H., & Chamberlain, J. S. (1996). Dlgh1, a mouse homolog of the drosophila discs-large gene, is located on chromosome 16. *Mammalian Genome*, 7, 623–624.

Campbell, J. G. (1952). Bicornuate uterus. *Canadian Medical Association Journal, 67*, 355–356.

Carroll, T. J., Park, J. S., Hayashi, S., Majumdar, A., & McMahon, A. P. (2005). Wnt9b plays a central role in the regulation of mesenchymal to epithelial transitions underlying organogenesis of the mammalian urogenital system. *Developmental Cell, 9*, 283–292.

Caruana, G., & Bernstein, A. (2001). Craniofacial dysmorphogenesis including cleft palate in mice with an insertional mutation in the discs large gene. *Molecular and Cellular Biology, 21*, 1475–1483.

Chan, Y. Y., Jayaprakasan, K., Zamora, J., Thornton, J. G., Raine-Fenning, N., & Coomarasamy, A. (2011). The prevalence of congenital uterine anomalies in unselected and high-risk populations: A systematic review. *Human Reproduction Update, 17*, 761–771.

Chen, N., Zhao, S., Jolly, A., Wang, L., Pan, H., Yuan, J., et al. (2021). Perturbations of genes essential for Mullerian duct and Wolffian duct development in Mayer-Rokitansky-Kuster-Hauser syndrome. *American Journal of Human Genetics, 108*, 337–345.

Cheroki C., Krepischi-Santos A.C.V., Szuhai K., Brenner V., Kim C.A.E., Otto P.A., et al., 2008. Genomic imbalances associated with mullerian aplasia, Journal of Medical Genetics, 45 (4), 228-232, 10.1136/jmg.2007.051839. Epub 2007 Nov 26.

Cretekos, C. J., Weatherbee, S. D., Chen, C. H., Badwaik, N. K., Niswander, L., Behringer, R. R., et al. (2005). Embryonic staging system for the short-tailed fruit bat, Carollia perspicillata, a model organism for the mammalian order Chiroptera, based upon timed pregnancies in captive-bred animals. *Developmental Dynamics, 233*, 721–738.

Cunha, G. R., Robboy, S. J., Kurita, T., Isaacson, D., Shen, J., Cao, M., et al. (2018). Development of the human female reproductive tract. *Differentiation, 103*, 46–65.

Cunha, G. R., Cooke, P. S., & Kurita, T. (2004). Role of stromal-epithelial interactions in hormonal responses. *Archives of Histology and Cytology, 67*, 417–434.

Davies, W. M., & Walpole, G. R. O. (1949). Bicornuate uterus. *The Medical Journal of Australia, 2*, 166–167.

De Arcangelis A, Mark M, Kreidberg J, Sorokin L, Georges-Labouesse E., 1999. Synergistic activities of alpha3 and alpha6 integrins are required during apical ectodermal ridge formation and organogenesis in the mouse. Development 126 (17), 3957-3968. doi: 10.1242/dev.126.17.3957. PMID: 10433923.

De Tomasi, L., David, P., Humbert, C., Silbermann, F., Arrondel, C., Tores, F., et al. (2017). Mutations in GREB1L cause bilateral kidney agenesis in humans and mice. *American Journal of Human Genetics, 101*, 803–814.

Deutscher, E., & Hung-Chang Yao, H. (2007). Essential roles of mesenchyme-derived beta-catenin in mouse Mullerian duct morphogenesis. *Developmental Biology, 307*, 227–236.

Dressler, G. R., Deutsch, U., Chowdhury, K., Nornes, H. O., & Gruss, P. (1990). Pax2, a new murine paired-box-containing gene and its expression in the developing excretory system. *Development, 109*, 787–795.

Eyaid W., Al-Qattan M.M., Al Abdulkareem I., Fetaini N., Al Balwi M., 2011. A novel homozygous missense mutation (c.610G>A, p.Gly204Ser) in the WNT7A gene causes tetra-amelia in two Saudi families. The American Journal of Medical Genetics, 155A (3):599-604. doi: 10.1002/ajmg.a.33717. Epub 2011 Feb 22.PMID: 21344627.

Fairfield, H., Gilbert, G. J., Barter, M., Corrigan, R. R., Curtain, M., Ding, Y., et al. (2011). Mutation discovery in mice by whole exome sequencing. *Genome Biology*, *12*, R86.

Fontana, L., Gentilin, B., Fedele, L., Gervasini, C., & Miozzo, M. (2017). Genetics of Mayer-Rokitansky-Kuster-Hauser (MRKH) syndrome. *Clinical Genetics*, *91*, 233–246.

Fujii, T., Pichel, J. G., Taira, M., Toyama, R., Dawid, I. B., & Westphal, H. (1994). Expression patterns of the murine LIM class homeobox gene lim1 in the developing brain and excretory system. *Developmental Dynamics*, *199*, 73–83.

Garavelli L, Wischmeijer A, Rosato S, Gelmini C, Reverberi S, Sassi S, et al., 2011. Al-Awadi-Raas-Rothschild (limb/pelvis/uterus-hypoplasia/aplasia) syndrome and WNT7A mutations: genetic homogeneity and nosological delineation. The American Journal of Medical Genetics. 155A (2):332-336. doi: 10.1002/ajmg.a.33793. Epub 2010 Dec 22.PMID: 21271649.

Ghyselinck NB, Dupé V, Dierich A, Messaddeq N, Garnier JM, Rochette-Egly C, et al., 1997. Role of the retinoic acid receptor beta (RARbeta) during mouse development. The International Journal of Developmental Biology 41 (3), 425-447. PMID: 9240560.

Goodman F.R., Bacchelli C., Brady A.F., Brueton L.A., Fryns J.P., Mortlock D.P., et al., 2000. Novel HOXA13 mutations and the phenotypic spectrum of hand-foot-genital syndrome. The American Journal of Human Genetics 67 (1), 197-202. doi: 10.1086/302961. Epub 2000 Jun 5.PMID: 10839976.

Grünwald, P. (1937). Zur Entwicklungsmechanik des Urogenitalsystems beim Huhn. *Wilhelm Roux' Archiv für Entwicklungsmechanik der Organismen*, *136*, 786–813.

Guioli, S., Sekido, R., & Lovell-Badge, R. (2007). The origin of the Mullerian duct in chick and mouse. *Developmental Biology*, *302*, 389–398.

Hafez, E. S. E. (1970). *Reproduction and breeding techniques for laboratory animals*. Philadelphia: Lea & Febiger.

Hayashi, K., Yoshioka, S., Reardon, S. N., Rucker, E. B., 3rd, Spencer, T. E., DeMayo, F. J., et al. (2011). WNTs in the neonatal mouse uterus: Potential regulation of endometrial gland development. *Biology of Reproduction*, *84*, 308–319.

Herlin, M. K., Le, V. Q., Hojland, A. T., Ernst, A., Okkels, H., Petersen, A. C., et al. (2019). Whole-exome sequencing identifies a GREB1L variant in a three-generation family with Mullerian and renal agenesis: A novel candidate gene in Mayer-Rokitansky-Kuster-Hauser (MRKH) syndrome. A case report. *Human Reproduction*, *34*, 1838–1846.

Huang, C. C., Orvis, G. D., Kwan, K. M., & Behringer, R. R. (2014). Lhx1 is required in Mullerian duct epithelium for uterine development. *Developmental Biology*, *389*, 124–136.

Iizuka-Kogo, A., Ishidao, T., Akiyama, T., & Senda, T. (2007). Abnormal development of urogenital organs in Dlgh1-deficient mice. *Development*, *134*, 1799–1807.

Innis, J. W. (1993). In M. P. Adam, H. H. Ardinger, R. A. Pagon, S. E. Wallace, L. J. H. Bean, G. Mirzaa, & A. Amemiya (Eds.), *Hand-foot-genital syndrome*. Seattle (WA): GeneReviews((R)).

Innis J.W., (2006). Hand-Foot-Genital Syndrome. In: Adam M.P., Ardinger H.H., Pagon R.A., Wallace S.E., Bean L.J.H., Gripp K.W., Mirzaa G.M., Amemiya A., (Eds.), GeneReviews® [Internet]. Seattle (WA): University of Washington, Seattle; 1993–2022. PMID: 20301596. [updated 2019 Aug 8].

Ishiyama, D., Nakamura, Y., Tanaka, T., Magata, F., Matsuda, F., & Maeda, K. I. (2019). Severe incomplete fusion of the Mullerian ducts influences reproduction in Holstein cattle. *Theriogenology*, *123*, 209–215.

Jacquinet, A., Boujemla, B., Fasquelle, C., Thiry, J., Josse, C., Lumaka, A., et al. (2020). GREB1L variants in familial and sporadic hereditary urogenital adysplasia and Mayer-Rokitansky-Kuster-Hauser syndrome. *Clinical Genetics, 98*, 126–137.

Kantaputra, P. N., Mundlos, S., & Sripathomsawat, W. (2010). A Novel homozygous Arg222Trp missense mutation in WNT7A in two sisters with severe Al-Awadi/Raas-Rothschild/Schinzel phocomelia syndrome. *The American Journal of Medical Genetics, 152A*(11), 2832–2837. https://doi.org/10.1002/ajmg.a.33673. PMID: 20949531

Karavanov, A. A., Karavanova, I., Perantoni, A., & Dawid, I. B. (1998). Expression pattern of the rat Lim-1 homeobox gene suggests a dual role during kidney development. *The International Journal of Developmental Biology, 42*, 61–66.

Kispert, A., Vainio, S., & McMahon, A. P. (1998). Wnt-4 is a mesenchymal signal for epithelial transformation of metanephric mesenchyme in the developing kidney. *Development, 125*, 4225–4234.

Kobayashi, A., & Behringer, R. R. (2003). Developmental genetics of the female reproductive tract in mammals. *Nature Reviews. Genetics, 4*, 969–980.

Kobayashi, A., Kwan, K. M., Carrol, T. J., McMahon, A. P., Mendelsohn, C. L., & Behringer, R. R. (2005). Distinct and sequential tissue-specific activities of the LIM-class. homeobox gene Lim1 for tubular morphogenesis during kidney development. *Development, 132*, 2809–2823.

Kobayashi, A., Shawlot, W., Kania, A., & Behringer, R. R. (2004). Requirement of Lim1 for female reproductive tract development. *Development, 131*, 539–549.

Koff, A. K. (1933). Development of the vagina in the human fetus. *Contributions to Embryology, 24*, 61–U13.

Krutzsch, P. H., & Crichton, E. G. (1987). Reproductive-biology of the male little mastiff bat, Mormopterus-Planiceps (Chiroptera, Molossidae), in Southeast Australia. *The American Journal of Anatomy, 178*, 352–368.

Laprise, P., Viel, A., & Rivard, N. (2004). Human homolog of disc-large is required for adherens junction assembly and differentiation of human intestinal epithelial cells. *The Journal of Biological Chemistry, 279*, 10157–10166.

Ledig, S., & Wieacker, P. (2018). Clinical and genetic aspects of Mayer-Rokitansky-Kuster-Hauser syndrome. *Medizinische Genetik: Mitteilungsblatt des Berufsverbandes Medizinische Genetik e.V, 30*, 3–11.

Ledig S., Brucker S., Barresi G., Schomburg J., Rall K., Wieacker P., 2012. Frame shift mutation of LHX1 is associated with Mayer-Rokitansky-Kuster-Hauser (MRKH) syndrome. Human Reproduction 27 (9), 2872-2875. doi: 10.1093/humrep/des206. Epub 2012 Jun 26.PMID: 22740494.

Liang H.I., Fu S.C., Yin C.H., Chang C.C., 2020. Herlyn-Werner-Wunderlich syndrome: An unusual case with presentation of menorrhagia. Taiwanese Journal of Obstetrics and Gynecology 59 (6), 948-951. doi: 10.1016/j.tjog.2020.09.026.PMID: 33218419.

Lindner, TH, Njolstad, PR, Horikawa, Y, Bostad, L, Bell, GI, & Sovik, O. (1999). A novel syndrome of diabetes mellitus, renal dysfunction and genital malformation associated with a partial deletion of the pseudo-POU domain of hepatocyte nuclear factor-1beta. *Human Molecular Genetics, 8*(11), 2001–2008. https://doi.org/10.1093/hmg/8.11.2001. PMID: 10484768.

Liszewska-Kaplon, M., Strozik, M., Kotarski, L., Baglaj, M., & Hirnle, L. (2020). Mayer-Rokitansky-Kuster-Hauser syndrome as an interdisciplinary problem. *Advances in Clinical and Experimental Medicine, 29*, 505–511.

Mericskay, M., Kitajewski, J., & Sassoon, D. (2004). Wnt5a is required for proper epithelial-mesenchymal interactions in the uterus. *Development, 131*, 2061–2072.

Miglino, M. A., dos Santos, T. C., Kanashiro, C., & dos Santos Ferraz, R. H. (2013). Morphology and reproductive physiology of female capybaras. In J. R. Moreira, K. M. P. M. B. Ferraz, E. A. Herrera, & D. W. Macdonald (Eds.), *Capybara: Biology, Use and Conservation of an Exceptional Neotropical Species* (pp. 131–146). New York, NY: Springer New York.

Mikhael, S., Dugar, S., Morton, M., Chorich, L. P., Tam, K. B., Lossie, A. C., et al. (2021). Genetics of agenesis/hypoplasia of the uterus and vagina: Narrowing down the number of candidate genes for Mayer-Rokitansky-Kuster-Hauser syndrome. *Human Genetics, 140*, 667–680.

Miller C, Sassoon DA., 1998. Wnt-7a maintains appropriate uterine patterning during the development of the mouse female reproductive tract. Development 125 (16), 3201-3211. doi: 10.1242/dev.125.16.3201.PMID: 9671592.

Miller, C., Pavlova, A., & Sassoon, D. A. (1998). Differential expression patterns of Wnt genes in the murine female reproductive tract during development and the estrous cycle. *Mechanisms of Development, 76*, 91–99.

Mishina, Y., Rey, R., Finegold, M. J., Matzuk, M. M., Josso, N., Cate, R. L., et al. (1996). Genetic analysis of the Mullerian-inhibiting substance signal transduction pathway in mammalian sexual differentiation. *Genes & Development, 10*, 2577–2587.

Mittal R.J., 2016. Herlyn-Werner-Wunderlich Syndrome, The Journal of Obstetrics and Gynecology of India (JOGI) 66 (2):128-130. doi: 10.1007/s13224-015-0707-8. Epub 2015 May 16.PMID: 27046969.

Miyamoto, N., Yoshida, M., Kuratani, S., Matsuo, I., & Aizawa, S. (1997). Defects of urogenital development in mice lacking Emx2. *Development, 124*, 1653–1664.

Morgan, L. M. (2004). A social biography of Carnegie embryo no. 836. *Anatomical Record. Part B, New Anatomist, 276*, 3–7.

Mortlock, D. P., & Innis, J. W. (1997). Mutation of HOXA13 in hand-foot-genital syndrome. *Nature Genetics, 15*, 179–180.

Mossman, H. W. (1987). *Vertebrate fetal membranes: Comparative ontogeny and morphology, evolution, phylogenetic significance, basic functions, research opportunities*. New Brunswick, N.J.: Rutgers University Press.

Mullen, R. D., & Behringer, R. R. (2014). Molecular genetics of Mullerian duct formation, regression and differentiation. *Sexual Development, 8*, 281–296.

Müller, J. (1830). *Bildungsgeschichte der Genitalien aus anatomischen Untersuchungen an Embryonen des Menschen und der Thiere; nebst einem Anhang über die chirurgische Behandlung der Hypospadia*. Düsseldorf: Arnz.

Nakayama, M., Nozu, K., Goto, Y., Kamei, K., Ito, S., Sato, H., et al. (2010). HNF1B alterations associated with congenital anomalies of the kidney and urinary tract. *The Journal of Pediatric Nephrology, 25*, 1073–1079.

O'Rahilly, R., Müller, F., & Streeter, G. L. (1987). *Developmental stages in human embryos: Including a revision of Streeter's "Horizons" and a survey of the Carnegie collection*. Washington, D.C.: Carnegie Institution of Washington.

Oram, R. A., Edghill, E. L., Blackman, J., Taylor, M. J., Kay, T., Flanagan, S. E., et al. (2010). Mutations in the hepatocyte nuclear factor-1beta (HNF1B) gene are common with combined uterine and renal malformations but are not found with isolated uterine malformations. *American Journal of Obstetrics and Gynecology, 203*(364), e361–e365.

Orazi, C., Lucchetti, M. C., Schingo, P. M., Marchetti, P., & Ferro, F. (2007). Herlyn-Werner-Wunderlich syndrome: Uterus didelphys, blind hemivagina and ipsilateral renal agenesis. Sonographic and MR findings in 11 cases. *Pediatric Radiology, 37*, 657–665.

O'Rahilly, R, & Müller, F. (1987). *Developmental stages in human embryos*. DC: Carnegie Institution of Washington.

Orvis, G. D., & Behringer, R. R. (2007). Cellular mechanisms of Mullerian duct formation in the mouse. *Developmental Biology, 306*, 493–504.

Parr, B. A., & McMahon, A. P. (1995). Dorsalizing signal Wnt-7a required for normal polarity of D-V and A-P axes of mouse limb. *Nature, 374*, 350–353.

Parr, B. A., & McMahon, A. P. (1998). Sexually dimorphic development of the mammalian reproductive tract requires Wnt-7a. *Nature, 395*, 707–710.

Patten, B. M. (1948). *Embryology of the pig*. New York: McGraw-Hill.

Pillet, J. (1967). Reconstruction du pelvis d'un embryon humain de 7, 5 mm (stade XVI de Streeter). *Comptes rendus de l'Association des anatomistes*, *52*, 1013–1023.
Purslow, C. E. (1922). A case of unilateral Hæmatokolpos, Hæmatometra and Hæmatosalpinx. *BJOG: An International Journal of Obstetrics & Gynaecology*, *29*, 643.
Raggio, I., Lefebvre, R. C., Poitras, P., Vaillancourt, D., & Champagne, J. P. (2006). Clinical diagnosis of uterine didelphia in Ayrshire heifer. *The Canadian Veterinary Journal*, *47*, 899–901.
Rasweiler, J. J. T., Oliveira, S. F., & Badwaik, N. K. (2002). An ultrastructural study of interstitial implantation in captive-bred, short-tailed fruit bats, Carollia perspicillata: Trophoblastic adhesion and penetration of the uterine epithelium. *Anatomy and Embryology (Berlin)*, *205*, 371–391.
Renfree, M. B. (1993). Ontogeny, genetic control, and phylogeny of female reproduction in monotreme and therian mammals. In F. S. Szalay, M. J. Novacek, & M. C. McKenna (Eds.), *Mammal phylogeny* (pp. 4–20). New York, NY: Springer US.
Roberts, S., Delury, C., & Marsh, E. (2012). The PDZ protein discs-large (DLG): The 'Jekyll and Hyde' of the epithelial polarity proteins. *The FEBS Journal*, *279*, 3549–3558.
Rodger, J. C. (1991). Fertilization of marsupials. In B. S. Dunbar, & M. G. O'Rand (Eds.), *A comparative overview of mammalian fertilization* (pp. 117–135). Boston, MA: Springer US.
Sanna-Cherchi, S., Khan, K., Westland, R., Krithivasan, P., Fievet, L., Rasouly, H. M., et al. (2017). Exome-wide association study identifies GREB1L mutations in congenital kidney malformations. *American Journal of Human Genetics*, *101*, 789–802.
Saravelos, S. H., Cocksedge, K. A., & Li, T. C. (2008). Prevalence and diagnosis of congenital uterine anomalies in women with reproductive failure: A critical appraisal. *Human Reproduction Update*, *14*, 415–429.
Schimming, B. C., Cesario, M. D., & Matheus, S. M. M. (2018). Morphology of the vaginal complex in the white-eared opossum (Didelphis albiventris): Gross anatomy and light microscopy. *Anatomia, Histologia, Embryologia*, *47*, 566–572.
Schrauwen, I., Kari, E., Mattox, J., Llaci, L., Smeeton, J., Naymik, M., et al. (2018). De novo variants in GREB1L are associated with non-syndromic inner ear malformations and deafness. *Human Genetics*, *137*, 459–470.
Schrauwen, I., Liaqat, K., Schatteman, I., Bharadwaj, T., Nasir, A., Acharya, A., et al. (2020). Autosomal dominantly inherited GREB1L variants in individuals with profound sensorineural hearing impairment. *Genes (Basel)*, *11*.
Shawlot, W., & Behringer, R. R. (1995). Requirement for Lim1 in head-organizer function. *Nature*, *374*, 425–430.
Spencer, T. E., Hayashi, K., Hu, J., & Carpenter, K. D. (2005). Comparative developmental biology of the mammalian uterus. *Current Topics in Developmental Biology*, *68*, 85–122.
Stark, K., Vainio, S., Vassileva, G., & McMahon, A. P. (1994). Epithelial transformation of metanephric mesenchyme in the developing kidney regulated by Wnt-4. *Nature*, *372*, 679–683.
Stern A.M., Gall Jr J.C., Perry B. L., Stimson C.W., 1969. Hand-foot-uterus syndrome, *Lancet* 1 (7594), 577, doi: 10.1016/s0140-6736(69)91994-1. PMID: 4179869.
Timurkaan, N., & Ozer, H. (2002). Uterus didelphys in a goat. *The Veterinary Record*, *151*, 217.
Torres, M., Gomez-Pardo, E., Dressler, G. R., & Gruss, P. (1995). Pax-2 controls multiple steps of urogenital development. *Development*, *121*, 4057–4065.
Vainio, S., Heikkila, M., Kispert, A., Chin, N., & McMahon, A. P. (1999). Female development in mammals is regulated by Wnt-4 signalling. *Nature*, *397*, 405–409.
Wick, R., & Kress, A. (2002). Ultrastructural changes in the uterine luminal and glandular epithelium during the oestrous cycle of the marsupial Monodelphis domestica (grey short-tailed opossum). *Cells, Tissues, Organs*, *170*, 111–131.

Winuthayanon, W., Hewitt, S. C., Orvis, G. D., Behringer, R. R., & Korach, K. S. (2010). Uterine epithelial estrogen receptor alpha is dispensable for proliferation but essential for complete biological and biochemical responses. *Proceedings of the National Academy of Sciences of the United States of America, 107*, 19272–19277.
Woods, D. F., & Bryant, P. J. (1991). The discs-large tumor suppressor gene of Drosophila encodes a guanylate kinase homolog localized at septate junctions. *Cell, 66*, 451–464.
Woods, C. G., Stricker, S., Seemann, P., Stern, R., Cox, J., Sherridan, E., et al. (2006). Mutations in WNT7A cause a range of limb malformations, including Fuhrmann syndrome and Al-Awadi/Raas-Rothschild/Schinzel phocomelia syndrome. *American Journal of Human Genetics, 79*, 402–408.
Woods, D. F., Hough, C., Peel, D., Callaini, G., & Bryant, P. J. (1996). Dlg protein is required for junction structure, cell polarity, and proliferation control in Drosophila epithelia. *The Journal of Cell Biology, 134*, 1469–1482.
Yamaguchi, T. P., Bradley, A., McMahon, A. P., & Jones, S. (1999). A Wnt5a pathway underlies outgrowth of multiple structures in the vertebrate embryo. *Development, 126*, 1211–1223.
Yu, J., Carroll, T. J., & McMahon, A. P. (2002). Sonic hedgehog regulates proliferation and differentiation of mesenchymal cells in the mouse metanephric kidney. *Development, 129*, 5301–5312.
Zhao, F., Zhou, J., Li, R., Dudley, E. A., & Ye, X. (2016). Novel function of LHFPL2 in female and male distal reproductive tract development. *Scientific Reports, 6*, 23037.

CHAPTER FOUR

Improving mouse models for the study of Alzheimer's disease

Alaina M. Reagan[a], Kristen D. Onos[a], Sarah E. Heuer[a,b], Michael Sasner[a], and Gareth R. Howell[a,b,c,*]
[a]The Jackson Laboratory, Bar Harbor, ME, United States
[b]Graduate Biomedical Sciences, Tufts University School of Medicine, Boston, MA, United States
[c]Graduate School of Biomedical Sciences and Engineering, University of Maine, Orono, ME, United States
*Corresponding author: e-mail address: gareth.howell@jax.org

Contents

Current Topics in Developmental Biology, Volume 148
ISSN 0070-2153
https://doi.org/10.1016/bs.ctdb.2021.12.005

Abstract

Alzheimer's disease (AD) is a complex neurodegenerative disease whose risk is influenced by genetic and environmental factors. Although a number of pathological hallmarks have been extensively studied over the last several decades, a complete picture of disease initiation and progression remains unclear. We now understand that numerous cell types and systems are involved in AD pathogenesis, and that this cellular profile may present differently for each individual, making the creation of relevant mouse models challenging. However, with increasingly diverse data made available by genome-wide association studies, we can identify and examine new genes and pathways involved in genetic risk for AD, many of which involve vascular health and inflammation. When developing mouse models, it is critical to assess (1) an aging timeline that represents onset and progression in humans, (2) genetic variants and context, (3) environmental factors present in human populations that result in both neuropathological and functional changes—themes that we address in this chapter.

1. Introduction

Alzheimer's disease (AD) is the most common form of age-related neurodegeneration, clinically characterized by a decline in cognitive function. It has previously been defined by the pathological "triad" of β-amyloid plaques, tau-positive neurofibrillary tangles, and neurodegeneration (ATN) but has now been expanded to a pathological "quad" to include vascular risk (ATNV) (Jack et al., 2018). Numerous genetic mutations and risk factors have been identified that are either causal or modify the disease progression, some of which relate to the buildup of amyloid plaques, and many that are associated with microglial function in the brain. AD has a strong genetic basis, with heritability estimates up to 80% (Gatz et al., 2006). However, genetic variants in well-established genes for AD, notably amyloid precursor protein (APP) and presenilin 1 and 2 (*PSEN1* and *PSEN2*) for the less prevalent early-onset AD (EOAD), and apolipoprotein E (*APOE*) for the more prevalent late-onset AD (LOAD), fail to explain the estimated heritability (Kamboh et al., 2012). It is now becoming clear that a combination of genetic factors beyond these likely contribute significantly to AD risk.

Well-known rare genetic and limited pathological features have long served as the basis for the creation and validation of mouse models of AD. Over the past 30 years, more than 100 genetically engineered mouse lines have been created that recapitulate specific aspects of clinical pathology (Jankowsky & Zheng, 2017). Many existing mouse models promote the development of amyloid plaques or neurofibrillary tangles (and a few exhibit signs of neurodegeneration), but all pathologies are typically not present within the same mouse. Although each feature is independently affiliated with human disease, some individuals with the same pathologies develop cognitive decline while others do not, once again underlining the complexity of pathogenesis. As such, few preclinical studies that used these current mouse models have been translated successfully to clinical therapies.

Genetic and genome-wide association studies have identified genes that modulate cholesterol metabolism, endocytosis, immune response and vascular health (DeTure & Dickson, 2019). By broadening our perception of which systems contribute to AD pathology, we can examine and integrate the roles of these more recently identified risk factors to provide deeper insight into mechanisms that drive disease. Data from the Mayo Clinic Brain Bank from 2007 to 2016 identified that most AD cases had coexisting non-Alzheimer pathologies, and comorbidities increased in frequency with age (DeTure & Dickson, 2019). This may indicate that in fact, many cases of AD should be considered as mixed dementias—a category that is likely under-diagnosed currently. It is important to examine the nature of coexisting pathologies, especially those that predate cognitive decline, not solely as secondary considerations, but as possible foundations upon which traditional AD pathology builds. As with many complex diseases of aging, environmental risk is closely intertwined with genetic risk. These include diet, physical activity, smoking and level of educational attainment (Livingston et al., 2020). Although it is difficult to accurately represent human diet and lifestyle choices within preclinical studies, efforts to model these risks in conjunction with genetics are extremely valuable, especially when genetic risk alone may not tip the scales toward pathology without an additional environmental stressor.

wGiven the failures of translation from preclinical to clinical studies, it is necessary to develop improved mouse models. To achieve this, it is important to not only study isolated genetic risk factors to understand basic mechanisms of disease pathophysiology, but also to include combinations of genetic and environmental risk factors. The historical focus ofboth clinical and preclinical research has been on neurons within the central nervous system (CNS); we now understand that myeloid-derived cells in the brain and complex cellular interactions between neurons, astrocytes, microglia and vascular cells in the brain (Henstridge, Hyman, & Spires-Jones, 2019) as well as comorbidities in peripheral systems such as diabetes and vascular compromise play a major role in most AD cases (Maciejewska, Czarnecka, & Szymański, 2021).

A number of existing reviews thoroughly examine classic AD mouse models (e.g., Jankowsky & Zheng, 2017; Sasaguri et al., 2017). In this chapter, we aim to highlight new strategies that focus on combinations of neuroinflammatory and vascular risk factors and the importance of genetic context as key contributors to AD/cognitive decline, including a discussion about the importance of humanized models. Finally, we will conclude with resources and tools that are currently under development, with the goal that they will become widely available to the new generation of AD researchers.

2. Overcoming limitations of existing mouse models relevant to early-onset AD

2.1 The limited translatability of familial AD models

As recently reviewed by us and others (Jankowsky & Zheng, 2017; Onos, Sukoff Rizzo, Howell, & Sasner, 2016; Sasaguri et al., 2017), numerous issues have been spotlighted with the mouse models for AD that have been available for decades. Unfortunately, these issues are key contributors to the lack of translatability of preclinical studies to the clinic, and have painted the mouse broadly as an ineffective model system for polygenic disease such as LOAD. Even while the majority of AD cases (>98%) are of the late-onset or sporadic type, there has been a dominance of the amyloid hypothesis (Selkoe & Hardy, 2016) in the field, simply due to the fact that familial or early-onset (EOAD) mutations have been known for decades. Because of this, almost all the AD rodent models in use today express EOAD mutations in *APP*, *PSEN1*, or *PSEN2*.

Models such as B6.*APP/PS1* (Jankowsky et al., 2001, 2004), B6.5XFAD (Oakley et al., 2006) and B6.hAPPJ20 (Mucke et al., 2000) have proven to be useful as models to study amyloidogenesis and neuroinflammation, but in most cases do not exhibit tau pathology or neurodegeneration. To model tau pathology, mutations in mouse *Mapt*, or humanized MAPT mice were developed such as hTau (Andorfer et al., 2003), but these do not necessarily drive tau pathology in a human-relevant fashion. One reason that results in these EOAD preclinical models have not translated to the clinic may be that they are not translationally relevant models of LOAD, and drugs tested in these models should only be moved into clinical trials for EOAD patients such as the Dominantly Inherited Alzheimer's Network (Salloway et al., 2021). It has been shown that drug development projects based on appropriate human genetics have a significantly greater success rate (Nelson et al., 2015).

Our understanding of late-onset genetics is more recent, and rapidly evolving. For models of LOAD, we need to consider the heterogeneity of disease, with underlying differences in genetic risk and disease etiology (Milind et al., 2020); we may need multiple models, each for a specific subtype of disease. It is also important to recognize that genetic risk factors vary by clinical population (Kunkle et al., 2021); for example, the risk conferred by the ε4 allele of APOE (Crean et al., 2011). Therefore, different clinical populations could require distinct animal models for preclinical testing. While most existing models express risk mutations in the pathway regulating amyloid metabolism in neurons, the majority of genetic risk for AD resides in genes expressed in non-neuronal cells, mainly in microglia. Most of the variants reside in noncoding regions, indicating perturbances of gene expression levels rather than changes in protein function (Novikova et al., 2021; Pimenova, Raj, & Goate, 2018).

The other significant concern with most existing AD mouse models is that they have been created using transgenic technology. These are dependent upon a heterologous promoter-expressed gene being inserted into the mouse genome, possibly affecting important noncoding regulatory regions. Practically, most models have been selected to have the most severe phenotype from a variety of transgenic founder lines, so that they strongly over-express the transgenic protein. In most cases the promoter used is not the homologous promoter, which can cause the protein to be expressed in cell types, tissues, and development/aging stages that differ from the endogenous gene, introducing artifactual phenotypes. Transgenic models are also more prone to phenotypic drift over time, possibly resulting in changes to transgenic protein expression due to loss of copy number or methylation of transgenic inserts. Many models were initially made on mixed genetic backgrounds, so phenotypes vary within a cohort and drift with time due to a lack of a fixed background.

Furthermore, transgenic strategies are not effective for modeling genetic risk from noncoding variants. We have recently shown that many of the widely used transgenic mouse models for Alzheimer's disease may disrupt endogenous gene expression at the site of transgene integration (Goodwin et al., 2019). Extensive effort may be required to ensure that the phenotype of a transgenic model results from the expressed transgene, rather than inadvertent loss of an unrelated gene (Gulbranson et al., 2021).

2.2 Improving utility of familial/early-onset AD models

With modern knock-in and CRISPR-based approaches, human coding sequences and/or genetic risk variants can be engineered into the endogenous locus in the mouse genome, leveraging promoter and regulatory regions to drive endogenous levels and sites of expression throughout the lifespan of the animal. A knock-in strategy was used to create the APP^{NL-G-F} model (Saito et al., 2014; Sasaguri et al., 2017), which expresses the Swedish, Iberian, and Arctic mutations in the context of a humanized Aβ1–42 sequence. This widely used model displays amyloidosis and neuroinflammation by 2 months of age, with synaptic loss and cognitive impairment by 6 months of age. While this model does not exhibit tau pathology or neuronal loss, it has proven very useful for studies on the biology of amyloid. It is important to note that models with just the Swedish and Iberian mutations (or the Swedish mutation alone) have very mild phenotypes even at advanced ages, while the Swedish mutation alone is a causative mutation in many cases of human EOAD. Thus, we need to consider species differences, most prominently the ~50-fold difference in lifespan, when comparing mouse models to human disease. It is also likely that pathology that occurs very early in the mouse lifespan will be disruptive to important developmental processes, muddying what results from typical aging vs disease-relevant features. More recently, the $hAPP^{SAA}$ model, which expresses humanized Aβ1–42 with the Swedish, Arctic and Austrian mutations, was created to provide an amyloid model that would be readily available for preclinical studies in the pharma and biotech community. This model exhibits extensive amyloid pathology, phosphorylated tau, a disease-associated microglia transcriptomic signature, and microglial metabolic disruption (Xia et al., 2021).

3. Developing mouse models of late-onset AD (LOAD) through the MODEL-AD consortium

3.1 Accelerating medicines partnerships: Alzheimer's disease

Over the past 10 years or more, federal funding to the National Institutes of Health (NIH) for Alzheimer's disease research has increased greatly, and currently more than $3 billion per year is committed from the NIH to finding a cure for AD. Much of the funding is administered by the National Institute on Aging (NIA). To maximize the impact of the increased funding, the NIA established AD summits, bringing together experts in the field to make recommendations for how to prevent or cure AD. These recommendations led to the creation of consortia centered around the concept of Open Science. Rather than all funding being distributed through the usual research grants, such as R01s, consortia were established through "U" (resource-based) funding mechanisms. Key consortia relevant to this review are the Accelerating Medicines Partnership: AD (AMP-AD), Model Organism Development and Evaluation for Late-onset AD (MODEL-AD), and TaRget Enablement to Accelerate Therapy Development for AD (TREAT-AD) (Fig. 1A). Critically, awardees were required to work together and release their data regularly through a publicly available platform hosted by Sage Bionetworks. Of the three, AMP-AD was the first consortium established that was charged with identifying novel targets for therapeutics, and developing novel biomarkers for early diagnoses. AMPs are a public-private partnership between the NIH, the US Food and Drug Administration, multiple biopharmaceutical and life sciences companies, and non-profit and other organizations to transform the model for developing new diagnostics and treatments. In addition to AD, AMP projects include autoimmune disorders, common metabolic diseases (CMD), Parkinson's disease (PD), Schizophrenia (SCZ) and Type 2 diabetes (T2D).

All AMP-AD data is being submitted to the AD Knowledge portal (formerly the AMP-AD Knowledge portal; https://adknowledgeportal.synapse.org). This led to a list of novel targets known as the "Wall of Targets," and included genes that had previously not been associated with AD, such as *VGF* (VGF nerve growth factor inducible (El Gaamouch et al., 2020)). These targets are now being assessed as potential treatment options for AD. Much of the work in AMP-AD focused on defining AD based on -omic data—particularly transcriptomics, proteomics and metabolomics (Baird et al., 2021; Sieberts et al., 2020). This has provided important novel insights into biological pathways, processes and networks that define AD, and provide a higher resolution picture of AD, moving beyond the somewhat simplistic ATNV definitions. For the transcriptomic analysis, all data (generated from multiple brain regions by members of the consortia) was standardized and put through a series of analyses to generate a set of "modules" that represent different biological processes (herein referred to as AMP-AD modules) (Fig. 1B) (Wan et al., 2020). Approximately 30 modules were clustered based on similarity of genes into 5 major groups that, through gene set enrichment, represented activity by major brain cell types including neurons, astrocytes, microglia, and oligodendrocytes. The AMP-AD modules provide a convenient way to use transcriptomic approaches to compare mouse models relevant to AD to human AD (Fig. 1C) (Preuss et al., 2020). Not surprisingly, the wealth of -omic data is allowing researchers to now identify subtypes of AD. In one recent study, transcriptomic data identified an inflammatory type of AD (termed subtype A) and a noninflammatory type (subtype B) (Milind et al., 2020). A second study

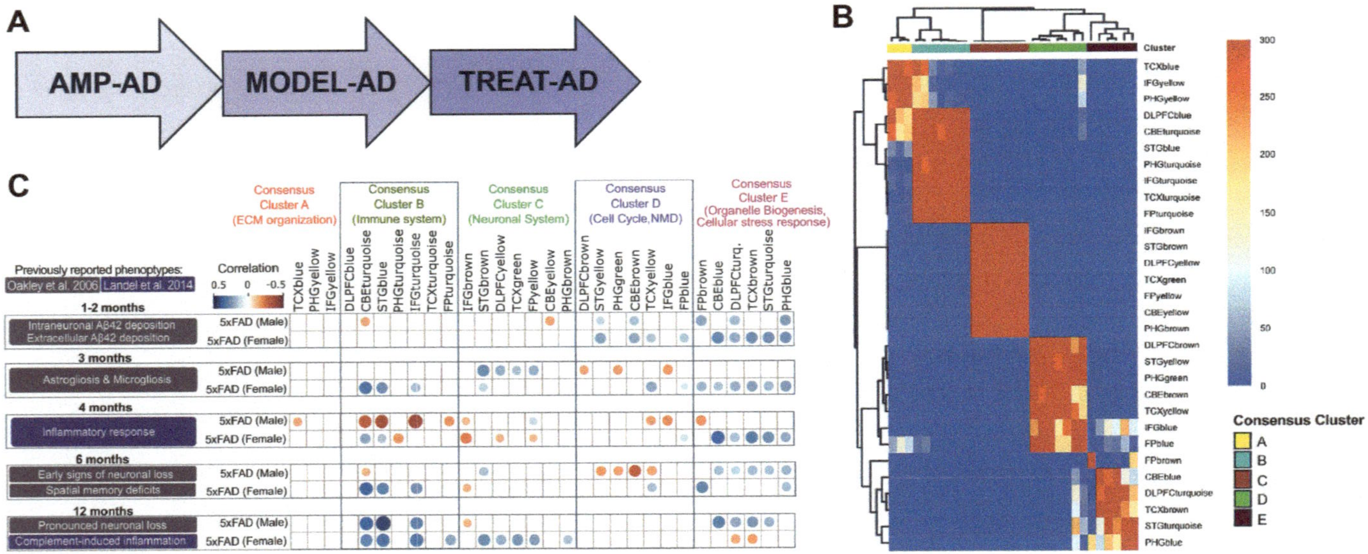

Fig. 1 (A) Outline of key NIH-funded consortia developed to accelerate collaborative research to better identify translational models and treatments for AD. (B) AMP-AD consortia results reported in Wan et al. (2020), the identification of 30 gene expression modules from human AD patient brain tissue that highlighted key pathways and cell-types enriched during AD. (C) Preuss et al., 2020, leverage AMP-AD data to demonstrate how human transcriptomics translate to current AD mouse models to evaluate face-validity of current models, and to identify where current models can be improved. For full details see original publications by Wan et al., 2020 (B) and Preuss et al., 2020 (C).

supported this finding, identifying three subtypes of which two were considered inflammatory, and the third noninflammatory (Neff et al., 2021). For those modeling AD in mice, it will be important to follow the fast-changing field of AD subtype characterization to ensure as close an alignment as possible between mouse models and human AD.

3.2 Model organism development and evaluation for late onset Alzheimer's disease (MODEL-AD)

At the 2015 NIA-led AD summit, the need for a more organized and systematic approach to modeling LOAD was recommended. There was a need to leverage the wealth of genetic and sequencing data that was being generated by many groups around the world to create mouse models based on the genetics of LOAD, so that the field could work to resolve those issues with current models mentioned above. Data was being generated by consortia including the International Genomics of Alzheimer's Project (IGAP), Alzheimer's Disease Neuroimaging Initiative (ADNI), Alzheimer's disease sequencing project (ADSP), as well as members of AMP-AD. Variations in more than 20 genes have been associated with LOAD, but their contribution to AD was not clear. Therefore, to develop models that more closely align to human LOAD, two programs were funded under the MODEL-AD umbrella (http://model-ad.org) (Oblak et al., 2020). The first program is a collaboration between Indiana University, The Jackson Laboratory (JAX), University of Pittsburgh, and Sage Bionetworks (IU/JAX/Pitt MODEL-AD)—the second is located at University of California, Irvine (UCI MODEL-AD). Both programs work together to create novel mouse models, phenotype them using methods aligned to human LOAD, and make all the mice and data available to the scientific community, including pharma, as early as possible. More than 40 mouse models have been created that are available through the JAX mouse repository; these models are listed here (http://model-ad.org). All protocols and data are available at the AD Knowledge Portal, which hosts experimental data from NIA's AD Translational Research Program, including AMP-AD, M2OVE-AD, MODEL-AD, Resilience-AD and Psych-AD. (https://adknowledgeportal.synapse.org/Explore/Programs/DetailsPage?Program=MODEL-AD).
Additionally, the MODEL-AD Explorer is an interactive interface that allows analysis of transcriptomic and neuropathology data as developed by the MODEL-AD Center: modeladexplorer.org.

Both MODEL-AD programs consist of an administrative core, bioinformatics and data management core (BDMC), and a disease modeling project

(DMP). The IU/JAX/Pitt MODEL-AD also established a preclinical testing core (PTC) charged with developing robust pipelines for performing preclinical studies in mice. Initially, the MODEL-AD consortium set out to standardize phenotypes in existing models of EOAD, focusing on the widely used *APP/PS1*, hTau, 3xTg and 5XFAD models. Phenotyping pipelines included traditional assays such as behavior and neuropathology, but also incorporated human-relevant biomarkers, magnetic resonance imaging (MRI), transcriptomics, proteomics and metabolomics, to align to human data generated by AMP-AD groups (Oblak et al., 2020). Following the establishment of the phenotyping pipelines, both groups set out to generate LOAD-relevant models, initially on the C57B/6J (B6) background. These included humanizing *APP* and *MAPT* genes, and creating allelic series for *APOE* (*e2, e3, e4*) (Foley et al., 2020) and *Trem2* (e.g., R47H).

The IU/JAX/Pitt MODEL-AD group then set about creating a series of "platform" strains to test the hypothesis that combinations of risk variants would drive AD-like phenotypes. The first model—termed LOAD1—was homozygous for *APOE4* ($APOE^{E4}$) and $Trem2^{R47H}$ (JAX stock no: 028709). Both male and female LOAD1 mice, as well as single mutants and B6 controls, were assessed using the deep phenotyping pipeline at 4, 8, 12 and 24 months (Kotredes et al., 2021). As expected, these mice did not develop amyloid and tau pathology, but *in vivo* positron emission tomography (PET)/MRI revealed genotype-specific regional alterations in glycolytic metabolism and vascular perfusion. Transcriptional profiling by RNA-Seq of brain hemispheres identified sex and age as the main sources of variation between genotypes, including age-specific enrichment of AD-related processes. Although this model does not develop sufficient AD-relevant changes to be suitable for preclinical testing, it does provide a potential "sensitizer model" to assess the effects of additional risk alleles and environmental risk factors such as dietary contributions. This led to the creation of a second model—termed LOAD2—that incorporates a humanized Aβ sequence on the LOAD1 background (JAX stock no: 030670). The humanized Aβ allele in LOAD2 is similar to an allele created by the UCI MODEL-AD group, with the difference being the UCI version also contains loxP sites flanking exon 14 (JAX stock no: 031050) (Baglietto-Vargas et al., 2021). hAβ mice were engineered to express human Aβ under the control of the mouse *App* locus. Simply changing three amino acids in the mouse Aβ peptide to its human counterpart led to the emergence of AD-relevant phenotypes, including deficits in cognition and synaptic plasticity. The IU/JAX/Pitt MODEL-AD group are currently phenotyping LOAD2 at 4, 12, 18 and 24 months of age, using similar approaches as described for LOAD1. Significant efforts are also underway to humanize the mouse *Mapt* locus, as it is expected that species differences between the human and mouse genes render the mouse MAPT protein less prone to form hyperphosphorylated Tau protein and neurofibrillary tangles. It should also be noted that the $Trem2^{R47H}$ mutation incorporated in to LOAD1 and LOAD2 mouse models reduces expression levels of mutant *Trem2*—an effect that appears mouse specific. Therefore, more recently, newer versions of $Trem2^{R47H}$ ($Trem2^{R47HHSS}$, JAX stock no: 027197; $Trem2^{R47HNSS}$, JAX stock no: 034036) have been created that preliminary data show do not affect mutant *Trem2* expression levels and are ideal for studying the biology of TREM2.

Both MODEL-AD programs are testing novel putative AD risk variants (both coding and noncoding) that were identified by the BDMC groups, or suggested from external studies. These have included variations in genes such as *ABCA7*, *BIN1*, *CR1*, *PLCG2*, *PICALM* and *MTHFR*. UCI MODEL-AD are crossing novel variants to 5XFAD as a primary screen to determine utility of a specific variant to modeling AD, while the IU/JAX/Pitt group are incorporating novel variants on either LOAD1 or LOAD2 backgrounds for primary screening and assessing relevance to AD by generating transcriptomic data to compare to the AMP-AD modules using the strategy described above (Fig. 1C). Promising variants are prioritized for deep phenotyping. A strain table of all models created by the MODEL-AD program is curated on the MODEL-AD website (https://www.model-ad.org/strain-table/). IU/JAX/Pitt MODEL-AD are also testing the potential of combining environmental with genetic factors using a 45% high fat/high sugar diet to increase alignment to AD. LOAD1, LOAD2, LOAD1.$Plcg2^{M28L}$ and LOAD1.$Mthfr^{C677T}$ are being provided a 45% high fat/high sugar that has been widely used to study the effects of poor diet and obesity. Future models are expected to contain combinations of genetic (based on polygenic risk analysis) and environmental risk factors based to alignment mouse data to human LOAD.

A primary goal of MODEL-AD is to provide LOAD models for performing preclinical testing of novel compounds. The PTC has established rigorous pipelines for testing novel compounds using Verubecestat (a beta secretase inhibitor) and Leviteracitam (a seizure medication), which are both in clinical trials for AD. Given their relevance to amyloid deposition, the PTC has used the 5XFAD mouse model for performing these studies but expect to incorporate MODEL-AD LOAD models in the future (Sukoff Rizzo et al., 2020). The PTC is also accepting nominations for preclinical screening of test compounds through the STOP-AD portal (https://stopadportal.synapse.org/#/). Submissions will be reviewed by a Steering Committee that evaluates and prioritizes nominations based on optimal drug-like properties and available experimental data that support the likelihood of success as a potential therapeutic for the treatment of Alzheimer's disease. Compounds selected for screening will be analyzed within the PTC labs at Indiana University and the University of Pittsburgh. On a larger scale than the PTC, TREAT-AD centers were recently created to improve, diversify, and reinvigorate the AD drug development pipelines. This includes the creation and distribution of tools that support target enabling packages (TEPs) for experimental validation of novel therapeutic targets, as well as initiation of early-stage drug discovery efforts against identified targets.

Collectively, NIA-funded consortia that are following the Open Science model have changed the way academia is contributing to the development of new therapeutic targets for AD. This includes the generation, phenotyping and dissemination of new mouse models for individual research labs and pharma alike to incorporate into their research and drug development programs.

4. Modeling vascular contributions

4.1 Emerging contribution of vascular dysfunction to AD

As stated in the introduction, multiple reviews exist that focus on hallmark characteristics of AD such as amyloid, tau and neuroinflammation (Jankowsky & Zheng, 2017; Onos et al., 2016; Sasaguri et al., 2017). However, it is becoming increasingly evident that cerebrovascular dysfunction contributes to dementia and AD (Nelson, Sweeney, Sagare, & Zlokovic, 2016; Snyder et al., 2015; Sweeney et al., 2019; Zlokovic, 2011), and a better understanding of mechanisms underlying vascular pathophysiology in AD and related dementias (ADRDs) have become a priority. Therefore, in this section we focus on the importance of vascular contributions to ADRDs, and emerging approaches to model this critical phenotype in mice.

There are two major ways to consider the source of cerebrovascular damage in the context of AD: damage that occurs prior to/may contribute to traditional AD pathology, and damage that occurs simultaneously or as a result of traditional AD pathology. However, overlap between the two is likely. Because a decline in vascular health is central to numerous systemic comorbidities, examining and recapitulating both genetic and environmental risk is critical to understanding vascular contributions to AD. Neurovascular dysfunction results in morphological as well as functional changes of the microvasculature (Kisler, Nelson, Montagne, & Zlokovic, 2017). Depending on the type and location of a vessel, the neurovascular unit (NVU) can be comprized of endothelial cells, pericytes, vascular smooth muscle cells (VSMCs), basement membrane, perivascular macrophages, microglia, astrocyte endfeet, and neuronal processes, which coordinate to form the protective blood–brain barrier (BBB) (Sweeney, Zhao, Montagne, Nelson, & Zlokovic, 2019). Disruption or damage to any or all of these components can result in leakage, reduced cerebral blood flow, ischemia, reduced cerebrovascular reactivity due to loss of intrinsic vascular tone or neurovascular coupling, each of which can contribute to neurodegeneration and cognitive decline (Kisler et al., 2017; Tarantini, Tran, Gordon, Ungvari, & Csiszar, 2017; Toth, Tarantini, Csiszar, & Ungvari, 2017). Few mouse models accurately reproduce the cerebrovascular phenotypes seen in human patients, in part due to differences between mouse and human exposure to environmental/lifestyle risk. However, variation in genes commonly associated with AD are reported to drive cerebrovascular dysfunction.

4.2 Vascular dysfunction in models of EOAD

Mice carrying mutations in APP and PSEN1 both present with cerebrovascular dysfunction. For example, early data from transgenic mice expressing APP^{swe} demonstrated that increased levels of Aβ results in reduced endothelial dilatory capabilities (Zhang, Eckman, Younkin, Hsiao, & Iadecola, 1997). Similarly, a separate study using 5XFAD mice showed reduced endothelial vasodilation and increased VSMC vasoconstriction in response to stimuli compared to controls (Iadecola et al., 1999), as well as attenuated neurovascular coupling (Niwa et al., 2000). Because these studies examined mice prior to pathological Aβ deposition, it suggests that low levels of soluble Aβ are sufficient to impair cerebrovascular reactivity. Aβ accumulation within the vessel wall causes cerebral amyloid angiopathy (CAA), which impairs vascular integrity and reactivity. CAA is observed in the majority of AD patients, and is also observed in transgenic mice with Dutch and Iowa Aβ mutations crossed with the Swedish APP mutants (TgSwDI APP) (Bell et al., 2009), as well as in older APP mice carrying Swedish and Indiana mutations (hAPPJ20) (Kimbrough, Robel, Roberson, & Sontheimer, 2015). Adult VSMCs can alter their differentiation status depending on their environment; this plasticity is beneficial during certain types of injury or stress. However, in AD patients and in TgSwDI APP and 5XFAD APP mice transcription factors that control VSMC differentiation are expressed at low levels, which results in increased expression of VSMC contractile proteins, including alpha smooth muscle actin, calponin and myosin heavy chain. The resulting hypercontractile VSMC phenotype, in combination with a reduction in endothelial dilatory capacity, attenuates cerebral blood flow (Chow et al., 2007). Attempting to reverse the effects of vascular depositions of Aβ in aged TgSwDI APP mice is not effective (Park et al., 2014), and many anti-amyloid immunotherapies have not been successful in AD patients because the treatment aggravates the existing vascular Aβ plaques (Wilcock & Colton, 2009). Thus, targeted intervention needs to occur early in disease.

In addition to extracellular Aβ pathology, Presenilin 1 (PSEN1) mutations also lead to significant cerebrovascular pathology in humans. BBB breakdown, pericyte and VSMC degeneration, and Aβ deposits in small cerebral arteries, arterioles and capillaries have been observed (Armstrong, 2008). In PSEN1 transgenic mouse models a similar profile is observed, with BBB breakdown, microhemorrhages, irregular microvessels, string vessels, basement membrane thickening, and reduced vascular density in the hippocampus (Gama Sosa et al., 2010; Wen et al., 2005).

4.3 Vascular dysfunction in genetic mouse models relevant to LOAD

The presence of one or two *APOE4* alleles is the strongest risk factor for the development of AD (Corder et al., 1993; Karch & Goate, 2015; Strittmatter et al., 1993). *APOE* plays a central role in cholesterol metabolism, and its disruption can promote BBB breakdown and CAA (Serrano-Pozo, Das, & Hyman, 2021). Even in *APOE4* carriers with no cognitive impairment, vascular dysfunction can be detected when compared to *APOE3* carriers, including reduced cerebrovascular reactivity, accelerated pericyte degeneration, and BBB breakdown (Halliday et al., 2016; Montagne et al., 2020). It is known that activation of the proinflammatory cyclophilin A-matrix metalloproteinase-9 pathway within the vasculature damages pericytes (Halliday et al., 2013). Compared to controls, transgenic *APOE4* mice show significantly increased levels of CypA in cerebral microvessels, resulting in pericyte degeneration, degradation of BBB tight junction and basement membrane proteins leading to BBB disruption (Bell et al., 2012). In contrast, transgenic mice expressing *APOE2* and *APOE3* are able to suppress the CypA-NFκB-MMP-9 pathway in pericytes, thus maintaining an intact BBB (Bell et al., 2012). In addition to regulating BBB maintenance, APOE plays a well-known role in the binding and clearance of Aβ, with the APOE2/Aβ and APOE3/Aβ complexes binding to low-density lipoprotein receptor-related protein 1 (LRP1), which is expressed in almost all cell types that comprise the NVU. These complexes are quickly moved across the BBB into the bloodstream. However, the APOE4-Aβ complexes have minimal interaction with LRP1, and are less effectively cleared from the brain by the very low density lipoprotein receptor (VLDLR) (Deane et al., 2008). It is now well understood that the reduced efficiency of Aβ clearance by APOE4 promotes Aβ accumulation in AD due to vascular dysfunction (Ma et al., 2018; Zhao, Nelson, Betsholtz, & Zlokovic, 2015), and is a target for clinical intervention.

Phosphatidylinositol binding clathrin assembly protein (*PICALM*) is a highly validated genetic risk factor for AD, and is significantly expressed in brain capillary endothelium with the primary function of endocytic and trafficking mechanisms (Thomas et al., 2016). Genome-wide association studies have linked up to 20 SNPs in the noncoding region of *PICALM* with AD (Harold et al., 2009; Lambert et al., 2013). Utilizing an *in vitro* model of the human BBB, it has been demonstrated that under homeostatic conditions, Aβ binds to LRP1 on the abluminal side of the cerebrovasculature, which recruits PICALM and initiates endocytosis of the Aβ-LRP1 complexes. However, in AD brain endothelium, PICALM levels are reduced, correlating with other notable pathology (Zhao, Sagare, et al., 2015). A mouse model deficient in brain endothelial *Picalm* demonstrates enhanced Aβ deposition in the brain as well as cognitive deficits, which can be rescued by endothelial-specific expression of *Picalm* (Zhao, Sagare, et al., 2015). Because it plays an important role in transporting and clearing Aβ across the BBB, including PICALM, risk variants within AD mouse models could improve our understanding of diverse vascular contributions to AD. To this end, UCI MODEL-AD used CRISPR to introduce the human-relevant H465R variant in the mouse *Picalm* gene (see strain table at MODEL-AD.org).

Clusterin (*CLU,* or apolipoprotein J, *APOJ*), is a glycoprotein that also plays a role in Aβ clearance (Bell et al., 2007). Genome-wide association studies have linked a SNP within CLU with late-onset AD (Harold et al., 2009; Lambert et al., 2009). The T allele is considered protective, while the C allele is considered a risk factor for AD (Karch & Goate, 2015; Lancaster et al., 2015). Using diffusion tensor imaging, at-risk individuals demonstrate reduced white matter integrity as well as hyperactivation of the prefrontal and limbic areas during working memory tasks, as seen by functional MRI (Lancaster et al., 2015). CLU is suggested to promote clearance/prevent aggregation of Aβ peptides across the BBB by binding to Aβ (Bell et al., 2007). While APOE uses LRP1 to traffic Aβ, Aβ42 bound to CLU can be rapidly cleared across the BBB via LRP2 as shown in guinea pigs and mice (Bell et al., 2007; Zlokovic et al., 1996). Variations in CLU are challenging to model in mice due to species differences between the human and mouse sequence, so humanizing approaches are required to faithfully recapitulate AD-relevant risk variants.

Mesenchyme homeobox gene 2 (MEOX2) is a transcription factor that regulates differentiation and remodeling in CNS vasculature (Candia et al., 1992; Gorski & Leal, 2003). A copy number variant in MEOX2 which reduces gene expression has been associated with AD risk (Rovelet-Lecrux et al., 2012). Low *Meox2* expression in brain endothelium results in vessel regression, as well as degradation of LRP1, which reduces Aβ clearance across the BBB. Increasing *Meox2* endothelial expression reverses these phenotypes (Wu et al., 2005). Deletion of *Meox2* is embryonically lethal, while heterozygous knockout mice with only one copy of *Meox2* show stunted endothelial cell development, and like human subjects, show reduced cerebral blood perfusion, reduced LRP1 expression, and impaired Aβ clearance (Wu et al., 2005). Further, *Meox2* haploinsufficiency increased neuronal cell loss in B6.*APP/PS1* mice (Soto et al., 2016), and accelerates axonal degeneration in a mouse model of glaucoma (Buchanan et al., 2019). To study the role of *MEOX2* in a more LOAD-relevant context, MODEL-AD has created LOAD1 mice heterozygous for a *Meox2* null mutation.

4.4 Comorbidities that likely contribute to cerebrovascular decline in dementia

The presence of additional vascular comorbidities promotes increased AD risk, so it is important to factor in these risks when developing mouse models. Using transverse aortic constriction or infusion of angiotensin II to model hypertension in mice, BBB impairment and Aβ deposition in the brain, as well as Aβ-induced neurovascular dysfunction and neurodegeneration were observed (Faraco et al., 2016;

Kruyer, Soplop, Strickland, & Norris, 2015). Type 2 diabetes mellitus is also considered to be a major vascular risk factor in the development of AD (Cheng, Huang, Deng, & Wang, 2012; Gudala, Bansal, Schifano, & Bhansali, 2013; Profenno, Porsteinsson, & Faraone, 2010). A number of experimental mouse models of Type 2 diabetes show abnormal vascular morphology and increased BBB permeability (Rom et al., 2019; Takeda et al., 2010). There is significant overlap between the mechanisms that promote vascular damage in diabetes and those that promote vascular damage in AD: vascular inflammation, oxidative stress, endoplasmic reticulum stress, elevated levels of advanced glycation end-products (AGEs) and their receptor (RAGE), and impaired insulin signaling/transport, all of which damage neurons, glia and cerebrovascular cells (De Felice & Lourenco, 2015; Srikanth et al., 2011; Tumminia, Vinciguerra, Parisi, & Frittitta, 2018). Atherosclerosis has long been associated with cholesterol-related pathology, diabetes, and aging, all of which are also risk factors for dementia and AD (Gupta & Iadecola, 2015; Iadecola, 2020). 5XFAD mice fed an atherogenic diet developed atherosclerosis in the aorta and showed increased amyloid deposition in the brain. The aortic atherosclerotic lesion area positively correlated with cerebral-amyloid deposits in transgenic mice, and behavioral assessments demonstrated that transgenic mice fed an atherogenic diet had more spatial learning impairment than those fed a normal diet, indicating a correlation between the diseases (Li, Cao, Garber, Kim, & Fukuchi, 2003). Hyperhomocysteinemia, or elevated plasma homocysteine, is known to induce vascular inflammation. Diet induced hyperhomocysteinemia in both wild type (Sudduth, Powell, Smith, Greenstein, & Wilcock, 2013) and B6.*APP/PS1* mice (Sudduth, Weekman, Brothers, Braun, & Wilcock, 2014) leads to memory deficits, microhemorrhages, neuroinflammation and increased MMP-2 and MMP-9 activity, which can result in breakdown of the ECM. Patients with a history of vascular disease often present with elevated homocysteine levels (Berwanger, Jeremy, & Stansby, 1995; Hankey & Eikelboom, 1999). Endothelial dysfunction, oxidative stress, hypomethylation, and thrombosis have been proposed as potential mechanisms to explain the relationship between hyperhomocysteinemia and cerebrovascular damage (Devlin et al., 2004). Variants in the methylenetetrahydrofolate (*MTHFR)* gene are known to cause hyperhomocysteinemia, and have been independently associated with AD (Hua, Zhao, Kong, & Ye, 2011; Rai, 2017; Stoccoro et al., 2017). A mouse model carrying the most common *MTHFR* risk variant has been created by our group; characterization is in progress.

5. Minimizing species differences to model inflammatory contributions to AD

5.1 Humanizing immune-relevant processes in mouse models

Increased inflammation in the brain (often referred to as neuroinflammation) is a central hallmark to aging and AD, observed by increased pro-inflammatory molecules. Recent genetic, pathological and omic-based studies have accumulated evidence that further implicates a dysregulated immune response in AD stemming from activated microglia (Efthymiou & Goate, 2017; Hopperton, Mohammad, Trépanier, Giuliano, & Bazinet, 2018; Kinney et al., 2018; Mostafavi et al., 2018; Salter & Stevens, 2017). GWAS implicates AD risk variants in and around genes that encode proteins important for innate immune function, including *CR1*, *CLU*, and *TREM2*, highlighting microglia as causative players in AD onset, progression and severity (Efthymiou & Goate, 2017; Jansen et al., 2019; Kunkle et al., 2019). However, studying the effects of these variants in AD mouse models poses challenges, since not all genetic loci for immune-related genes are conserved between humans and mice, requiring the need to "humanize" immune-related AD risk loci. One such locus is the Regulators of Complement Activation (RCA) cluster that encodes proteins in the complement cascade that is part of the innate immune response to microbial infection. Genetic variation between mice and humans in the CR1 locus within the RCA causes the complete absence of CR1 in mice (Farries & Atkinson, 1991; Jacobson & Weis, 2008; Nonaka, 2014). CR1 has multiple allotypes in humans that differ in size through the presence of long homologous repeats (LHRs), and certain allotypes may be responsible for modifying risk to developing AD (Biffi et al., 2010, 2012; Hazrati et al., 2012). To study the effects of CR1 allotype in causing disease risk, we engineered a mouse model to replace mouse *Cr2* with human *CR1* and *CR2*, as well as utilizing a Flp/FRT system to excise one LHR in the humanized CR1 gene to produce the $CR1^{long}$ and $CR1^{short}$ allotypes that are thought to be the most relevant to human disease, which was not possible using the mouse genetic orthologs (Jackson et al., 2020).

5.2 Deciphering peripheral and central immune contributions using iPSC-derived microglia

Since most genetic risk for LOAD is expressed in cells of the myeloid lineage, there is a current need for model systems to allow the study of the role of risk variants in these cell types.

It is well established that mouse and human immune systems are significantly different, making the generation of humanized mouse models using immunological loci that confer AD risk imperative to the study of how immunological factors influence AD susceptibility (Mestas & Hughes, 2004; Shay et al., 2013). However, a difficulty with studying immune-relevant AD pathways is deciphering central vs peripheral immune responses, especially since peripheral immune cells express many of the same genes and proteins as resident CNS microglia. Additionally, mice and human microglia show differential transcriptional and cellular responses to aging and AD, which could stem from the important genetic differences in immunological loci outside of those with known AD risk (Galatro et al., 2017; Geirsdottir et al., 2019; Mancuso et al., 2019; Masuda et al., 2019).

New discoveries in the generation of patient-derived induced pluripotent stem cells (iPSCs), as well as a more concrete understanding of the developmental underpinnings of microglia from early-life hematopoietic progenitors, has facilitated new opportunities to study the relationship between immune-relevant genetic risk and AD, as well as the microglia-specific contributions to AD neuropathology. Several protocols have been recently released that allow for rapid and efficient production of microglia-like (iMGL) cells from mouse and human iPSCs (Abud et al., 2017; McQuade et al., 2018; Pandya et al., 2017). Functional *in vitro* experiments such as pro-inflammatory cytokine secretion, ADP-induced calcium mobilization and migration, as well as phagocytosis of purified synaptosomes, have validated that these microglia behave similarly to their *in vivo* counterparts (Abud et al., 2017). However, caveats to studying microglia solely *in vitro* are that immune cells show an extreme sensitivity to environmental changes, and they do not transcriptionally recapitulate purified human microglia when cultured in isolation (Bohlen et al., 2017; Gosselin et al., 2017; Hasselmann et al., 2019; He, Taylor, Yao, & Bhattacharya, 2021).

To continue studying genetic influences on microglia function without relying on *in vitro* manipulations and mouse/human genetic differences, chimeric mouse models have been developed in which human iPSCs are differentiated into induced hematopoietic progenitor cells (iHPCs), and subsequently xenotransplanted into microglia-deficient mice using cranial injections (Hasselmann et al., 2019; Mancuso et al., 2019). These microglia show greater transcriptomic conservation to isolated human microglia, as well as functional responses to focal injury, neuronal phagocytosis and inflammatory challenge (Hasselmann et al., 2019). The transplanted microglia can be generated from human iPS cells that have either been selected to express LOAD risk variants or have been genetically engineered by CRISPR to express a specific LOAD variant. The latter approach enables the comparison of responses in systems that are genetically identical, other than the variant of interest. Additionally, transplanted microglia also show a robust cytokine-mediated inflammatory response to injected oligomeric amyloid, making future study of microglia using chimeric mouse models more relevant to evaluating contributions to AD pathology (Mancuso et al., 2019). A key to successfully studying microglia-specific contributions to AD is developing and leveraging the proper immunodeficient host mouse models such as the $hCSF1^{KI}$ mouse that supports engraftment of human microglia (Hasselmann et al., 2019; Mancuso et al., 2019; Rongvaux et al., 2014). Furthermore, new mouse models, such as the $Csf1r^{FIRE/FIRE}$ line, have been developed in which resident macrophages are selectively depleted without altering peripheral immune cells, allowing for the disentanglement of peripheral versus central contributions to AD (Rojo et al., 2019).

6. Mouse models at single-cell or cell-state resolution

As the study of AD and other dementias evolves away from a focus on neurons to other cell types, the field has begun to develop mouse models that express reporter molecules in specific cell types (e.g., microglia) (Kaiser & Feng, 2019; McKinsey et al., 2020), and is now moving toward reporters for specific cell states (e.g., damage or disease-associated microglia). These models will enable not only cell imaging but also cell-state-specific sorting, such that various ‘omics approaches (transcriptomics, proteomics, lipidomics) can be used to understand these cell states at the molecular level. Another experimental approach is to generate mouse lines that carry a floxed gene of interest to be combined with another model (or viral vector) expressing *Cre* recombinase in these cell types and states, to enable specific ablation of key genes only in a specific cell type (Kaiser & Feng, 2019; Masuda et al., 2020; McKinsey et al., 2020). This strategy has been used to show that key AD-related genes such as APOE (Huynh et al., 2019; Liu et al., 2017), BIN1 (De Rossi et al., 2020; McAvoy et al., 2019), and C1QA (Fonseca et al., 2017) act in a cell-autonomous manner.

7. Integrating genetic context into mouse models to better represent the patient

The mouse resources described thus far in this chapter utilize a limited number of genetic backgrounds. In humans, genetic context, or the genetics of an individual, has been shown to modulate age-related cognitive decline and the age-of-onset or progression of AD, strongly suggesting the existence of multiple genetic mechanisms that underlie vulnerability or resilience. Genetic context, in combination with environmental exposures, also exerts a strong effect on the rate of aging, lifespan and/or healthspan, which is also the greatest risk factor for AD. This likely means that one-size-fits-all therapies for patients will not be successful due to our genetically heterogenous population. This is highlighted by the number of failed clinical trials, and until recently, the lack of new therapeutics made available. With the advancement and decreased cost of genome sequencing, there is greater access to more patient transcriptomics and proteomics data than ever before. As discussed above, this rich dataset has allowed researchers to stratify these patients into "subtypes" of risk for AD (Milind et al., 2020; Neff et al., 2021; Wan et al., 2020). Therefore, embracing—rather than avoiding—genetic diversity in our approaches to modeling AD is imperative to begin to untangle the various ways gene interactions modulate individual trajectories during aging and in complex disease.

Until recently, the mouse background strain for models of specific gene function or disease-related variants has been limited due to efforts to maintain consistency and reproducibility across labs around the world. Classical laboratory strains are an admixture of *Mus musculus* subspecies, and were primarily derived from domesticated mouse stocks. They were often selected for a specific trait, such as fecundity, ease of handling, and most often, successful survival of the inbreeding process. While these strains, such as B6 and 129/SvlmJ, have proved to be successful models for Mendelian genetics in which a single mutation produces a strong effect, this has likely limited the recapitulation of aspects of more complex processes, such as aging and neurodegenerative disease.

A number of mouse resources that have varying levels of genetic diversity are now available to the research community that move beyond previously used laboratory strains. One approach has been the development of wild-derived inbred strains originating from stocks trapped in disparate regions of the world, that are representative of specific house mouse species: *Mus musculus domesticus, Mus musculus musculus,* and *Mus musculus castaneus* (Fig. 2A). As these strains were caught and then inbred in the lab much more recently, their genomes are closer to recapitulating combinations of genetic variants that would exist in the natural world. Natural genetic variation may be a critical missing component in understanding how differences, such as in immune function, contribute, interact and ultimately fail in the context of neurodegenerative diseases. Furthermore, wild-derived strains have been shown to differ in onset of sexual maturity, lifespan (Yuan et al., 2020), and processes of aging (Selman & Swindell, 2018). Much of the current work across disciplines has focused on three specific wild-derived strains: PWK/PhJ, WSB/EiJ and CAST/EiJ, as these were used as founders, in combination with classic laboratory strains, for development of diversity genome mapping resources such as the Collaborative Cross

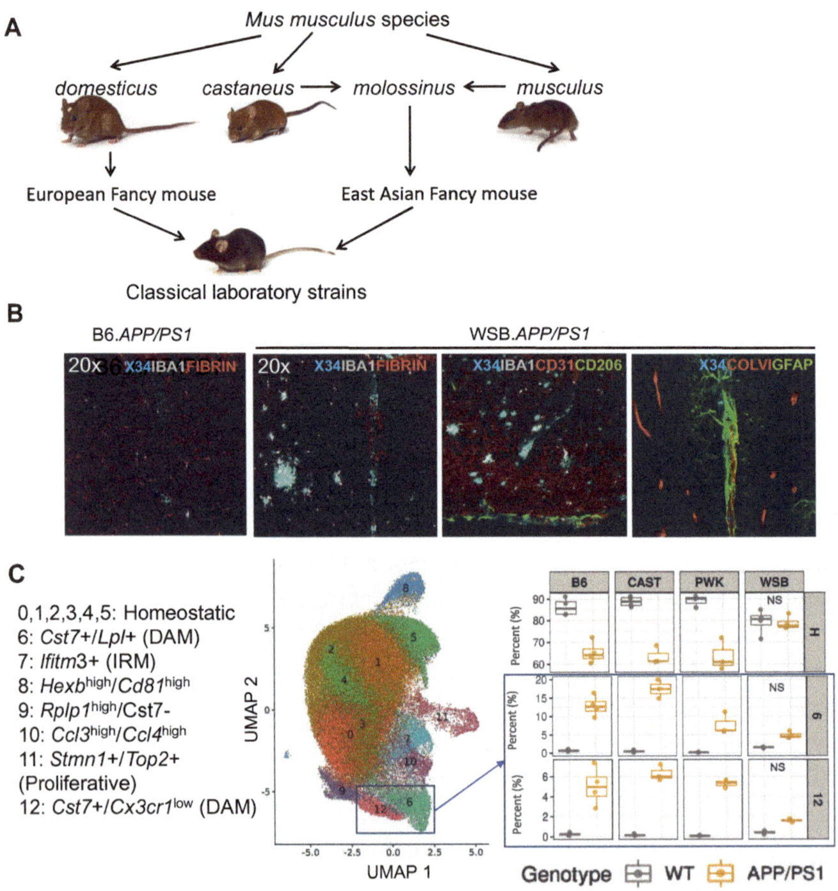

Fig. 2 (A) Outline of key NIH-funded consortia developed to accelerate collaborative research to better identify translational models and treatments for AD. (B) B6. *APP/PS1* rarely show accumulation of vascular amyloid, while WSB.*APP/PS1* exhibit robust CAA in cortical and leptomeningeal vessels with associated leakage as indicated by fibrin. Co-staining of X34 with IBA1 reveals little recruitment of microglia to CAA, and instead highlights astrocytes wrapped around areas of collagen loss. (C) Single-cell microglia sequencing identifies many microglia states. Unlike other wild-derived strains, there is no difference in the abundance of homeostatic microglia, or a significant increase in DAM in WSB.*APP/PS1* compared with WSB.

(Phillippi et al., 2014; Threadgill, Miller, Churchill, & de Villena, 2011) and Diversity Outbred (Chesler, 2014) populations. While the genomes of these populations are a random combination of founder-contributed alleles, the wild-derived contributed alleles are often the defining difference for traits studied. For a complete review on the differences that underlie isogenic, recombinant and outbred mouse strains and mapping populations, please refer to these reviews (Li & Auwerx, 2020; Saul, Philip, Reinholdt, & Chesler, 2019). There has also been renewed interest in the development of wild-derived inbred lines, and groups such as the Michael Nachman (Phifer-Rixey et al., 2018) and Beth Dumont labs are in the process of inbreeding and importing new wild-derived strains to make freely available to the research community. These strains are all *Mus musculus domesticus*, but were caught in extremely different environments. This will help to further tease apart the impact of environment on genetic variation underlying specific biological processes.

It has already been well-established in prior sections that currently used mouse models of AD are limited in their representation of human-relevant pathologies. While these models may be successful in depositing large amounts of parenchymal plaques and glial activation, they may fail to demonstrate other important mixed pathologies such as CAA, vascular leakage or neuronal loss. While this lack can be attributed to a number of factors, such as differing lifespans and/or differences between human and mouse proteins, this can also be due to a narrow examination of genetic context. Our lab recently employed wild-derived strains to examine differences in amyloid pathology and immune function by introduction of *APP/PS1* (Onos et al., 2019). In comparison to B6, one of these strains, WSB.*APP/PS1* exhibited lower numbers of parenchymal amyloid plaques, but significantly more CAA partnered with robust fibrin leakage—a marker of blood–brain barrier compromise (Fig. 2B). Cortical and hippocampal neuron counts revealed moderate cell loss in layer I and II of the parietal cortex and CA1 in female WSB.*APP/PS1* at 8 months. While mutations in *APP* (e.g., Danish, Iowa or Dutch; rare in human cases) have been shown to drive CAA in some mouse models, CAA has not been reported in B6 mice carrying the *APP/PS1* transgenes without other interventions. The combination of both parenchymal and vascular plaque accumulation is an important component to modeling of human AD, as an estimated 85–95% of individuals with AD (Attems, 2005; Biffi & Greenberg, 2011) present with CAA. Furthermore, WSB.*APP/PS1* also exhibited unique neuroinflammatory responses compared to other AD strains assessed, with WSB.

APP/PS1 failing to demonstrate a significant increase in disease-associated microglia (DAM) (Fig. 2C) (Yang et al., 2021), but a strong association of astrocytes to both cortical and leptomeningeal vascular deposits of amyloid. Overall, this suggests that the WSB genetic context may be representative of patients in which interactions between cerebrovascular dysfunction, CAA and neuroinflammation contribute to the risk of neurodegeneration. This work led directly to the incorporation of multiple genetic contexts into the MODEL-AD consortium. Ultimately, the use of genetically diverse mouse models is likely to be the best and most clinically relevant approach to modeling different patient populations at the preclinical level.

8. Training resources for use of animal models of Alzheimer's disease

The Jackson Laboratory hosts an annual course entitled *Principles and Techniques for Improving Preclinical Translation in Alzheimer's Disease.* This is a led by the staff of the MODEL-AD Center, and includes both classroom instruction and practical training sessions. Lectures focus on the selection of appropriate animal models for preclinical testing; experimental design and analysis; pharmacokinetic, pharmacodynamic, and functional testing; transcriptomics, proteomics, biomarkers, neuropathology, and in vivo imaging as clinically relevant measures; and the ARRIVE (Animal Research: Reporting of In Vivo Experiments) guidelines. Practical sessions include hands-on laboratory training in relevant methods, such as compound dosing, serial blood and CSF collection, and brain tissue dissections, and experimental design and biostatistical practicums.

9. Conclusions

Although much is known about the overarching pathological features of AD, it is clear that there are significantly varied combinations of genetic and environmental risks that can result in these diagnosable features. Through the use of extensive -omics data in recent years, we have been able to uncover a broad scope of factors that contribute to risk. We are now able to translate this foundational data into more realistic mouse models, which we can use to explore the complex biology that leads to the classic AD pathology. Because early clinical intervention is the ultimate goal, understanding which systems and pathways prime the CNS for future damage is critical. This profile will likely present differently based on age, sex, race, ethnicity, environment and social history, so we will need to leverage mouse diversity as well as human diversity in our studies. To further amplify the usefulness of more diverse mouse models, we aim to make data accessibility and communication a priority for current and future projects.

References

Abud, E. M., Ramirez, R. N., Martinez, E. S., Healy, L. M., Nguyen, C. H. H., Newman, S. A., et al. (2017). iPSC-derived human microglia-like cells to study neurological diseases. *Neuron, 94*(2), 278–293.e279. https://doi.org/10.1016/j.neuron.2017.03.042.

Andorfer, C., Kress, Y., Espinoza, M., de Silva, R., Tucker, K. L., Barde, Y. A., et al. (2003). Hyperphosphorylation and aggregation of tau in mice expressing normal human tau isoforms. *Journal of Neurochemistry, 86*(3), 582–590. https://doi.org/10.1046/j.1471-4159.2003.01879.x.

Armstrong, R. A. (2008). Spatial correlations between beta-amyloid (Abeta) deposits and blood vessels in familial Alzheimer's disease. *Folia Neuropathologica, 46*(4), 241–248.

Attems, J. (2005). Sporadic cerebral amyloid angiopathy: Pathology, clinical implications, and possible pathomechanisms. *Acta Neuropathologica, 110*(4), 345–359. https://doi.org/10.1007/s00401-005-1074-9.

Baglietto-Vargas, D., Forner, S., Cai, L., Martini, A. C., Trujillo-Estrada, L., Swarup, V., et al. (2021). Generation of a humanized Aβ expressing mouse demonstrating aspects of Alzheimer's disease-like pathology. *Nature Communications, 12*(1), 2421. https://doi.org/10.1038/s41467-021-22624-z.

Baird, D. A., Liu, J. Z., Zheng, J., Sieberts, S. K., Perumal, T., Elsworth, B., et al. (2021). Identifying drug targets for neurological and psychiatric disease via genetics and the brain transcriptome. *PLoS Genetics, 17*(1), e1009224. https://doi.org/10.1371/journal.pgen.1009224.

Bell, R. D., Deane, R., Chow, N., Long, X., Sagare, A., Singh, I., et al. (2009). SRF and myocardin regulate LRP-mediated amyloid-beta clearance in brain vascular cells. *Nature Cell Biology, 11*(2), 143–153. https://doi.org/10.1038/ncb1819.

Bell, R. D., Sagare, A. P., Friedman, A. E., Bedi, G. S., Holtzman, D. M., Deane, R., et al. (2007). Transport pathways for clearance of human Alzheimer's amyloid beta-peptide and apolipoproteins E and J in the mouse central nervous system. *Journal of Cerebral Blood Flow and Metabolism, 27*(5), 909–918. https://doi.org/10.1038/sj.jcbfm.9600419.

Bell, R. D., Winkler, E. A., Singh, I., Sagare, A. P., Deane, R., Wu, Z., et al. (2012). Apolipoprotein E controls cerebrovascular integrity via cyclophilin A. *Nature, 485*(7399), 512–516. https://doi.org/10.1038/nature11087.

Berwanger, C. S., Jeremy, J. Y., & Stansby, G. (1995). Homocysteine and vascular disease. *The British Journal of Surgery, 82*(6), 726–731. https://doi.org/10.1002/bjs.1800820604.

Biffi, A., Anderson, C. D., Desikan, R. S., Sabuncu, M., Cortellini, L., Schmansky, N., et al. (2010). Genetic variation and neuroimaging measures in Alzheimer disease. *Archives of Neurology, 67*(6), 677–685. https://doi.org/10.1001/archneurol.2010.108.

Biffi, A., & Greenberg, S. M. (2011). Cerebral amyloid angiopathy: A systematic review. *Journal of Clinical Neurology*, 7(1), 1–9. https://doi.org/10.3988/jcn.2011.7.1.1.

Biffi, A., Shulman, J. M., Jagiella, J. M., Cortellini, L., Ayres, A. M., Schwab, K., et al. (2012). Genetic variation at CR1 increases risk of cerebral amyloid angiopathy. *Neurology, 78*(5), 334–341. https://doi.org/10.1212/WNL.0b013e3182452b40.

Bohlen, C. J., Bennett, F. C., Tucker, A. F., Collins, H. Y., Mulinyawe, S. B., & Barres, B. A. (2017). Diverse requirements for microglial survival, specification, and function revealed by defined-medium cultures. *Neuron, 94*(4), 759–773.e758. https://doi.org/10.1016/j.neuron.2017.04.043.

Buchanan, R. A., Foley, K. E., Pepper, K. W., Reagan, A. M., Keezer, K. J., Hewes, A. A., et al. (2019). Meox2 haploinsufficiency accelerates axonal degeneration in DBA/2J glaucoma. *Investigative Ophthalmology & Visual Science*, *60*(10), 3283–3296. https://doi.org/10.1167/iovs.18-26126.

Candia, A. F., Hu, J., Crosby, J., Lalley, P. A., Noden, D., Nadeau, J. H., et al. (1992). Mox-1 and Mox-2 define a novel homeobox gene subfamily and are differentially expressed during early mesodermal patterning in mouse embryos. *Development*, *116*(4), 1123–1136.

Cheng, G., Huang, C., Deng, H., & Wang, H. (2012). Diabetes as a risk factor for dementia and mild cognitive impairment: A meta-analysis of longitudinal studies. *Internal Medicine Journal*, *42*(5), 484–491. https://doi.org/10.1111/j.1445-5994.2012.02758.x.

Chesler, E. J. (2014). Out of the bottleneck: The diversity outcross and collaborative cross mouse populations in behavioral genetics research. *Mammalian Genome*, *25*(1–2), 3–11. https://doi.org/10.1007/s00335-013-9492-9.

Chow, N., Bell, R. D., Deane, R., Streb, J. W., Chen, J., Brooks, A., et al. (2007). Serum response factor and myocardin mediate arterial hypercontractility and cerebral blood flow dysregulation in Alzheimer's phenotype. *Proceedings of the National Academy of Sciences of the United States of America*, *104*(3), 823–828. https://doi.org/10.1073/pnas.0608251104.

Corder, E. H., Saunders, A. M., Strittmatter, W. J., Schmechel, D. E., Gaskell, P. C., Small, G. W., et al. (1993). Gene dose of apolipoprotein E type 4 allele and the risk of Alzheimer's disease in late onset families. *Science*, *261*(5123), 921–923. https://doi.org/10.1126/science.8346443.

Crean, S., Ward, A., Mercaldi, C. J., Collins, J. M., Cook, M. N., Baker, N. L., et al. (2011). Apolipoprotein E epsilon4 prevalence in Alzheimer's disease patients varies across global populations: A systematic literature review and meta-analysis. *Dementia and Geriatric Cognitive Disorders*, *31*(1), 20–30. https://doi.org/10.1159/000321984.

De Felice, F. G., & Lourenco, M. V. (2015). Brain metabolic stress and neuroinflammation at the basis of cognitive impairment in Alzheimer's disease. *Frontiers in Aging Neuroscience*, 7, 94. https://doi.org/10.3389/fnagi.2015.00094.

De Rossi, P., Nomura, T., Andrew, R. J., Masse, N. Y., Sampathkumar, V., Musial, T. F., et al. (2020). Neuronal BIN1 regulates presynaptic neurotransmitter release and memory consolidation. *Cell Reports*, *30*(10), 3520–3535 e3527. https://doi.org/10.1016/j.celrep.2020.02.026.

Deane, R., Sagare, A., Hamm, K., Parisi, M., Lane, S., Finn, M. B., et al. (2008). apoE isoform-specific disruption of amyloid beta peptide clearance from mouse brain. *The Journal of Clinical Investigation*, *118*(12), 4002–4013. https://doi.org/10.1172/jci36663.

DeTure, M. A., & Dickson, D. W. (2019). The neuropathological diagnosis of Alzheimer's disease. *Molecular Neurodegeneration*, *14*(1), 32. https://doi.org/10.1186/s13024-019-0333-5.

Devlin, A. M., Arning, E., Bottiglieri, T., Faraci, F. M., Rozen, R., & Lentz, S. R. (2004). Effect of Mthfr genotype on diet-induced hyperhomocysteinemia and vascular function in mice. *Blood*, *103*(7), 2624–2629. https://doi.org/10.1182/blood-2003-09-3078.

Efthymiou, A. G., & Goate, A. M. (2017). Late onset Alzheimer's disease genetics implicates microglial pathways in disease risk. *Molecular Neurodegeneration*, *12*(1), 43. https://doi.org/10.1186/s13024-017-0184-x.

El Gaamouch, F., Audrain, M., Lin, W. J., Beckmann, N., Jiang, C., Hariharan, S., et al. (2020). VGF-derived peptide TLQP-21 modulates microglial function through C3aR1 signaling pathways and reduces neuropathology in 5xFAD mice. *Molecular Neurodegeneration*, *15*(1), 4. https://doi.org/10.1186/s13024-020-0357-x.

Faraco, G., Park, L., Zhou, P., Luo, W., Paul, S. M., Anrather, J., et al. (2016). Hypertension enhances Aβ-induced neurovascular dysfunction, promotes β-secretase activity, and leads to amyloidogenic processing of APP. *Journal of Cerebral Blood Flow and Metabolism*, *36*(1), 241–252. https://doi.org/10.1038/jcbfm.2015.79.

Farries, T. C., & Atkinson, J. P. (1991). Evolution of the complement system. *Immunology Today*, *12*(9), 295–300. https://doi.org/10.1016/0167-5699(91)90002-b.

Foley, K. E., Garceau, D. T., Kotredes, K. P., Carter, G. W., Sasner, M., & Howell, G. R. (2020). APOEε3/ε4 and APOEε4/ε4 genotypes drive unique gene signatures in the cortex of young mice. *bioRxiv*. https://doi.org/10.1101/2020.10.28.359422. 2020.2010.2028.359422.

Fonseca, M. I., Chu, S. H., Hernandez, M. X., Fang, M. J., Modarresi, L., Selvan, P., et al. (2017). Cell-specific deletion of C1qa identifies microglia as the dominant source of C1q in mouse brain. *Journal of Neuroinflammation*, *14*(1), 48. https://doi.org/10.1186/s12974-017-0814-9.

Galatro, T. F., Holtman, I. R., Lerario, A. M., Vainchtein, I. D., Brouwer, N., Sola, P. R., et al. (2017). Transcriptomic analysis of purified human cortical microglia reveals age-associated changes. *Nature Neuroscience*, *20*(8), 1162–1171. https://doi.org/10.1038/nn.4597.

Gama Sosa, M. A., Gasperi, R. D., Rocher, A. B., Wang, A. C., Janssen, W. G., Flores, T., et al. (2010). Age-related vascular pathology in transgenic mice expressing presenilin 1-associated familial Alzheimer's disease mutations. *The American Journal of Pathology*, *176*(1), 353–368. https://doi.org/10.2353/ajpath.2010.090482.

Gatz, M., Reynolds, C. A., Fratiglioni, L., Johansson, B., Mortimer, J. A., Berg, S., et al. (2006). Role of genes and environments for explaining Alzheimer disease. *Archives of General Psychiatry*, *63*(2), 168–174. https://doi.org/10.1001/archpsyc.63.2.168.

Geirsdottir, L., David, E., Keren-Shaul, H., Weiner, A., Bohlen, S. C., Neuber, J., et al. (2019). Cross-species single-cell analysis reveals divergence of the primate microglia program. *Cell*, *179*(7), 1609–1622.e1616. https://doi.org/10.1016/j.cell.2019.11.010.

Goodwin, L. O., Splinter, E., Davis, T. L., Urban, R., He, H., Braun, R. E., et al. (2019). Large-scale discovery of mouse transgenic integration sites reveals frequent structural variation and insertional mutagenesis. *Genome Research*, *29*(3), 494–505. https://doi.org/10.1101/gr.233866.117.

Gorski, D. H., & Leal, A. J. (2003). Inhibition of endothelial cell activation by the homeobox gene Gax. *The Journal of Surgical Research*, *111*(1), 91–99. https://doi.org/10.1016/s0022-4804(03)00042-8.

Gosselin, D., Skola, D., Coufal, N. G., Holtman, I. R., Schlachetzki, J. C. M., Sajti, E., et al. (2017). An environment-dependent transcriptional network specifies human microglia identity. *Science*, *356*(6344). https://doi.org/10.1126/science.aal3222.

Gudala, K., Bansal, D., Schifano, F., & Bhansali, A. (2013). Diabetes mellitus and risk of dementia: A meta-analysis of prospective observational studies. *Journal of Diabetes Investigation*, *4*(6), 640–650. https://doi.org/10.1111/jdi.12087.

Gulbranson, D. R., Ho, K., Yu, G. Q., Yu, X., Das, M., Shao, E., et al. (2021). Phenotypic differences between the Alzheimer's disease-related hAPP-J20 model and heterozygous Zbtb20 knockout mice. *eNeuro*. https://doi.org/10.1523/ENEURO.0089-21.2021.

Gupta, A., & Iadecola, C. (2015). Impaired Aβ clearance: A potential link between atherosclerosis and Alzheimer's disease. *Frontiers in Aging Neuroscience*, 7, 115. https://doi.org/10.3389/fnagi.2015.00115.

Halliday, M. R., Pomara, N., Sagare, A. P., Mack, W. J., Frangione, B., & Zlokovic, B. V. (2013). Relationship between cyclophilin a levels and matrix metalloproteinase 9 activity in cerebrospinal fluid of cognitively normal apolipoprotein e4 carriers and blood-brain barrier breakdown. *JAMA Neurology*, *70*(9), 1198–1200. https://doi.org/10.1001/jamaneurol.2013.3841.

Halliday, M. R., Rege, S. V., Ma, Q., Zhao, Z., Miller, C. A., Winkler, E. A., et al. (2016). Accelerated pericyte degeneration and blood-brain barrier breakdown in apolipoprotein E4 carriers with Alzheimer's disease. *Journal of Cerebral Blood Flow and Metabolism*, *36*(1), 216–227. https://doi.org/10.1038/jcbfm.2015.44.

Hankey, G. J., & Eikelboom, J. W. (1999). Homocysteine and vascular disease. *Lancet*, *354*(9176), 407–413. https://doi.org/10.1016/s0140-6736(98)11058-9.

Harold, D., Abraham, R., Hollingworth, P., Sims, R., Gerrish, A., Hamshere, M. L., et al. (2009). Genome-wide association study identifies variants at CLU and PICALM associated with Alzheimer's disease. *Nature Genetics*, *41*(10), 1088–1093. https://doi.org/10.1038/ng.440.

Hasselmann, J., Coburn, M. A., England, W., Figueroa Velez, D. X., Kiani Shabestari, S., Tu, C. H., et al. (2019). Development of a chimeric model to study and manipulate human microglia in vivo. *Neuron*, *103*(6), 1016–1033.e1010. https://doi.org/10.1016/j.neuron.2019.07.002.

Hazrati, L. N., Van Cauwenberghe, C., Brooks, P. L., Brouwers, N., Ghani, M., Sato, C., et al. (2012). Genetic association of CR1 with Alzheimer's disease: A tentative disease mechanism. *Neurobiology of Aging*, *33*(12). https://doi.org/10.1016/j.neurobiolaging.2012.07.001. 2949.e2945-2949.e2912.

He, Y., Taylor, N., Yao, X., & Bhattacharya, A. (2021). Mouse primary microglia respond differently to LPS and poly(I:C) in vitro. *Scientific Reports*, *11*(1), 10447. https://doi.org/10.1038/s41598-021-89777-1.

Henstridge, C. M., Hyman, B. T., & Spires-Jones, T. L. (2019). Beyond the neuron-cellular interactions early in Alzheimer disease pathogenesis. *Nature Reviews. Neuroscience*, *20*(2), 94–108. https://doi.org/10.1038/s41583-018-0113-1.

Hopperton, K. E., Mohammad, D., Trépanier, M. O., Giuliano, V., & Bazinet, R. P. (2018). Markers of microglia in post-mortem brain samples from patients with Alzheimer's disease: A systematic review. *Molecular Psychiatry*, *23*(2), 177–198. https://doi.org/10.1038/mp.2017.246.

Hua, Y., Zhao, H., Kong, Y., & Ye, M. (2011). Association between the MTHFR gene and Alzheimer's disease: A meta-analysis. *The International Journal of Neuroscience*, *121*(8), 462–471. https://doi.org/10.3109/00207454.2011.578778.

Huynh, T. V., Wang, C., Tran, A. C., Tabor, G. T., Mahan, T. E., Francis, C. M., et al. (2019). Lack of hepatic apoE does not influence early Abeta deposition: Observations from a new APOE knock-in model. *Molecular Neurodegeneration*, *14*(1), 37. https://doi.org/10.1186/s13024-019-0337-1.

Iadecola, C. (2020). Revisiting atherosclerosis and dementia. *Nature Neuroscience*, *23*(6), 691–692. https://doi.org/10.1038/s41593-020-0626-6.

Iadecola, C., Zhang, F., Niwa, K., Eckman, C., Turner, S. K., Fischer, E., et al. (1999). SOD1 rescues cerebral endothelial dysfunction in mice overexpressing amyloid precursor protein. *Nature Neuroscience*, *2*(2), 157–161. https://doi.org/10.1038/5715.

Jack, C. R., Jr., Bennett, D. A., Blennow, K., Carrillo, M. C., Dunn, B., Haeberlein, S. B., et al. (2018). NIA-AA research framework: Toward a biological definition of Alzheimer's disease. *Alzheimers Dement*, *14*(4), 535–562. https://doi.org/10.1016/j.jalz.2018.02.018.

Jackson, H. M., Foley, K. E., O'Rourke, R., Stearns, T. M., Fathalla, D., Morgan, B. P., et al. (2020). A novel mouse model expressing human forms for complement receptors CR1 and CR2. *BMC Genetics*, *21*(1), 101. https://doi.org/10.1186/s12863-020-00893-9.

Jacobson, A. C., & Weis, J. H. (2008). Comparative functional evolution of human and mouse CR1 and CR2. *Journal of Immunology*, *181*(5), 2953–2959. https://doi.org/10.4049/jimmunol.181.5.2953.

Jankowsky, J. L., Fadale, D. J., Anderson, J., Xu, G. M., Gonzales, V., Jenkins, N. A., et al. (2004). Mutant presenilins specifically elevate the levels of the 42 residue beta-amyloid peptide in vivo: Evidence for augmentation of a 42-specific gamma secretase. *Human Molecular Genetics*, *13*(2), 159–170. https://doi.org/10.1093/hmg/ddh019.

Jankowsky, J. L., Slunt, H. H., Ratovitski, T., Jenkins, N. A., Copeland, N. G., & Borchelt, D. R. (2001). Co-expression of multiple transgenes in mouse CNS: A comparison of strategies. *Biomolecular Engineering*, *17*(6), 157–165. https://doi.org/10.1016/s1389-0344(01)00067-3.

Jankowsky, J. L., & Zheng, H. (2017). Practical considerations for choosing a mouse model of Alzheimer's disease. *Molecular Neurodegeneration*, *12*(1), 89. https://doi.org/10.1186/s13024-017-0231-7.

Jansen, I. E., Savage, J. E., Watanabe, K., Bryois, J., Williams, D. M., Steinberg, S., et al. (2019). Genome-wide meta-analysis identifies new loci and functional pathways influencing Alzheimer's disease risk. *Nature Genetics*, *51*(3), 404–413. https://doi.org/10.1038/s41588-018-0311-9.

Kaiser, T., & Feng, G. (2019). Tmem119-EGFP and Tmem119-CreERT2 transgenic mice for labeling and manipulating microglia. *eNeuro*, *6*(4). https://doi.org/10.1523/ENEURO.0448-18.2019.

Kamboh, M. I., Demirci, F. Y., Wang, X., Minster, R. L., Carrasquillo, M. M., Pankratz, V. S., et al. (2012). Genome-wide association study of Alzheimer's disease. *Translational Psychiatry*, *2*(5), e117. https://doi.org/10.1038/tp.2012.45.

Karch, C. M., & Goate, A. M. (2015). Alzheimer's disease risk genes and mechanisms of disease pathogenesis. *Biological Psychiatry*, 77(1), 43–51. https://doi.org/10.1016/j.biopsych.2014.05.006.

Kimbrough, I. F., Robel, S., Roberson, E. D., & Sontheimer, H. (2015). Vascular amyloidosis impairs the gliovascular unit in a mouse model of Alzheimer's disease. *Brain*, *138*(Pt 12), 3716–3733. https://doi.org/10.1093/brain/awv327.

Kinney, J. W., Bemiller, S. M., Murtishaw, A. S., Leisgang, A. M., Salazar, A. M., & Lamb, B. T. (2018). Inflammation as a central mechanism in Alzheimer's disease. *Alzheimers Dement (N Y)*, *4*, 575–590. https://doi.org/10.1016/j.trci.2018.06.014.

Kisler, K., Nelson, A. R., Montagne, A., & Zlokovic, B. V. (2017). Cerebral blood flow regulation and neurovascular dysfunction in Alzheimer disease. *Nature Reviews. Neuroscience*, *18*(7), 419–434. https://doi.org/10.1038/nrn.2017.48.

Kotredes, K., Oblak, A., Pandey, R., Lin, P., Garceau, D., Williams, H., et al. (2021). *A multi-discipline phenotyping platform for late-onset Alzheimer's disease employed on a novel, humanized APOEe4. Trem2*R47H mouse model. Research Square*. https://doi.org/10.21203/rs.3.rs-135213/v1.

Kruyer, A., Soplop, N., Strickland, S., & Norris, E. H. (2015). Chronic hypertension leads to neurodegeneration in the TgSwDI mouse model of Alzheimer's disease. *Hypertension*, *66*(1), 175–182. https://doi.org/10.1161/hypertensionaha.115.05524.

Kunkle, B. W., Grenier-Boley, B., Sims, R., Bis, J. C., Damotte, V., Naj, A. C., et al. (2019). Genetic meta-analysis of diagnosed Alzheimer's disease identifies new risk loci and implicates Aβ, tau, immunity and lipid processing. *Nature Genetics*, *51*(3), 414–430. https://doi.org/10.1038/s41588-019-0358-2.

Kunkle, B. W., Schmidt, M., Klein, H. U., Naj, A. C., Hamilton-Nelson, K. L., Larson, E. B., et al. (2021). Novel Alzheimer disease risk loci and pathways in African American individuals using the African genome resources panel: A meta-analysis. *JAMA Neurology*, *78*(1), 102–113. https://doi.org/10.1001/jamaneurol.2020.3536.

Lambert, J. C., Heath, S., Even, G., Campion, D., Sleegers, K., Hiltunen, M., et al. (2009). Genome-wide association study identifies variants at CLU and CR1 associated with Alzheimer's disease. *Nature Genetics*, *41*(10), 1094–1099. https://doi.org/10.1038/ng.439.

Lambert, J. C., Ibrahim-Verbaas, C. A., Harold, D., Naj, A. C., Sims, R., Bellenguez, C., et al. (2013). Meta-analysis of 74,046 individuals identifies 11 new susceptibility loci for Alzheimer's disease. *Nature Genetics*, *45*(12), 1452–1458. https://doi.org/10.1038/ng.2802.

Lancaster, T. M., Brindley, L. M., Tansey, K. E., Sims, R. C., Mantripragada, K., Owen, M. J., et al. (2015). Alzheimer's disease risk variant in CLU is associated with neural inefficiency in healthy individuals. *Alzheimers Dement, 11*(10), 1144–1152. https://doi.org/10.1016/j.jalz.2014.10.012.
Li, H., & Auwerx, J. (2020). Mouse systems genetics as a prelude to precision medicine. *Trends in Genetics, 36*(4), 259–272. https://doi.org/10.1016/j.tig.2020.01.004.
Li, L., Cao, D., Garber, D. W., Kim, H., & Fukuchi, K. (2003). Association of aortic atherosclerosis with cerebral beta-amyloidosis and learning deficits in a mouse model of Alzheimer's disease. *The American Journal of Pathology, 163*(6), 2155–2164. https://doi.org/10.1016/s0002-9440(10)63572-9.
Liu, C. C., Zhao, N., Fu, Y., Wang, N., Linares, C., Tsai, C. W., et al. (2017). ApoE4 accelerates early seeding of amyloid pathology. *Neuron, 96*(5), 1024–1032 e1023. https://doi.org/10.1016/j.neuron.2017.11.013.
Livingston, G., Huntley, J., Sommerlad, A., Ames, D., Ballard, C., Banerjee, S., et al. (2020). Dementia prevention, intervention, and care: 2020 report of the lancet commission. *Lancet, 396*(10248), 413–446. https://doi.org/10.1016/s0140-6736(20)30367-6.
Ma, Q., Zhao, Z., Sagare, A. P., Wu, Y., Wang, M., Owens, N. C., et al. (2018). Blood-brain barrier-associated pericytes internalize and clear aggregated amyloid-β42 by LRP1-dependent apolipoprotein E isoform-specific mechanism. *Molecular Neurodegeneration, 13*(1), 57. https://doi.org/10.1186/s13024-018-0286-0.
Maciejewska, K., Czarnecka, K., & Szymański, P. (2021). A review of the mechanisms underlying selected comorbidities in Alzheimer's disease. *Pharmacological Reports*. https://doi.org/10.1007/s43440-021-00293-5.
Mancuso, R., Van Den Daele, J., Fattorelli, N., Wolfs, L., Balusu, S., Burton, O., et al. (2019). Stem-cell-derived human microglia transplanted in mouse brain to study human disease. *Nature Neuroscience, 22*(12), 2111–2116. https://doi.org/10.1038/s41593-019-0525-x.
Masuda, T., Amann, L., Sankowski, R., Staszewski, O., Lenz, M., d'Errico, P., et al. (2020). Novel Hexb-based tools for studying microglia in the CNS. *Nature Immunology, 21*(7), 802–815. https://doi.org/10.1038/s41590-020-0707-4.
Masuda, T., Sankowski, R., Staszewski, O., Böttcher, C., Amann, L., Sagar, et al. (2019). Spatial and temporal heterogeneity of mouse and human microglia at single-cell resolution. *Nature, 566*(7744), 388–392. https://doi.org/10.1038/s41586-019-0924-x.
McAvoy, K. M., Rajamohamed Sait, H., Marsh, G., Peterson, M., Reynolds, T. L., Gagnon, J., et al. (2019). Cell-autonomous and non-cell autonomous effects of neuronal BIN1 loss in vivo. *PLoS One, 14*(8), e0220125. https://doi.org/10.1371/journal.pone.0220125.
McKinsey, G. L., Lizama, C. O., Keown-Lang, A. E., Niu, A., Santander, N., Larpthaveesarp, A., et al. (2020). A new genetic strategy for targeting microglia in development and disease. *eLife, 9*. https://doi.org/10.7554/eLife.54590.
McQuade, A., Coburn, M., Tu, C. H., Hasselmann, J., Davtyan, H., & Blurton-Jones, M. (2018). Development and validation of a simplified method to generate human microglia from pluripotent stem cells. *Molecular Neurodegeneration, 13*(1), 67. https://doi.org/10.1186/s13024-018-0297-x.
Mestas, J., & Hughes, C. C. (2004). Of mice and not men: Differences between mouse and human immunology. *Journal of Immunology, 172*(5), 2731–2738. https://doi.org/10.4049/jimmunol.172.5.2731.
Milind, N., Preuss, C., Haber, A., Ananda, G., Mukherjee, S., John, C., et al. (2020). Transcriptomic stratification of late-onset Alzheimer's cases reveals novel genetic modifiers of disease pathology. *PLoS Genetics, 16*(6), e1008775. https://doi.org/10.1371/journal.pgen.1008775.

Montagne, A., Nation, D. A., Sagare, A. P., Barisano, G., Sweeney, M. D., Chakhoyan, A., et al. (2020). APOE4 leads to blood-brain barrier dysfunction predicting cognitive decline. *Nature*, *581*(7806), 71–76. https://doi.org/10.1038/s41586-020-2247-3.

Mostafavi, S., Gaiteri, C., Sullivan, S. E., White, C. C., Tasaki, S., Xu, J., et al. (2018). A molecular network of the aging human brain provides insights into the pathology and cognitive decline of Alzheimer's disease. *Nature Neuroscience*, *21*(6), 811–819. https://doi.org/10.1038/s41593-018-0154-9.

Mucke, L., Masliah, E., Yu, G. Q., Mallory, M., Rockenstein, E. M., Tatsuno, G., et al. (2000). High-level neuronal expression of abeta 1-42 in wild-type human amyloid protein precursor transgenic mice: Synaptotoxicity without plaque formation. *The Journal of Neuroscience*, *20*(11), 4050–4058.

Neff, R. A., Wang, M., Vatansever, S., Guo, L., Ming, C., Wang, Q., et al. (2021). Molecular subtyping of Alzheimer's disease using RNA sequencing data reveals novel mechanisms and targets. *Science Advances*, 7(2). https://doi.org/10.1126/sciadv.abb5398.

Nelson, A. R., Sweeney, M. D., Sagare, A. P., & Zlokovic, B. V. (2016). Neurovascular dysfunction and neurodegeneration in dementia and Alzheimer's disease. *Biochimica et Biophysica Acta*, *1862*(5), 887–900. https://doi.org/10.1016/j.bbadis.2015.12.016.

Nelson, M. R., Tipney, H., Painter, J. L., Shen, J., Nicoletti, P., Shen, Y., et al. (2015). The support of human genetic evidence for approved drug indications. *Nature Genetics*, *47*(8), 856–860. https://doi.org/10.1038/ng.3314.

Niwa, K., Younkin, L., Ebeling, C., Turner, S. K., Westaway, D., Younkin, S., et al. (2000). Abeta 1-40-related reduction in functional hyperemia in mouse neocortex during somatosensory activation. *Proceedings of the National Academy of Sciences of the United States of America*, *97*(17), 9735–9740. https://doi.org/10.1073/pnas.97.17.9735.

Nonaka, M. (2014). Evolution of the complement system. *Sub-Cellular Biochemistry*, *80*, 31–43. https://doi.org/10.1007/978-94-017-8881-6_3.

Novikova, G., Kapoor, M., Tcw, J., Abud, E. M., Efthymiou, A. G., Chen, S. X., et al. (2021). Integration of Alzheimer's disease genetics and myeloid genomics identifies disease risk regulatory elements and genes. *Nature Communications*, *12*(1), 1610. https://doi.org/10.1038/s41467-021-21823-y.

Oakley, H., Cole, S. L., Logan, S., Maus, E., Shao, P., Craft, J., et al. (2006). Intraneuronal beta-amyloid aggregates, neurodegeneration, and neuron loss in transgenic mice with five familial Alzheimer's disease mutations: Potential factors in amyloid plaque formation. *The Journal of Neuroscience*, *26*(40), 10129–10140. https://doi.org/10.1523/JNEUROSCI.1202-06.2006.

Oblak, A. L., Forner, S., Territo, P. R., Sasner, M., Carter, G. W., Howell, G. R., et al. (2020). Model organism development and evaluation for late-onset Alzheimer's disease: MODEL-AD. *Alzheimers Dement (N Y)*, *6*(1), e12110. https://doi.org/10.1002/trc2.12110.

Onos, K. D., Sukoff Rizzo, S. J., Howell, G. R., & Sasner, M. (2016). Toward more predictive genetic mouse models of Alzheimer's disease. *Brain Research Bulletin*, *122*, 1–11. https://doi.org/10.1016/j.brainresbull.2015.12.003.

Onos, K. D., Uyar, A., Keezer, K. J., Jackson, H. M., Preuss, C., Acklin, C. J., et al. (2019). Enhancing face validity of mouse models of Alzheimer's disease with natural genetic variation. *PLoS Genetics*, *15*(5), e1008155. https://doi.org/10.1371/journal.pgen.1008155.

Pandya, H., Shen, M. J., Ichikawa, D. M., Sedlock, A. B., Choi, Y., Johnson, K. R., et al. (2017). Differentiation of human and murine induced pluripotent stem cells to microglia-like cells. *Nature Neuroscience*, *20*(5), 753–759. https://doi.org/10.1038/nn.4534.

Park, L., Koizumi, K., El Jamal, S., Zhou, P., Previti, M. L., Van Nostrand, W. E., et al. (2014). Age-dependent neurovascular dysfunction and damage in a mouse model of cerebral amyloid angiopathy. *Stroke*, *45*(6), 1815–1821. https://doi.org/10.1161/strokeaha.114.005179.

Phifer-Rixey, M., Bi, K., Ferris, K. G., Sheehan, M. J., Lin, D., Mack, K. L., et al. (2018). The genomic basis of environmental adaptation in house mice. *PLoS Genetics*, *14*(9), e1007672. https://doi.org/10.1371/journal.pgen.1007672.

Phillippi, J., Xie, Y., Miller, D. R., Bell, T. A., Zhang, Z., Lenarcic, A. B., et al. (2014). Using the emerging collaborative cross to probe the immune system. *Genes and Immunity*, *15*(1), 38–46. https://doi.org/10.1038/gene.2013.59.

Pimenova, A. A., Raj, T., & Goate, A. M. (2018). Untangling genetic risk for Alzheimer's disease. *Biological Psychiatry*, *83*(4), 300–310. https://doi.org/10.1016/j.biopsych.2017.05.014.

Preuss, C., Pandey, R., Piazza, E., Fine, A., Uyar, A., Perumal, T., et al. (2020). A novel systems biology approach to evaluate mouse models of late-onset Alzheimer's disease. *Molecular Neurodegeneration*, *15*(1), 67. https://doi.org/10.1186/s13024-020-00412-5.

Profenno, L. A., Porsteinsson, A. P., & Faraone, S. V. (2010). Meta-analysis of Alzheimer's disease risk with obesity, diabetes, and related disorders. *Biological Psychiatry*, *67*(6), 505–512. https://doi.org/10.1016/j.biopsych.2009.02.013.

Rai, V. (2017). Methylenetetrahydrofolate reductase (MTHFR) C677T polymorphism and Alzheimer disease risk: A meta-analysis. *Molecular Neurobiology*, *54*(2), 1173–1186. https://doi.org/10.1007/s12035-016-9722-8.

Rojo, R., Raper, A., Ozdemir, D. D., Lefevre, L., Grabert, K., Wollscheid-Lengeling, E., et al. (2019). Deletion of a Csf1r enhancer selectively impacts CSF1R expression and development of tissue macrophage populations. *Nature Communications*, *10*(1), 3215. https://doi.org/10.1038/s41467-019-11053-8.

Rom, S., Zuluaga-Ramirez, V., Gajghate, S., Seliga, A., Winfield, M., Heldt, N. A., et al. (2019). Hyperglycemia-driven neuroinflammation compromises BBB leading to memory loss in both diabetes mellitus (DM) type 1 and type 2 mouse models. *Molecular Neurobiology*, *56*(3), 1883–1896. https://doi.org/10.1007/s12035-018-1195-5.

Rongvaux, A., Willinger, T., Martinek, J., Strowig, T., Gearty, S. V., Teichmann, L. L., et al. (2014). Development and function of human innate immune cells in a humanized mouse model. *Nature Biotechnology*, *32*(4), 364–372. https://doi.org/10.1038/nbt.2858.

Rovelet-Lecrux, A., Legallic, S., Wallon, D., Flaman, J. M., Martinaud, O., Bombois, S., et al. (2012). A genome-wide study reveals rare CNVs exclusive to extreme phenotypes of Alzheimer disease. *European Journal of Human Genetics*, *20*(6), 613–617. https://doi.org/10.1038/ejhg.2011.225.

Saito, T., Matsuba, Y., Mihira, N., Takano, J., Nilsson, P., Itohara, S., et al. (2014). Single app knock-in mouse models of Alzheimer's disease. *Nature Neuroscience*, *17*(5), 661–663. https://doi.org/10.1038/nn.3697.

Salloway, S., Farlow, M., McDade, E., Clifford, D. B., Wang, G., Llibre-Guerra, J. J., et al. (2021). A trial of gantenerumab or solanezumab in dominantly inherited Alzheimer's disease. *Nature Medicine*, *27*(7), 1187–1196. https://doi.org/10.1038/s41591-021-01369-8.

Salter, M. W., & Stevens, B. (2017). Microglia emerge as central players in brain disease. *Nature Medicine*, *23*(9), 1018–1027. https://doi.org/10.1038/nm.4397.

Sasaguri, H., Nilsson, P., Hashimoto, S., Nagata, K., Saito, T., De Strooper, B., et al. (2017). APP mouse models for Alzheimer's disease preclinical studies. *The EMBO Journal*, *36*(17), 2473–2487. https://doi.org/10.15252/embj.201797397.

Saul, M. C., Philip, V. M., Reinholdt, L. G., & Chesler, E. J. (2019). High-diversity mouse populations for complex traits. *Trends in Genetics*, *35*(7), 501–514. https://doi.org/10.1016/j.tig.2019.04.003.

Selkoe, D. J., & Hardy, J. (2016). The amyloid hypothesis of Alzheimer's disease at 25 years. *EMBO Molecular Medicine*, *8*(6), 595–608. https://doi.org/10.15252/emmm.201606210.

Selman, C., & Swindell, W. R. (2018). Putting a strain on diversity. *The EMBO Journal*, *37*(22). https://doi.org/10.15252/embj.2018100862.

Serrano-Pozo, A., Das, S., & Hyman, B. T. (2021). APOE and Alzheimer's disease: Advances in genetics, pathophysiology, and therapeutic approaches. *Lancet Neurology, 20*(1), 68–80. https://doi.org/10.1016/s1474-4422(20)30412-9.
Shay, T., Jojic, V., Zuk, O., Rothamel, K., Puyraimond-Zemmour, D., Feng, T., et al. (2013). Conservation and divergence in the transcriptional programs of the human and mouse immune systems. *Proceedings of the National Academy of Sciences of the United States of America, 110*(8), 2946–2951. https://doi.org/10.1073/pnas.1222738110.
Sieberts, S. K., Perumal, T. M., Carrasquillo, M. M., Allen, M., Reddy, J. S., Hoffman, G. E., et al. (2020). Large eQTL meta-analysis reveals differing patterns between cerebral cortical and cerebellar brain regions. *Scientific Data, 7*(1), 340. https://doi.org/10.1038/s41597-020-00642-8.
Snyder, H. M., Corriveau, R. A., Craft, S., Faber, J. E., Greenberg, S. M., Knopman, D., et al. (2015). Vascular contributions to cognitive impairment and dementia including Alzheimer's disease. *Alzheimers Dement, 11*(6), 710–717. https://doi.org/10.1016/j.jalz.2014.10.008.
Soto, I., Grabowska, W. A., Onos, K. D., Graham, L. C., Jackson, H. M., Simeone, S. N., et al. (2016). Meox2 haploinsufficiency increases neuronal cell loss in a mouse model of Alzheimer's disease. *Neurobiology of Aging, 42*, 50–60. https://doi.org/10.1016/j.neurobiolaging.2016.02.025.
Srikanth, V., Maczurek, A., Phan, T., Steele, M., Westcott, B., Juskiw, D., et al. (2011). Advanced glycation endproducts and their receptor RAGE in Alzheimer's disease. *Neurobiology of Aging, 32*(5), 763–777. https://doi.org/10.1016/j.neurobiolaging.2009.04.016.
Stoccoro, A., Tannorella, P., Salluzzo, M. G., Ferri, R., Romano, C., Nacmias, B., et al. (2017). The methylenetetrahydrofolate reductase C677T polymorphism and risk for late-onset Alzheimer's disease: Further evidence in an Italian multicenter study. *Journal of Alzheimer's Disease, 56*(4), 1451–1457. https://doi.org/10.3233/jad-161081.
Strittmatter, W. J., Weisgraber, K. H., Huang, D. Y., Dong, L. M., Salvesen, G. S., Pericak-Vance, M., et al. (1993). Binding of human apolipoprotein E to synthetic amyloid beta peptide: Isoform-specific effects and implications for late-onset Alzheimer disease. *Proceedings of the National Academy of Sciences of the United States of America, 90*(17), 8098–8102. https://doi.org/10.1073/pnas.90.17.8098.
Sudduth, T. L., Powell, D. K., Smith, C. D., Greenstein, A., & Wilcock, D. M. (2013). Induction of hyperhomocysteinemia models vascular dementia by induction of cerebral microhemorrhages and neuroinflammation. *Journal of Cerebral Blood Flow and Metabolism, 33*(5), 708–715. https://doi.org/10.1038/jcbfm.2013.1.
Sudduth, T. L., Weekman, E. M., Brothers, H. M., Braun, K., & Wilcock, D. M. (2014). β-Amyloid deposition is shifted to the vasculature and memory impairment is exacerbated when hyperhomocysteinemia is induced in APP/PS1 transgenic mice. *Alzheimer's Research & Therapy, 6*(3), 32. https://doi.org/10.1186/alzrt262.
Sukoff Rizzo, S. J., Masters, A., Onos, K. D., Quinney, S., Sasner, M., Oblak, A., et al. (2020). Improving preclinical to clinical translation in Alzheimer's disease research. *Alzheimers Dement (N Y), 6*(1), e12038. https://doi.org/10.1002/trc2.12038.
Sweeney, M. D., Montagne, A., Sagare, A. P., Nation, D. A., Schneider, L. S., Chui, H. C., et al. (2019). Vascular dysfunction-the disregarded partner of Alzheimer's disease. *Alzheimers Dement, 15*(1), 158–167. https://doi.org/10.1016/j.jalz.2018.07.222.
Sweeney, M. D., Zhao, Z., Montagne, A., Nelson, A. R., & Zlokovic, B. V. (2019). Blood-brain barrier: From physiology to disease and Back. *Physiological Reviews, 99*(1), 21–78. https://doi.org/10.1152/physrev.00050.2017.
Takeda, S., Sato, N., Uchio-Yamada, K., Sawada, K., Kunieda, T., Takeuchi, D., et al. (2010). Diabetes-accelerated memory dysfunction via cerebrovascular inflammation

and Abeta deposition in an Alzheimer mouse model with diabetes. *Proceedings of the National Academy of Sciences of the United States of America, 107*(15), 7036–7041. https://doi.org/10.1073/pnas.1000645107.
Tarantini, S., Tran, C. H. T., Gordon, G. R., Ungvari, Z., & Csiszar, A. (2017). Impaired neurovascular coupling in aging and Alzheimer's disease: Contribution of astrocyte dysfunction and endothelial impairment to cognitive decline. *Experimental Gerontology, 94*, 52–58. https://doi.org/10.1016/j.exger.2016.11.004.
Thomas, R. S., Henson, A., Gerrish, A., Jones, L., Williams, J., & Kidd, E. J. (2016). Decreasing the expression of PICALM reduces endocytosis and the activity of β-secretase: Implications for Alzheimer's disease. *BMC Neuroscience, 17*(1), 50. https://doi.org/10.1186/s12868-016-0288-1.
Threadgill, D. W., Miller, D. R., Churchill, G. A., & de Villena, F. P. (2011). The collaborative cross: A recombinant inbred mouse population for the systems genetic era. *ILAR Journal, 52*(1), 24–31. https://doi.org/10.1093/ilar.52.1.24.
Toth, P., Tarantini, S., Csiszar, A., & Ungvari, Z. (2017). Functional vascular contributions to cognitive impairment and dementia: Mechanisms and consequences of cerebral autoregulatory dysfunction, endothelial impairment, and neurovascular uncoupling in aging. *American Journal of Physiology. Heart and Circulatory Physiology, 312*(1), H1–h20. https://doi.org/10.1152/ajpheart.00581.2016.
Tumminia, A., Vinciguerra, F., Parisi, M., & Frittitta, L. (2018). Type 2 diabetes mellitus and Alzheimer's disease: Role of insulin Signalling and therapeutic implications. *International Journal of Molecular Sciences, 19*(11). https://doi.org/10.3390/ijms19113306.
Wan, Y. W., Al-Ouran, R., Mangleburg, C. G., Perumal, T. M., Lee, T. V., Allison, K., et al. (2020). Meta-analysis of the Alzheimer's disease human brain transcriptome and functional dissection in mouse models. *Cell Reports, 32*(2), 107908. https://doi.org/10.1016/j.celrep.2020.107908.
Wen, P. H., De Gasperi, R., Sosa, M. A., Rocher, A. B., Friedrich, V. L., Jr., Hof, P. R., et al. (2005). Selective expression of presenilin 1 in neural progenitor cells rescues the cerebral hemorrhages and cortical lamination defects in presenilin 1-null mutant mice. *Development, 132*(17), 3873–3883. https://doi.org/10.1242/dev.01946.
Wilcock, D. M., & Colton, C. A. (2009). Immunotherapy, vascular pathology, and microhemorrhages in transgenic mice. *CNS & Neurological Disorders Drug Targets, 8*(1), 50–64. https://doi.org/10.2174/187152709787601858.
Wu, Z., Guo, H., Chow, N., Sallstrom, J., Bell, R. D., Deane, R., et al. (2005). Role of the MEOX2 homeobox gene in neurovascular dysfunction in Alzheimer disease. *Nature Medicine, 11*(9), 959–965. https://doi.org/10.1038/nm1287.
Xia, D., Lianoglou, S., Sandmann, T., Calvert, M., Suh, J. H., Thomsen, E., et al. (2021). Fibrillar Aβ causes profound microglial metabolic perturbations in a novel APP knock-in mouse model. *bioRxiv*. https://doi.org/10.1101/2021.01.19.426731. 2021.2001.2019.426731.
Yang, H. S., Onos, K. D., Choi, K., Keezer, K. J., Skelly, D. A., Carter, G. W., et al. (2021). Natural genetic variation determines microglia heterogeneity in wild-derived mouse models of Alzheimer's disease. *Cell Reports, 34*(6), 108739. https://doi.org/10.1016/j.celrep.2021.108739.
Yuan, R., Musters, C. J. M., Zhu, Y., Evans, T. R., Sun, Y., Chesler, E. J., et al. (2020). Genetic differences and longevity-related phenotypes influence lifespan and lifespan variation in a sex-specific manner in mice. *Aging Cell, 19*(11), e13263. https://doi.org/10.1111/acel.13263.
Zhang, F., Eckman, C., Younkin, S., Hsiao, K. K., & Iadecola, C. (1997). Increased susceptibility to ischemic brain damage in transgenic mice overexpressing the amyloid precursor protein. *The Journal of Neuroscience, 17*(20), 7655–7661. https://doi.org/10.1523/jneurosci.17-20-07655.1997.

Zhao, Z., Nelson, A. R., Betsholtz, C., & Zlokovic, B. V. (2015). Establishment and dysfunction of the blood-brain barrier. *Cell*, *163*(5), 1064–1078. https://doi.org/10.1016/j.cell.2015.10.067.

Zhao, Z., Sagare, A. P., Ma, Q., Halliday, M. R., Kong, P., Kisler, K., et al. (2015). Central role for PICALM in amyloid-β blood-brain barrier transcytosis and clearance. *Nature Neuroscience*, *18*(7), 978–987. https://doi.org/10.1038/nn.4025.

Zlokovic, B. V. (2011). Neurovascular pathways to neurodegeneration in Alzheimer's disease and other disorders. *Nature Reviews. Neuroscience*, *12*(12), 723–738. https://doi.org/10.1038/nrn3114.

Zlokovic, B. V., Martel, C. L., Matsubara, E., McComb, J. G., Zheng, G., McCluskey, R. T., et al. (1996). Glycoprotein 330/megalin: Probable role in receptor-mediated transport of apolipoprotein J alone and in a complex with Alzheimer disease amyloid beta at the blood-brain and blood-cerebrospinal fluid barriers. *Proceedings of the National Academy of Sciences of the United States of America*, *93*(9), 4229–4234. https://doi.org/10.1073/pnas.93.9.4229.

CHAPTER FIVE

Mouse models for the study of clustered protocadherins

Cathy M. McLeod[a] and Andrew M. Garrett[a,b,*]

[a]Department of Pharmacology, Wayne State University School of Medicine, Detroit, MI, United States
[b]Department of Ophthalmology, Visual, and Anatomical Sciences, Wayne State University School of Medicine, Detroit, MI, United States
*Corresponding author: e-mail address: andrew.garrett@wayne.edu

Contents

Abstract

Since their first description, the clustered protocadherins (cPcdhs) have sparked interest for their potential to generate diverse cell-surface recognition cues and their widespread expression in the nervous system. Through the use of mouse models, we have learned a great deal about the functions served by cPcdhs, and how their molecular diversity is regulated. cPcdhs are essential contributors to a host of processes during neural circuit formation, including neuronal survival, dendritic and axonal branching, self-avoidance and targeting, and synapse formation. Their expression is controlled by the interplay of epigenetic marks with proximal and distal elements involving high order DNA looping, regulating transcription factor binding. Here, we will review various mouse models targeting the cPcdh locus and how they have been instructive in uncovering the regulation and function of the cPcdhs.

Current Topics in Developmental Biology, Volume 148
ISSN 0070-2153
https://doi.org/10.1016/bs.ctdb.2021.12.006

1. Introduction

1.1 Clustered protocadherins

The clustered protocadherins (cPcdhs) are cell adhesion molecules (CAMs) of the cadherin superfamily named for their striking genetic organization in three tandem gene clusters. The first members were discovered in a yeast two hybrid screen for novel binding partners of Fyn tyrosine kinase (Kohmura et al., 1998). Two clones isolated from this screen bore a marked resemblance to cadherins and were found to be predominately expressed in the central nervous system (CNS), and were thus originally named cadherin related neuronal receptors or, CNR1 and CNR2 (Kohmura et al., 1998). The C-termini of the CNRs were identical in both amino acid and cDNA sequence, and Southern blot analyses revealed that they were members of a larger family with a shared 3' sequence and divergent 5' sequence (Kohmura et al., 1998). Sequence analyses identified three distinct gene clusters on human chromosome 5q31 and mouse chromosome 18 occupying nearly 1 MB and encoding ~60 protein isoforms classified in the protocadherin branch of the larger cadherin superfamily (Sugino et al., 2000; Wu & Maniatis, 1999, 2000). The three clusters *Pcdha, Pcdhb,* and *Pcdhg,* encode the α-, β-, and γ-protocadherins, respectively (Fig. 1A).

In mouse, the *Pcdha* and *Pcdhg* clusters are comprised of large "variable" exons (14 and 22 respectively) and three smaller "constant" exons. The variable exons encode six extracellular cadherin repeats (EC1–EC6), a transmembrane domain, and short variable cytoplasmic domain while the constant exons encode a cluster-specific common cytoplasmic C-terminus (Fig. 1). The middle cluster, *Pcdhb,* contains 22 variable exons without any constant exons (Sugino et al., 2000; Wu & Maniatis, 1999, 2000). Isoforms are further divided within the cluster by sequence homology. The *Pcdhg* cluster contains A, B, and C subgroups (i.e., γA, γB, γC, Fig. 1B). The *Pcdha* cluster includes the original CNRs and two additional C-type isoforms (αC, Fig. 1A) (Sugino et al., 2000; Wu & Maniatis, 1999). The five C-type isoforms (two in α, three in γ) are more closely related to each other than to other isoforms within the same cluster. In both the α and γ cluster, the C-type exons are the closest variable exons to the constant exons.

The exonic structure of the cPcdhs suggests that variable extracellular interactions could link to cytoplasmic interactions common within clusters (mediated by cluster-specific C-termini) as well as others distinct between isoforms (mediated by variable cytoplasmic domains). These are features

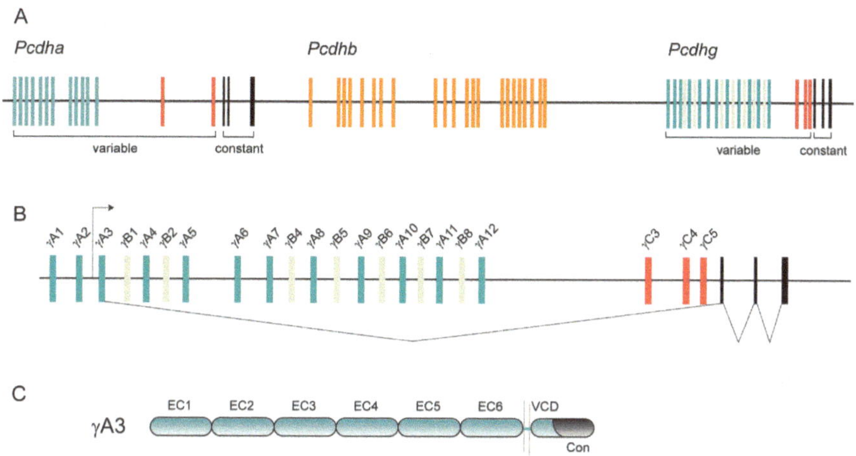

Fig. 1 The cPcdhs. (A) The cPcdh locus is comprised of three gene clusters: *Pcdha*(α), *Pcdhb*(β) and *Pcdhg*(γ), which are arranged in tandem on mouse chromosome 18. *Pcdha* and *Pcdhg* have 14 and 22 variable exons respectively, which splice to three constant exons. They include five total C type variable exons which are more homologous to each other than to other variable exons within the cluster. *Pcdhb* does not splice to constant exons. (B) The *Pcdhg* variable exons are divided into A, B and C types based on sequence homology. Like the other cPcdhs, *Pcdhg* variable exons are each preceded by their own promoter. (C) Each variable exon encodes most of the protein, including six extracellular cadherin repeats, a transmembrane domain and a short variable cytoplasmic region. The constant exons encode a common intracellular region within a cluster.

one might expect in molecules that facilitate neural circuit formation, as developing neurons use cell surface recognition molecules to sample the extracellular environment in order to achieve their appropriate morphology and form specific synaptic connections amongst a myriad of choices. Consistent with a role in neural development, sequence changes or epigenetic disruptions of the cPcdhs have been identified in disorders including dyslexia, major depression, Down syndrome, schizophrenia, and autism (El Hajj et al., 2016; Iossifov et al., 2014; Naskar et al., 2018; Shao et al., 2019; Vadodaria et al., 2019).

In this chapter, we will summarize how the use of mouse models has expanded our knowledge of cPcdh regulation and function as part of the complex mechanisms that govern neural circuitry. Mice are a particularly advantageous model as the organization of the mouse cPcdh locus is conserved with humans and nearly all isoforms are orthologous. Many mouse models targeting the cPcdh locus have been generated over the last 20 years, including those targeting individual clusters (Figs. 2 and 3) and multiple

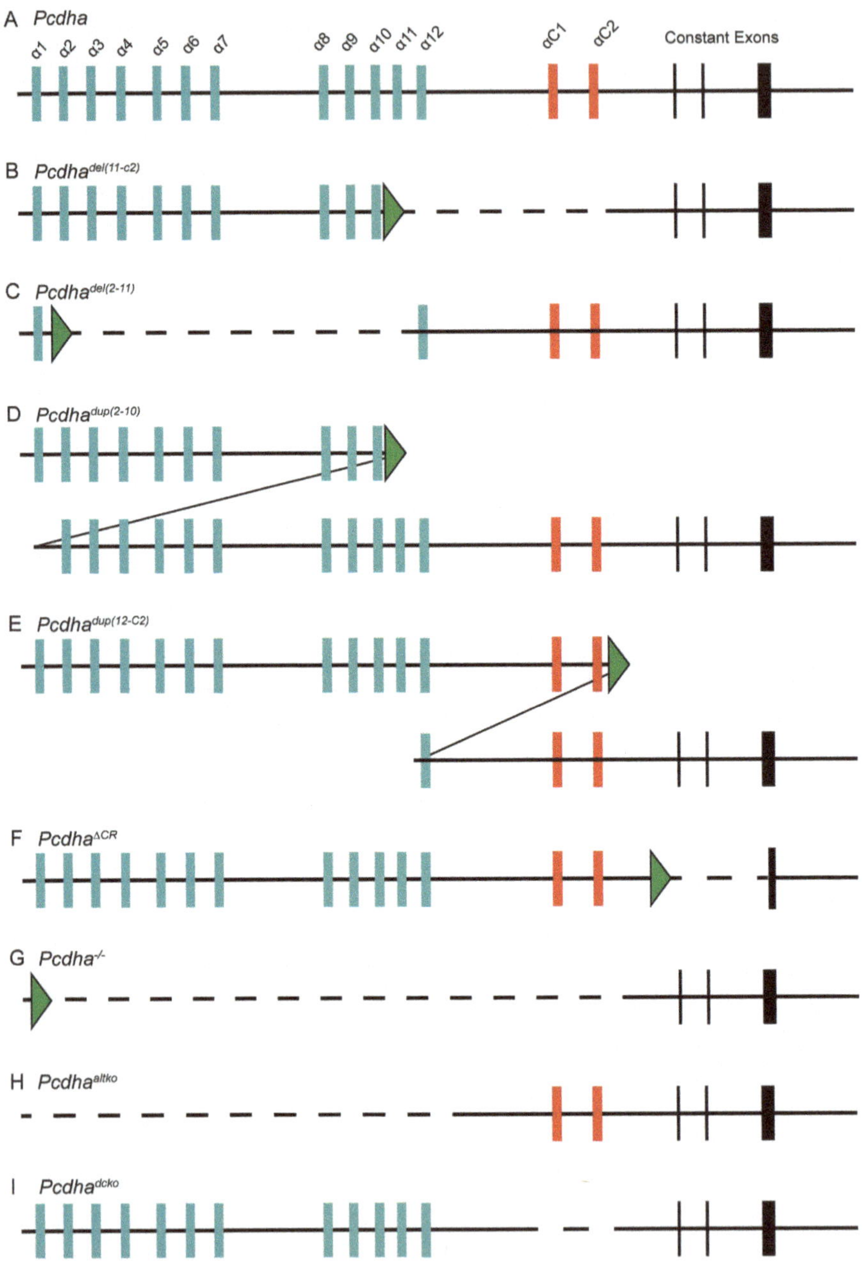

Fig. 2 Models targeting the *Pcdha* locus. (A) The *Pcdha* locus is comprised of 14 variable exons spliced to three constant exons. (B, C) Deletions and (D, E) duplications of the 5′ and 3′ Pcdha variable exons were achieved by the TAMERE method of trans-allelic recombination (Noguchi et al., 2009). (F) To disrupt the entire cluster, the constant exons of *Pcdha* were disrupted using TAMERE (Hasegawa et al., 2008). (G) In a separate allele disrupting the entire locus, the variable exons were deleted using serial loxP targeting followed by Cre recombination (Chen et al., 2017). (H) The alternate (non-C type) variable exons of *Pcdha*, or (I) the C type variable exons were removed using Crispr/Cas9 (Iossifov et al., 2014). Deletions are indicated by the dashed line, and loxP sites are denoted by green triangles.

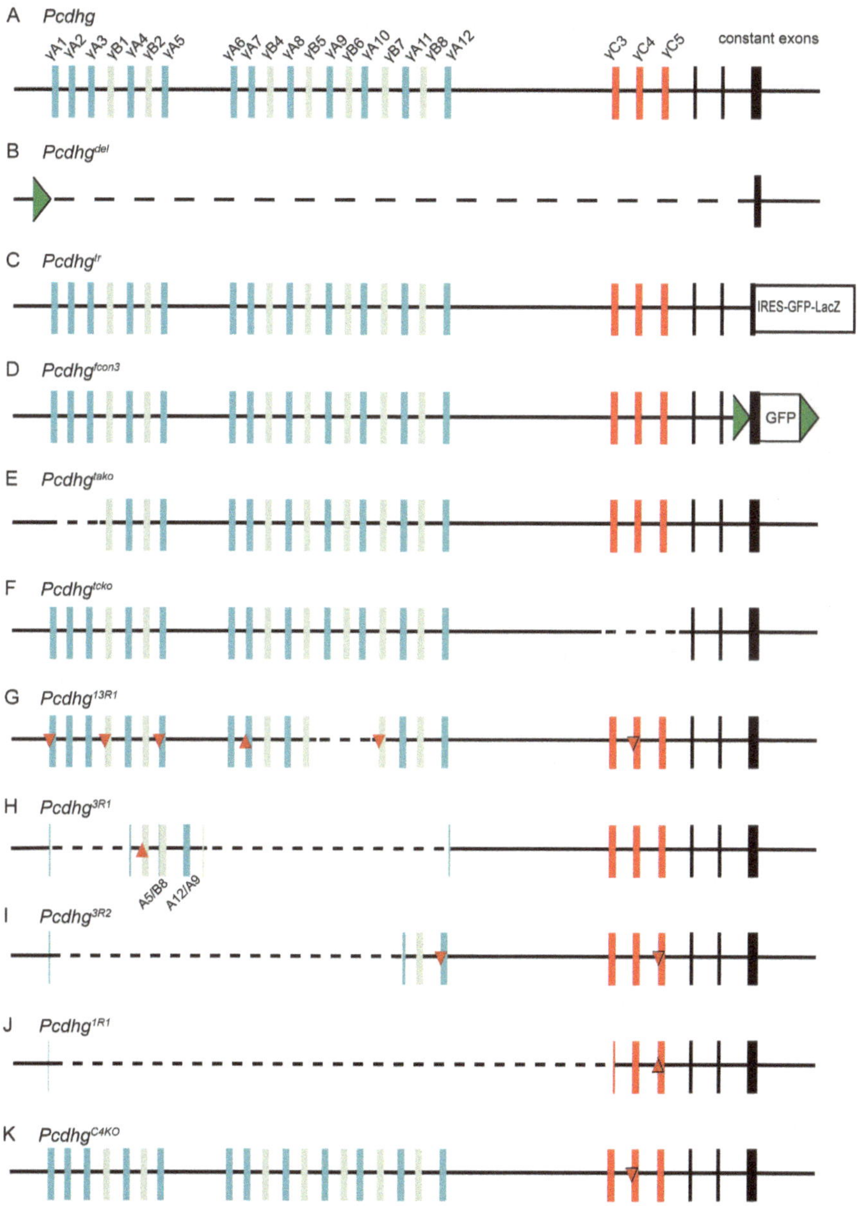

Fig. 3 Models targeting the *Pcdhg* locus. (A) The *Pcdhg* locus includes 22 variable exons spliced to three constant exons. There are A, B and C type variable exons based on sequence homology. Three mutants were made to disrupt the entire cluster: (B) *Pcdhg*^{del} was made by sequential targeting to insert loxP sites upstream of the variable exons and within the third constant exons, followed by Cre excision of intervening

(Continued)

clusters together (Fig. 4). While we will not attempt to provide an exhaustive list, we will highlight models that have proven particularly useful for studying the cPcdhs at the cluster and individual isoform levels, and to begin answering questions that would be otherwise unfeasible to address.

1.2 The structure and binding specificity of cPcdhs

EC domains mediate calcium-dependent adhesion and are the definitional characteristic of the cadherin superfamily. Classical cadherins rely on their EC1 domain (N-terminal and most distal from the membrane) for homophilic specificity, with a large hydrophobic pocket serving as the adhesion site (Morishita et al., 2006). As cPcdhs are a member of the cadherin superfamily it was not unreasonable to suppose that they would engage in homophilic *trans* interactions, as classical cadherins do. However, crystallographic studies of cPcdh EC1 revealed that this hydrophobic pocket was not present (Morishita et al., 2006). Further structural analyses of larger portions of the cPcdh ectodomain identified that homophilic interactions were mediated by the EC2:EC3 interactions. The cPcdhs engage their EC domains in a head to tail fashion where EC1:EC4 of each protein will contact each other, as well as EC2:EC3 (Goodman, Rubinstein, Thu, Bahna, et al., 2016, Goodman, Rubinstein, Thu, Mannepalli, et al., 2016; Rubinstein et al., 2015). These EC2 and EC3 regions are most diverse among the cPcdh ectodomains, which is consistent with their role in strict *trans* specificity.

While *trans* interactions are strictly homophilic, the cPcdhs form a variety of diverse multimers in *cis* via covalent and noncovalent interactions, some of which are required for cell surface localization (Schreiner & Weiner, 2010).

Fig. 3—Cont'd sequence (Wang et al., 2002), (C) *Pcdhg*^{tr}, a hypomorphic truncation mutant was made by inserting an IRES-GFP-LacZ cassette into the third constant exon (Weiner et al., 2005), and (D) *Pcdhg*^{fcon3}, a conditional loss of function allele, was constructed by inserting loxP sites flanking the third constant exons with an in-frame GFP fusion (Lefebvre et al., 2008; Prasad et al., 2008). Several distinct approaches have been taken to reduce isoform diversity. (E, F) Traditional homologous recombination was used to delete the first three A type variable exons (*Pcdhg* ^{tako}, E) or the three C type variable exons (*Pcdhg*^{tcko}, F) (Chen et al., 2012). (G–J) A shotgun CRISPR/Cas9 strategy was used to make an unbiased allelic series of mutants including *Pcdhg*^{13R1}, which stands for 13 isoforms remaining (G), *Pcdhg*^{3R1} (H) and *Pcdhg*^{3R2} (I) with only three intact isoforms, and *Pcdhg*^{1R1} (J) with only one intact isoform (Garrett et al., 2019). (K) *Pcdhg*^{C4KO} was created using CRISPR/Cas9 to test the importance of the γC4 isoform (Morishita et al., 2006). Red downfacing triangles denote frame shifting small deletions while upward facing triangles stand for frame shifting small insertions. Large deletions are indicated by the dashed line, and loxP sites are denoted by green triangles.

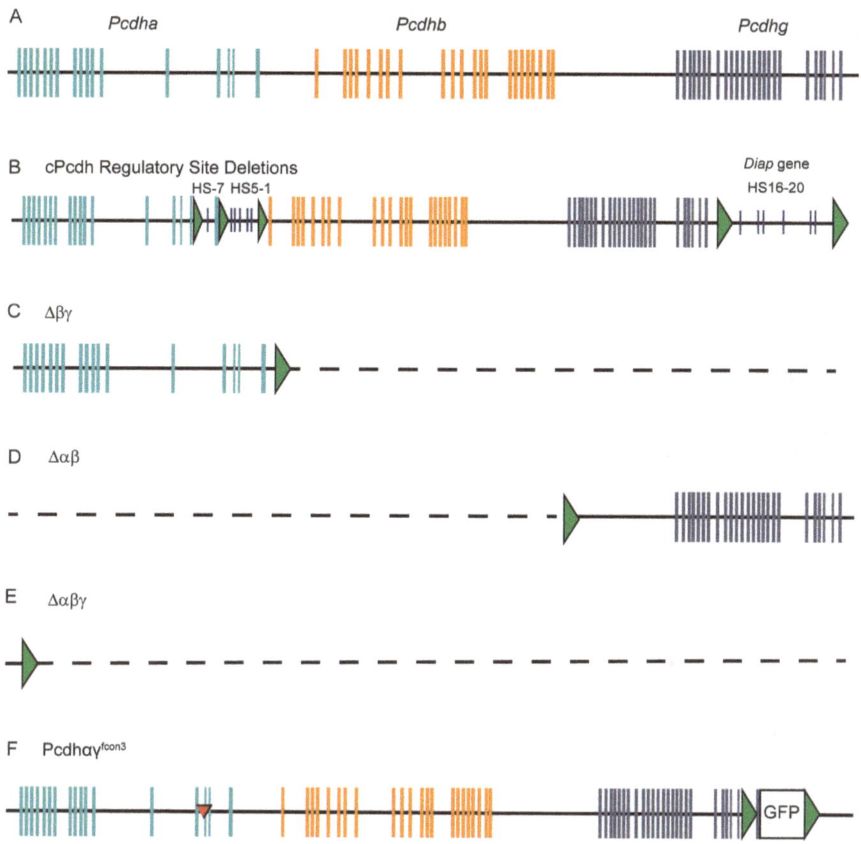

Fig. 4 Multi-cluster models. (A) The WT cPcdh locus. (B) Enhancer regions that regulate expression from the cPcdh locus were discovered by DNAse hypersensitivity assay. These regions were deleted by inserting flanking LoxP sites via gene targeting followed by Cre recombination. Multiple sites are indicated, but models were created by deleting them individually (Kehayova et al., 2011; Yokota et al., 2011). (C–E) Multi-cluster deletions were generated using different techniques by different groups, including distinct methods of trans-allelic recombination and multiple rounds of targeting by homologous recombination in embryonic stem cells (Hasegawa et al., 2016; Mountoufaris et al., 2017; Wu et al., 2007). (F) To disrupt both the *Pcdha* and *Pcdhg* clusters while circumventing the neonatal lethality of *Pcdhg* deletion, the conditional loss of function *Pcdhg*^{fcon3} allele was targeted by CRISPR/Cas9 to disrupt the *Pcdha* constant region (Ing-Esteves et al., 2018). The red downfacing triangle denotes a frame shifting small deletions, large deletions are indicated by the dashed line, and loxP sites are denoted by green triangles.

The specificity of *cis* interactions lies in the EC6 domain which has been shown as essential to ferry the non C-type α-Pcdhs and γC4-Pcdh isoforms to the membrane, as these isoforms require interaction with either a β or non-γC4 isoform for membrane transport and surface localization

(Murata, Hamada, Morishita, Mutoh, & Yagi, 2004; Thu et al., 2014). Importantly, these *cis* interactions regulate *trans* specificity. In a cell based adhesion assay, homophilic interactions were only observed when all isoforms were in common; even one mismatched isoform among opposing cell membranes was enough to prohibit interaction in a process of "interference," and even changing the relative ratio of the same isoforms reduced interaction (Schreiner & Weiner, 2010; Thu et al., 2014). The model for multimer formation based on these structures is a lattice of dimers (Rubinstein et al., 2015). Dimers in *cis* assemble on the membrane and bind with homophilic isoforms in *trans* on the opposing cell surfaces, the alternating *cis* and *trans* interactions lead to the formation of a zipper-like structure so long as isoforms match between membranes—a single mismatch will end the zipper formation (Rubinstein et al., 2015). A recent study using cryo-electron tomography has provided experimental evidence for this model (Brasch et al., 2019).

1.3 The evolution of the cPcdhs

The processes of gene conversion and adaptive variation are hypothesized to be major drivers of concerted evolution among cPcdh genes in humans, mice, rats and teleost fish (Noonan, Grimwood, Schmutz, Dickson, & Myers, 2004; Taguchi, Koide, Shiroishi, & Yagi, 2005; Yu, Yew, Rajasegaran, & Venkatesh, 2007). Gene conversion is suggested by comparative studies of the zebrafish and mammalian cPcdh genes. Genes within a species are more similar than their orthologs in another species, indicating that specialized functions may be generated by gene homogenization (Noonan et al., 2004). Some regions of the coding region, however, are protected from such conversion events, such as the EC3 coding region which is essential for homophilic specificity, suggesting conversion is prohibited as it would perturb proper homophilic recognition (Morgan, 2008; Noonan et al., 2004; Taguchi et al., 2005). Conversely, the EC6 domain seems to be the most prone to conversion and homogenization events within a species, which could account for the promiscuous *cis* interactions in which this domain engages (Noonan et al., 2004; Taguchi et al., 2005). The evolution of this cluster of genes is especially interesting given that these duplication events could contribute to the complexity of brain development. For example, zebrafish lack the beta cluster altogether, as well as the A type and B type gamma genes that are present in mammals. These additional genes allow for a greater repertoire of diverse molecules for potential use in neural circuits in mammals compared with zebrafish (Noonan et al., 2004). While previously thought to be only a vertebrate innovation, the cPcdhs have separately arisen in cephalopods which also boast a sophisticated CNS, supporting the idea that the cPcdh repertoire contributes to CNS complexity (Albertin et al., 2015; Styfhals, Seuntjens, Simakov, Sanges, & Fiorito, 2018; Yu et al., 2008).

2. Models to study isoform expression

The mechanisms of isoform choice at the expression level have important implications for how the cPcdhs contribute to neural development. Early studies using *in situ* hybridization found that individual isoforms were expressed in scattered cells in the CNS (Kohmura et al., 1998; Wang et al., 2002). Sequence analysis and 5' RACE identified individual promoters at the 5' end of each variable exon (Tasic et al., 2002). This left two possibilities for splicing of the variable exon to the constant exon: trans-splicing from the variable exon of one chromosome to a constant exon from the homologous chromosome; or cis splicing to a constant exon on the same chromosome. To determine which method of splicing occurred, Tasic and colleagues took advantage of strain-specific polymorphisms between C57BL/6 (B6) and Spret/Ei mice. By sequencing cDNA obtained from an F1 cross of these two strains, they found that splicing only occurred within but not between chromosomes, concluding that *Pcdha* and *Pcdhg* variable exons do not trans splice between homologous chromosomes (Tasic et al., 2002).

To ask if isoform choice was regulated at the single cell level, Takeshi Yagi's lab performed a series of single cell analyses on cerebellar Purkinje neurons (Esumi et al., 2005; Hirano et al., 2012; Kaneko et al., 2006). Purkinje neurons are large, and provided starting material for nested PCR amplification and single isoform detection. They found that individual neurons expressed all five of the C-type isoforms, ~2 of the other α isoforms, ~4 β isoforms, and ~4 of the other γ isoforms (Esumi et al., 2005; Hirano et al., 2012; Kaneko et al., 2006). These non-C-type isoforms appeared to be chosen stochastically, and are referred to as V-type isoforms, for variable expression. This pattern of expression suggests that rather than regulating cell-type-specific recognition events during neural development, cPcdhs endow each neuron with a unique identity. In addition to assaying isoform choice, these studies also analyzed allelic choice, asking if one or both alleles of a given isoform were expressed (Esumi et al., 2005; Hirano et al., 2012; Kaneko et al., 2006). To do so, they again leveraged strain polymorphisms in an F1 cross: Expression of the *Pcdha* isoforms was analyzed using polymorphisms between the B6 and MSM strains, while in separate experiments *Pcdhb* and *Pcdhg* alleles were distinguished using polymorphisms between B6 and JF1. These studies revealed strict monoallelic expression of cPcdh isoforms, with no preference for the maternal or paternal allele. Interestingly, in a given cell, while expression of each variable exon was monoallelic, both cPcdh alleles were active. All five constitutively expressed C-type isoforms displayed biallelic expression in each Purkinje cell (Kaneko et al., 2006). This scattered "salt-and-pepper" distribution of isoform expression has been verified by *in situ* hybridization and isoform-specific antibody staining in varying cell types within the CNS (Hirano et al., 2012; Kohmura et al., 1998; Lobas et al., 2012; Wang et al., 2002), but it does not appear to be universal: Single cell RNA sequencing of serotonergic neurons in the brain revealed that they predominantly expressed the αC2 isoform to the near exclusion of all others, although the other C isoforms were also present at lower levels (Chen et al., 2017). Single cell expression analyses based on sequencing rely on amplification and are prone to false positive and false negative errors. Recent advances in highly multiplex fluorescent *in situ* hybridization techniques could provide a clearer picture of isoform choice in different cell types.

Several different genetic models have been used to analyze how the cPcdhs are regulated at the cluster and supracluster levels. A series of *Pcdha* mutants were used to analyze cluster level regulation. The first of these were designed using the TAMERE method, or *ta*rgeted *me*iotic *re*combination. LoxP sites were inserted at different locations within the variable region of the *Pcdha* cluster. Breeding these mice together, along with Cre recombinase expressed in the testis, resulted in recombination between homologous chromosomes, generating a series of complementary deletions and duplications: $Pcdha^{del(11\text{-}c2)}$ with a $Pcdha^{dup(12\text{-}c2)}$, and $Pcdha^{del(2\text{-}11)}$ with $Pcdha^{dup(2\text{-}10)}$ (Fig. 2B–E) (Noguchi et al., 2009). The level of constant exon expression in Purkinje neurons was unchanged in these mutants, indicating compensation in deletion mutants and downregulation of each individual isoform in duplication mutants, together suggesting isoform expression is regulated at the level of the whole cluster (Noguchi et al., 2009). Further, the V-type isoform α10 acquired constitutive expression in del(11-c2) mutants in which it became the 3' most isoform, suggesting that the position of C-type variable exons near the constant exons may play a role in their constitutive expression (Noguchi et al., 2009). Two deletion models targeting putative enhancer sites (identified in DNase hypersensitivity assays) were used to identify the regulatory elements controlling *Pcdha* expression: HS5-1 KO and HS7 KO (Fig. 4B) (Kehayova, Monahan, Chen, & Maniatis, 2011). The deletion of HS5-1 had a distinctly more detrimental effect on the expression of genes to the 3' end of the Pcdha cluster, whereas the 5' end was more moderately affected, suggesting that physical distance from the enhancer itself influences the level of expression. The only exception to this appeared to be PcdhaC2, which was not reliant on HS5-1, but saw a major decrease in expression in the HS7 KO mutant (Kehayova et al., 2011). There was a distinct split in the effect of HS5-1 deletion on expression levels between *Pcdha5* and *Pcdha6* exons, which suggests that these are two different domains within the α cluster reliant on different enhancer elements for expression. There was also a decrease in binding of the transcription factor CTCF to *Pcdha* promoters in the HS5-1 KO, suggesting that at least for the highly affected isoforms (*Pcdha6-Pcdha12*), CTCF regulates expression by binding both the enhancer and promoter. This HS5-1 KO also resulted in inappropriate *Pcdha* expression in cultured kidney and glia cells, suggesting a role for this enhancer in regulating neuronal-specific expression (Kehayova et al., 2011). Interestingly, deletion of this regulatory element had an inverse effect on the *Pcdhb* genes closest to the enhancer, which were upregulated in HS5-1's absence (Yokota et al., 2011). An additional sequence element responsible for *Pcdhb* regulation was identified downstream of the *Pcdhg* cluster by DNase I hypersensitivity assays, and named HS16-20 (Fig. 4B). Yokota and colleagues generated both a transgenic reporter mouse to confirm that this region had *Pcdhb* enhancer activity, as well as deletion mutants (Yokota et al., 2011). These latter animals were analyzed by qPCR, demonstrating diminished *Pcdhg* expression and drastically reduced *Pcdhb* expression, establishing HS16-20 as a cluster control region (CCR) for *Pcdhb* and *Pcdhg* (Yokota et al., 2011). To determine if CCR acts in *cis* or in *trans*, heterozygous +/HS16-20 (mutated C57BL/6 allele) mice were crossed to the JF1 strain to generate F1 heterozygous mutants. Only JF1 sequences were seen in the +/HS16-20 animals, indicating that the CCR acts only in *cis* (Yokota et al., 2011).

To determine how individual isoforms are "selected" for expression, Kawaguchi and colleagues analyzed two mouse cell lines, C1300 and M3, which express distinct cohorts of *Pcdha* isoforms. Bisulfite sequencing revealed that promoter methylation negatively correlated with the expression level of a given isoform, where hypermethylation of a *Pcdha* promoter CpG site repressed transcription. The promoters of C type isoforms (αC1 and αC2) however were hypomethylated, consistent with their constitutive and biallelic expression (Kawaguchi et al., 2008). Consistent with this observation, the de novo methyltransferase Dnmt3b has been shown to regulate the methylation of cPcdh promoters. Dnmt3b KO animals exhibited branching errors in their Purkinje neurons, where an increased number of cPcdh isoforms were expressed. Interestingly, the C type isoforms were not affected by Dnmt3b loss, suggesting that their constitutive expression is regulated separately from that of the rest of the cluster (Toyoda et al., 2014).

The HS5-1 study described above implicated CTCF in the regulation of the cPcdhs. CTCF conditional knockouts (CTCF-cKO) were used to functionally test this, using the Nex-Cre line (Goebbels et al., 2006) to limit CTCF loss of function to postmitotic cortical and hippocampal projection neurons. Quantitative RT-PCR demonstrated that expression of all cPcdh stochastic isoforms except *Pcdhb1* were drastically decreased in the absence of CTCF, but the expression of the C-types except *Pcdhac1* was upregulated. Thus, CTCF is necessary for proper expression of the stochastically expressed Pcdhs in neurons of the cortex and hippocampus, but not necessarily the constitutive ones (Hirayama, Tarusawa, Yoshimura, Galjart, & Yagi, 2012). Further studies in neuroblastoma cell lines demonstrated that CTCF can bind to cPcdh promoter regions along with Rad21, a surrogate for the whole cohesin complex (Monahan et al., 2012). This binding is hypothesized to facilitate DNA loops joining the distal enhancer regions to individual variable exon promoters, dependent on methylation status, to express individual isoforms.

In addition to these loops between enhancers and promoters to regulate individual exons, large chromatin looping is an important structural regulator of the entire cPcdh locus. Large topological associated domains (TADs) insulate the cPcdh locus (Jiang et al., 2017). The methyltransferase Setdb1 is essential for establishing these TADs. In Setdb1 conditional loss of function mutants specific to the postnatal forebrain, TADs were disrupted resulting in hypomethylation that left the cPcdhs open to CTCF, leading to improper expression of the cPcdhs and dysfunctional dendritic morphology in mutant mice (Jiang et al., 2017).

3. Models to study function

3.1 Single cluster models

The initial analyses of cPcdh function came at the level of individual clusters with ablation mutants, beginning with the *Pcdhg* cluster. A complete deletion mutant ($Pcdhg^{del}$) was developed using two rounds of embryonic stem cell (ESC) targeting to insert loxP sites flanking the entire cluster: the first loxP cassette was inserted within the third constant exon using homologous recombination, and positive clones were subjected to another insertion upstream of the first variable exon (Wang et al., 2002). Transient Cre recombinase expression in double-positive clones resulted in excision of the entire cluster. The resulting mice were dubbed *Pcdhg*^{del} (Fig. 3B). Homozygous mutants exhibited a hunched posture, tremor, lack of stimulus response, and shallow breathing, followed by death soon after birth. This was accompanied by widespread interneuron cell death in the spinal cord and a significant reduction in synapse number (Wang et al., 2002). Weiner and colleagues took two approaches to test if this synaptic reduction was secondary to the cell death: they crossed *Pcdhg*^{del} mice with mutants for the pro-apoptotic gene *Bax* which prevented apoptosis, and they analyzed a hypomorphic mutant, *Pcdhg*^{tr}, which had no excessive cell death. In both cases, significantly fewer synapses were found in the spinal cord, demonstrating that γ-Pcdhs functioned in synaptic development independently from cell death (Weiner, Wang, Tapia, & Sanes, 2005). The *Pcdhg*^{tr} mutant was truncated 57 amino acids from the C terminus by the insertion of an IRES-GFP-LacZ cassette in the third constant exon of the *Pcdhg* cluster (Fig. 3C).

A conditional allele, *Pcdhg*^{fcon3}, was developed to circumvent the early postnatal lethality, allowing analysis of later developing tissue. This conditional mutant was generated by the insertion of loxP sites on either side of the third constant exon, with an in-frame GFP fusion (Fig. 3D). Cre recombinase expression in the target tissue will excise this exon, disrupting the entire *Pcdhg* cluster (Lefebvre, Zhang, Meister, Wang, & Sanes, 2008; Prasad, Wang, Gray, & Weiner, 2008). This allele has been used to great effect in teasing out the cell autonomy of the previously described phenotypes of neuronal cell death (Carriere et al., 2020; Lefebvre et al., 2008; Mancia Leon et al., 2020; Prasad et al., 2008; Su et al., 2010), and synapse malformation (Garrett & Weiner, 2009; Molumby et al., 2017; Prasad & Weiner, 2011), and has been instrumental in the discovery of two functions related to dendrite morphology: promotion of dendrite arborization in projection neurons in the cerebral cortex (Garrett, Schreiner, Lobas, & Weiner, 2012), and promotion of dendrite self-avoidance in Purkinje neurons in the cerebellum and starburst amacrine cells (SACs) in the retina (Kostadinov & Sanes, 2015; Lefebvre, Kostadinov, Chen, Maniatis, & Sanes, 2012).

Mutants disrupting the *Pcdha* cluster were generated by targeting the constant exons. An allele lacking the entire constant region, *Pcdha*$^{\Delta CR}$, was created by first generating two separate alleles, one with a loxP site before the first constant exon, and one with a loxP site within the third constant exon. These were combined through the TAMERE method of trans-allelic recombination with the intervening sequence excised (Fig. 2F) (Hasegawa et al., 2008). Unlike Pcdhg mutants, homozygous *Pcdha*$^{\Delta CR/\Delta CR}$ animals survived through adulthood with normal fertility. These mutants exhibited axonal targeting defects in the olfactory bulb, where olfactory sensory neurons expressing a single odorant receptor failed to coalesce their axons onto a single glomerulus (Hasegawa et al., 2008). These mice also revealed a function for α-Pcdhs in the axon development of serotonergic neurons in the raphe nucleus, as these projections in the null animals failed to diffusely innervate the brain, but formed clusters (Katori et al., 2009). This phenotype, which is likely related to the self-avoidance function described for γ-Pcdhs, was also verified in a separately developed mouse model (*Pcdha*$^{-/-}$) in which all *Pcdha* variable exons were excised using serial loxP targeting and Cre excision, similarly to *Pcdhg*^{del} (Fig. 2G) (Chen et al., 2017).

In addition to the large variable exons, the *Pcdha* cluster also generates variable transcripts from the constant region through alternative splicing within constant exon 3: A-type transcripts result from normal transcription through exon 3, while B-type transcripts involve an additional splice event within the exon. The predicted proteins differ at the C-terminus (Sugino et al., 2000). To isolate the functional relevance of these variants, two more models were made using standard homologous recombination to insert stop codons to disrupt the A or B transcripts (*Pcdha*$^{\Delta A}$ and *Pcdha*$^{\Delta B}$, respectively (Fukuda et al., 2008)). The A-type mutants, but not B-type, showed enhanced fear memory and dysfunctional targeting of serotonergic fibers (Fukuda et al., 2008; Katori et al., 2009). B-type specific functions have not been described.

The *Pcdhb* cluster has been targeted various times, both individually and in combination with the *Pcdha* or *Pcdhg* cluster (Chen et al., 2017; Hasegawa et al., 2016; Wu, Ying, Wu, & Capecchi, 2007). These multi-cluster models will be discussed below. As of yet, no *Pcdhb*-specific functions have been described.

3.2 Multi-cluster models

Since the three tandem cPcdh clusters produce proteins that can interact promiscuously in *cis* to generate large lattices, they are likely to have overlapping functions and to compensate for each other to some extent.

One way to address these cooperative functions is to create multi-cluster mutants (Hasegawa et al., 2008; Ing-Esteves et al., 2018; Wu et al., 2007). The first multi-cluster deletion encompassed *Pcdha* and *Pcdhb*, and was made through trans-allelic recombination using strong Cre expression during mitosis (rather than meiosis in the TAMERE method)(Wu et al., 2007). Hasegawa et. al. made a set of three multi-cluster mutants, $\Delta\alpha\beta$, $\Delta\beta\gamma$, and triple mutant $\Delta\alpha\beta\gamma$ using the TAMERE method (Fig. 4C–E). A third set of mutants, including a tri-cluster $\Delta\alpha\beta\gamma$, was generated through multiple rounds of targeting in ESCs to insert loxP sites, followed by Cre excision (Fig. 4E). Here, hypoxanthine phosphoribosyl transferase (HPRT) minigene cassettes were inserted to allow selection for rare long-range deletions by HPRT complementation (Mountoufaris et al., 2017).

Analysis of these large deletions revealed specific and shared roles. All mutations disrupting *Pcdhg* resulted in neonatal lethality and excessive cell death, while all mutants with *Pcdhg* intact survived with normal cell numbers. However, $\Delta\beta\gamma$ and $\Delta\alpha\beta\gamma$ mutants exhibited more severe cell loss than $\Delta\gamma$ alone, suggesting that γ-Pcdhs alone are sufficient for normal neuronal survival, but some cooperation occurs (Hasegawa et al., 2016). The olfactory targeting phenotype was more severe in multi-cluster mutants than those targeting *Pcdha* only, suggesting cPcdh diversity is an important contributor to this function (Hasegawa et al., 2016; Mountoufaris et al., 2017). Further, $\Delta\beta\gamma$ mutants exhibited lung abnormalities not found in single-cluster mutants, suggesting an overlapping function between the two clusters (Hasegawa et al., 2016).

The cPcdhs share their ~1 Mb genomic domain with other genes, and deletions encompassing *Pcdhb* and *Pcdhg* also removed the *Slc25a2* and *Taf7* genes. Further, any disruption of *Pcdhg* resulted in early postnatal lethality, precluding analysis of later developing phenotypes such as self-avoidance. Ing-Esteves and colleagues approached the locus with a lighter touch to overcome both challenges in generating multi-cluster mutants. Starting with the $Pcdhg^{fcon3}$ allele, they used CRISPR/Cas9 to disrupt the first constant exon of *Pcdha* (Ing-Esteves et al., 2018). The resulting mice ($Pcdhag^{f}$, Fig. 4F) when homozygous were functionally null for *Pcdha* with conditional control of *Pcdhg* disruption. Using tissue specific Cre lines, they found that both neuronal survival and self-avoidance were more severely disrupted in double mutants than in *Pcdhg* mutants alone. Interestingly, the effects of *Pcdha* on survival was dose dependent. This was shown by crossing $Pcdhag^{f}$ to $Pcdhg^{fcon3}$ or $Pcdha^{-/-}$ with and without Cre, resulting in gene dosages in the following configurations: a^2g^2, a^1g^1, a^0g^2, a^0g^1, a^2g^0, a^1g^0, and a^0g^0. Some cell loss was detectable in a^0g^1 mice, and became more severe when *Pcdha* dosage was reduced in the complete absence of *Pcdhg* (Ing-Esteves et al., 2018). Thus, in agreement with previous studies, these models show that there are complex combinatorial and redundant functions between the cPcdhs.

3.3 Isoform specific models

There is increasing evidence for isoform-specific functions. In the *Pcdhg* cluster, deletions were made removing the first three A-type isoforms (γA1-γA3, *Pcdhg*^{tako} for *t*riple *A* *k*nock*o*ut, Fig. 3E) and others removing the three C-type isoforms (γC3-γC5, *Pcdhg*^{tcko} Fig. 3F). Deletion of the C isoforms with *Pcdhg*^{tcko} resulted in apoptosis comparable to *Pcdhg*^{del}, while the *Pcdhg*^{tako} animals were unaffected, indicating an essential role for γC3-C5 in neuronal survival (Chen et al., 2012). Neither mutant had disrupted self-avoidance in SACs in the retina (Lefebvre et al., 2012). γC5 was found to associate with the GABA receptor γ2 subunit, and hippocampal cultures from Pcdhg^{tcko} mutants had an overall decrease in GABAergic synapses. shRNAs specifically targeting γC5 for knockdown resulted in a similar reduction (Li et al., 2012). Using a cell line model, it was demonstrated that γC3, but not other isoforms, can inhibit Wnt signaling by binding Axin1 and stabilizing its localization to the cell membrane (Mah, Houston, & Weiner, 2016). Newly developed models will shed light on the functional importance of this and other isoforms (Garrett et al., 2019).

A recent study generated an array of new alleles reducing the diversity of Pcdhg isoforms using a CRISPR/Cas9 approach (Garrett et al., 2019). sgRNA guides against all 22 variable exons were pooled and microinjected into pronuclei along with mRNA encoding Cas9. The resulting founder mice were bred to wild type animals to generate G1 heterozygous mice. More than 90 G1 animals were screened for heterozygous mutations and potential off-target sites using a custom amplicon sequencing array, and interesting strains were cryopreserved. Around 25 unique strains were cryopreserved with between 1 and 21 variable exons disrupted. Mutations included discrete insertion/deletion frameshifting events and large deletions or rearrangements between guide sites. Every isoform was disrupted in at least one strain. Three strains were initially chosen to breed to homozygosity: two with 3 isoforms remaining (*Pcdhg*^{3R1} and *Pcdhg*^{3R2}) and one with 13 isoforms remaining (*Pcdhg*^{13R1}) (Fig. 3G–I). Analysis of these strains suggested that γC4 was particularly important for postnatal viability and neuronal survival, prompting the analysis of two additional strains: *Pcdhg*^{1R1} in which only γC4 was left intact, and *Pcdhg*^{C4KO}, a strain generated by only targeting γC4 (Fig. 3J–K). *Pcdhg*$^{1R1/1R1}$ homozygous mutants were viable and fertile, while *Pcdhg*$^{C4KO/C4KO}$ animals died shortly after birth with neuronal apoptosis comparable to *Pcdhg*^{del} mutants, indicating an independent role for γC4 in neuronal survival (Garrett et al., 2019). Together with the multi-cluster mutants discussed above, this suggests a model where γC4 alone is essential for normal neuronal survival, but other isoforms in other clusters can compensate in its absence in a limited way to promote survival.

Deletions within the *Pcdha* cluster were described above (Fig. 2B and C), and have been used to investigate isoform-specific functions. To determine which isoforms are responsible for their importance in serotonergic projections, two different deletion mutants were compared: *Pcdha*$^{del(2\text{-}11)}$ where isoforms a2-a11 were deleted and *Pcdha*$^{del(11\text{-}C2)}$ in which a11-c2 were removed. Normal projections were maintained in *Pcdha*$^{del(2\text{-}11)}$, but *Pcdha*$^{del(11\text{-}C2)}$ revealed an irregular serotonergic branch pattern. In situ hybridization confirmed that only αC2 was highly expressed in serotonergic neurons (Katori et al., 2017). The importance of αC2 was demonstrated concomitantly in a study using a CRISPR/Cas9 approach to generate mice lacking all *Pcdha* isoforms except the two αC isoforms, *Pcdha*^{altko}, and another strain lacking only the αC isoforms, *Pcdha*^{dcko} (Fig. 2H and I), which demonstrated that only αC isoforms were required for serotonergic targeting. They also made the case based on expression that αC2 was the crucial isoform (Chen et al., 2017).

3.4 Overexpression models

Overexpression of specific *Pcdhg* isoforms has not demonstrated isoform-specific functions *per se*, but it has provided insight into the role of homophilic interaction and isoform diversity. Pcdhg A1-mCherry and C3-mCherry overexpression (OE) animals were generated by knock-in to the ROSA26 locus with Cre-dependent expression driven by a CAG promoter (Fig. 5) (Lefebvre et al., 2012). When crossed with Chat-Cre to drive expression in cholinergic cells (limited to SACs in the retina), overexpression of either isoform was able to rescue self-avoidance defects in *Pcdhg* loss of function mutants. However, SACs in these mice had fewer synaptic connections with neighboring SACs measured by morphological overlap and paired electrophysiological recordings (Kostadinov & Sanes,

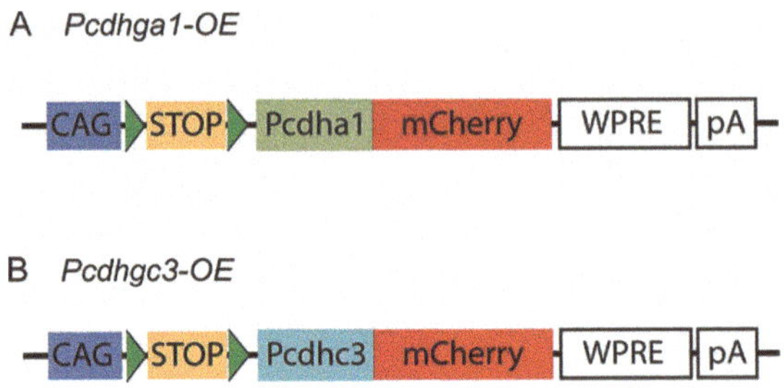

Fig. 5 Overexpression models. Cre-dependent overexpression of either *Pcdhga1* or *Pcdhgc3* was achieved by knock-in into the ROSA26 locus (Lefebvre et al., 2012). Each knock-in cassette included a CAG promoter, a multi-polyA stop sequence flanked by loxP sites, cDNA for each isoform fused with mCherry, and WPRE (woodchuck hepatitis virus posttranscriptional regulatory element) and polyA sequence.

2015; Lefebvre et al., 2012). This suggests a model in which cPcdhs facilitate self/non-self recognition. That is, individual neurons stochastically express a subset of isoforms that differs from their neighbors, allowing sister dendrites from the same neuron to recognize and repel each other, while still overlapping and forming reciprocal connections with neighboring SACs that express a different repertoire of isoforms.

These same overexpression lines were used to investigate the role of homophilic interaction in dendrite arborization, with nearly opposite results to those found in SACs. In a study from Molumby and colleagues, when homophilic matching between a neuron and neighboring cells was increased by driving expression throughout the cortex (Emx-1-Cre) or in astrocytes in the cortex (GFAP-Cre), arborization increased. This increase could even be limited to dendrites passing through a specific cortical layer where matching was locally increased (Sim1-Cre). Conversely, if matching was reduced by driving expression of the rarely expressed γA1 isoform in scattered cells, arborization decreased (Molumby, Keeler, & Weiner, 2016). In this context, dendrite arborization was enhanced by homophilic matching between neighbors, not just between sister neurites of a single neuron.

4. Future directions

While much has been learned about the cPcdhs through the use of mouse models over the last 20 years, there are outstanding questions. One of these is highlighted by the disparate effects of homophilic γ-Pcdh matching: matching between sister neurites but not between neighboring cells promotes self-avoidance while allowing overlap with neighbors in the retina (Kostadinov & Sanes, 2015; Lefebvre et al., 2012), but homophilic matching between neighbors promotes dendrite arborization in cortical neurons (Molumby et al., 2016). The basic question of adhesion (consistent with branching phenotypes) or repulsion (consistent with self-avoidance phenotypes) is unresolved, and could point towards a co-receptor or cell-type-specific signaling modulator.

A second set of questions is raised by the isoform diversity. There are at least three possibilities: individual isoforms could be uniquely responsible for individual functions, individual isoforms could be redundant with each other, or diversity *per se* could be required for any given function. As we have seen, there is evidence for each of these models for different functions. γC4 is necessary and sufficient among the *Pcdhg* isoforms for neuronal survival and postnatal viability, providing clear evidence for an isoform-specific function (Garrett et al., 2019), yet cell death phenotypes were more severe in multi-cluster mutants, suggesting some level of isoform redundancy (Hasegawa et al., 2016; Ing-Esteves et al., 2018). Similarly, phenotypes in the olfactory system are consistent with either redundancy or a role for diversity *per se* (Mountoufaris et al., 2017). No single isoform was required for self-avoidance (Lefebvre et al., 2012), and this function along with that of dendrite arborization could depend on isoform diversity, as the levels of isoform matching could balance the contacts between neighboring cells to promote self/non-self discrimination and dendritic arbor complexity. The recently generated allelic series of mouse mutants reducing *Pcdhg* isoform diversity will likely illuminate the level of diversity required for these functions (Garrett et al., 2019).

Intracellular binding partners and molecular mechanisms for some functions have been described (Chen et al., 2009; Garrett et al., 2012; Han, Lin, Meng, & Wang, 2010; Keeler, Schreiner, & Weiner, 2015; Molumby et al., 2017), while mechanisms for other functions remain elusive. For example, mechanisms unique to γC4 for promoting neuronal survival have not been uncovered, and the molecular pathways of self-avoidance remain unknown. We expect that mouse models will be essential for addressing these questions moving forward.

References

Albertin, C. B., Simakov, O., Mitros, T., Wang, Z. Y., Pungor, J. R., Edsinger-Gonzales, E., et al. (2015). The octopus genome and the evolution of cephalopod neural and morphological novelties. *Nature*, *524*, 220–224.

Brasch, J., Goodman, K. M., Noble, A. J., Rapp, M., Mannepalli, S., Bahna, F., et al. (2019). Visualization of clustered protocadherin neuronal self-recognition complexes. *Nature*, *569*, 280–283.

Carriere, C. H., Wang, W. X., Sing, A. D., Fekete, A., Jones, B. E., Yee, Y., et al. (2020). The gamma-protocadherins regulate the survival of GABAergic interneurons during developmental cell death. *Journal of Neuroscience*, *40*, 8652–8668.
Chen, W. V., Alvarez, F. J., Lefebvre, J. L., Friedman, B., Nwakeze, C., Geiman, E., et al. (2012). Functional significance of isoform diversification in the protocadherin gamma gene cluster. *Neuron*, *75*, 402–409.
Chen, J., Lu, Y., Meng, S., Han, M. H., Lin, C., & Wang, X. (2009). Alpha- and gamma-protocadherins negatively regulate PYK2. *The Journal of Biological Chemistry*, *284*, 2880–2890.
Chen, W. V., Nwakeze, C. L., Denny, C. A., O'Keeffe, S., Rieger, M. A., Mountoufaris, G., et al. (2017). Pcdhalphac2 is required for axonal tiling and assembly of serotonergic circuitries in mice. *Science*, *356*, 406–411.
El Hajj, N., Dittrich, M., Böck, J., Kraus, T. F., Nanda, I., Müller, T., et al. (2016). Epigenetic dysregulation in the developing Down syndrome cortex. *Epigenetics*, *11*, 563–578.
Esumi, S., Kakazu, N., Taguchi, Y., Hirayama, T., Sasaki, A., Hirabayashi, T., et al. (2005). Monoallelic yet combinatorial expression of variable exons of the protocadherin-alpha gene cluster in single neurons. *Nature Genetics*, *37*, 171–176.
Fukuda, E., Hamada, S., Hasegawa, S., Katori, S., Sanbo, M., Miyakawa, T., et al. (2008). Down-regulation of protocadherin-alpha A isoforms in mice changes contextual fear conditioning and spatial working memory. *European Journal of Neuroscience*, *28*, 1362–1376.
Garrett, A. M., Bosch, P. J., Steffen, D. M., Fuller, L. C., Marcucci, C. G., Koch, A. A., et al. (2019). CRISPR/Cas9 interrogation of the mouse Pcdhg gene cluster reveals a crucial isoform-specific role for Pcdhgc4. *PLoS Genetics*, *15*, e1008554.
Garrett, A. M., Schreiner, D., Lobas, M. A., & Weiner, J. A. (2012). gamma-protocadherins control cortical dendrite arborization by regulating the activity of a FAK/PKC/MARCKS signaling pathway. *Neuron*, *74*, 269–276.
Garrett, A. M., & Weiner, J. A. (2009). Control of CNS synapse development by {gamma}-protocadherin-mediated astrocyte-neuron contact. *Journal of Neuroscience*, *29*, 11723–11731.
Goebbels, S., Bormuth, I., Bode, U., Hermanson, O., Schwab, M. H., & Nave, K. A. (2006). Genetic targeting of principal neurons in neocortex and hippocampus of NEX-Cre mice. *Genesis*, *44*, 611–621.
Goodman, K. M., Rubinstein, R., Thu, C. A., Bahna, F., Mannepalli, S., Ahlsén, G., et al. (2016). Structural basis of diverse homophilic recognition by clustered α- and β-protocadherins. *Neuron*, *90*, 709–723.
Goodman, K. M., Rubinstein, R., Thu, C. A., Mannepalli, S., Bahna, F., Ahlsén, G., et al. (2016). γ-Protocadherin structural diversity and functional implications. *Elife*, *5*.
Han, M. H., Lin, C., Meng, S., & Wang, X. (2010). Proteomics analysis reveals overlapping functions of clustered protocadherins. *Molecular & Cellular Proteomics*, *9*, 71–83.
Hasegawa, S., Hamada, S., Kumode, Y., Esumi, S., Katori, S., Fukuda, E., et al. (2008). The protocadherin-alpha family is involved in axonal coalescence of olfactory sensory neurons into glomeruli of the olfactory bulb in mouse. *Molecular and Cellular Neurosciences*, *38*, 66–79.
Hasegawa, S., Kumagai, M., Hagihara, M., Nishimaru, H., Hirano, K., Kaneko, R., et al. (2016). Distinct and cooperative functions for the protocadherin-alpha, -beta and -gamma clusters in neuronal survival and axon targeting. *Frontiers in Molecular Neuroscience*, *9*, 155.
Hirano, K., Kaneko, R., Izawa, T., Kawaguchi, M., Kitsukawa, T., & Yagi, T. (2012). Single-neuron diversity generated by protocadherin-beta cluster in mouse central and peripheral nervous systems. *Frontiers in Molecular Neuroscience*, *5*, 90.

Hirayama, T., Tarusawa, E., Yoshimura, Y., Galjart, N., & Yagi, T. (2012). CTCF is required for neural development and stochastic expression of clustered Pcdh genes in neurons. *Cell Reports*, *2*, 345–357.

Ing-Esteves, S., Kostadinov, D., Marocha, J., Sing, A. D., Joseph, K. S., Laboulaye, M. A., et al. (2018). Combinatorial effects of alpha- and gamma-protocadherins on neuronal survival and dendritic self-avoidance. *The Journal of Neuroscience*, *38*, 2713–2729.

Iossifov, I., O'Roak, B. J., Sanders, S. J., Ronemus, M., Krumm, N., Levy, D., et al. (2014). The contribution of de novo coding mutations to autism spectrum disorder. *Nature*, *515*, 216–221.

Jiang, Y., Loh, Y. E., Rajarajan, P., Hirayama, T., Liao, W., Kassim, B. S., et al. (2017). The methyltransferase SETDB1 regulates a large neuron-specific topological chromatin domain. *Nature Genetics*, *49*, 1239–1250.

Kaneko, R., Kato, H., Kawamura, Y., Esumi, S., Hirayama, T., Hirabayashi, T., et al. (2006). Allelic gene regulation of Pcdh-alpha and Pcdh-gamma clusters involving both monoallelic and biallelic expression in single Purkinje cells. *The Journal of Biological Chemistry*, *281*, 30551–30560.

Katori, S., Hamada, S., Noguchi, Y., Fukuda, E., Yamamoto, T., Yamamoto, H., et al. (2009). Protocadherin-alpha family is required for serotonergic projections to appropriately innervate target brain areas. *Journal of Neuroscience*, *29*, 9137–9147.

Katori, S., Noguchi-Katori, Y., Okayama, A., Kawamura, Y., Luo, W., Sakimura, K., et al. (2017). Protocadherin-αC2 is required for diffuse projections of serotonergic axons. *Science Reports*, *7*, 15908.

Kawaguchi, M., Toyama, T., Kaneko, R., Hirayama, T., Kawamura, Y., & Yagi, T. (2008). Relationship between DNA methylation states and transcription of individual isoforms encoded by the protocadherin-alpha gene cluster. *Journal of Biological Chemistry*, *283*, 12064–12075.

Keeler, A. B., Schreiner, D., & Weiner, J. A. (2015). Protein kinase C phosphorylation of a gamma-protocadherin C-terminal lipid binding domain regulates focal adhesion kinase inhibition and dendrite arborization. *Journal of Biological Chemistry*, *290*, 20674–20686.

Kehayova, P., Monahan, K., Chen, W., & Maniatis, T. (2011). Regulatory elements required for the activation and repression of the protocadherin-alpha gene cluster. *Proceedings of the National Academy of Sciences of the United States of America*, *108*, 17195–17200.

Kohmura, N., Senzaki, K., Hamada, S., Kai, N., Yasuda, R., Watanabe, M., et al. (1998). Diversity revealed by a novel family of cadherins expressed in neurons at a synaptic complex. *Neuron*, *20*, 1137–1151.

Kostadinov, D., & Sanes, J. R. (2015). Protocadherin-dependent dendritic self-avoidance regulates neural connectivity and circuit function. *Elife*, *4*.

Lefebvre, J. L., Kostadinov, D., Chen, W. V., Maniatis, T., & Sanes, J. R. (2012). Protocadherins mediate dendritic self-avoidance in the mammalian nervous system. *Nature*, *488*, 517–521.

Lefebvre, J. L., Zhang, Y., Meister, M., Wang, X., & Sanes, J. R. (2008). gamma-Protocadherins regulate neuronal survival but are dispensable for circuit formation in retina. *Development*, *135*, 4141–4151.

Li, Y., Xiao, H., Chiou, T. T., Jin, H., Bonhomme, B., Miralles, C. P., et al. (2012). Molecular and functional interaction between protocadherin-γC5 and GABAA receptors. *Journal of Neuroscience*, *32*, 11780–11797.

Lobas, M. A., Helsper, L., Vernon, C. G., Schreiner, D., Zhang, Y., Holtzman, M. J., et al. (2012). Molecular heterogeneity in the choroid plexus epithelium: The 22-member gamma-protocadherin family is differentially expressed, apically localized, and implicated in CSF regulation. *Journal of Neurochemistry*, *120*, 913–927.

Mah, K. M., Houston, D. W., & Weiner, J. A. (2016). The γ-Protocadherin-C3 isoform inhibits canonical Wnt signalling by binding to and stabilizing Axin1 at the membrane. *Science Reports*, *6*, 31665.

Mancia Leon, W. R., Spatazza, J., Rakela, B., Chatterjee, A., Pande, V., Maniatis, T., et al. (2020). Clustered gamma-protocadherins regulate cortical interneuron programmed cell death. *Elife*, *9*.

Molumby, M. J., Anderson, R. M., Newbold, D. J., Koblesky, N. K., Garrett, A. M., Schreiner, D., et al. (2017). gamma-Protocadherins interact with neuroligin-1 and negatively regulate dendritic spine morphogenesis. *Cell Reports*, *18*, 2702–2714.

Molumby, M. J., Keeler, A. B., & Weiner, J. A. (2016). Homophilic protocadherin cell-cell interactions promote dendrite complexity. *Cell Reports*, *15*, 1037–1050.

Monahan, K., Rudnick, N. D., Kehayova, P. D., Pauli, F., Newberry, K. M., Myers, R. M., et al. (2012). Role of CCCTC binding factor (CTCF) and cohesin in the generation of single-cell diversity of protocadherin-α gene expression. *Proceedings of the National Academy of Sciences of the United States of America*, *109*, 9125–9130.

Morgan, M. (2008). Models for the recent evolution of protocadherin gene clusters. *Biocell*, *32*, 9–26.

Morishita, H., Umitsu, M., Murata, Y., Shibata, N., Udaka, K., Higuchi, Y., et al. (2006). Structure of the cadherin-related neuronal receptor/protocadherin-alpha first extracellular cadherin domain reveals diversity across cadherin families. *Journal of Biological Chemistry*, *281*, 33650–33663.

Mountoufaris, G., Chen, W. V., Hirabayashi, Y., O'Keeffe, S., Chevee, M., Nwakeze, C. L., et al. (2017). Multicluster Pcdh diversity is required for mouse olfactory neural circuit assembly. *Science*, *356*, 411–414.

Murata, Y., Hamada, S., Morishita, H., Mutoh, T., & Yagi, T. (2004). Interaction with protocadherin-gamma regulates the cell surface expression of protocadherin-alpha. *Journal of Biological Chemistry*, *279*, 49508–49516.

Naskar, T., Faruq, M., Banerjee, P., Khan, M., Midha, R., Kumari, R., et al. (2018). Ancestral variations of the PCDHG gene cluster predispose to dyslexia in a multiplex family. *EBioMedicine*, *28*, 168–179.

Noguchi, Y., Hirabayashi, T., Katori, S., Kawamura, Y., Sanbo, M., Hirabayashi, M., et al. (2009). Total expression and dual gene-regulatory mechanisms maintained in deletions and duplications of the Pcdha cluster. *Journal of Biological Chemistry*, *284*, 32002–32014.

Noonan, J. P., Grimwood, J., Schmutz, J., Dickson, M., & Myers, R. M. (2004). Gene conversion and the evolution of protocadherin gene cluster diversity. *Genome Research*, *14*, 354–366.

Prasad, T., Wang, X., Gray, P. A., & Weiner, J. A. (2008). A differential developmental pattern of spinal interneuron apoptosis during synaptogenesis: Insights from genetic analyses of the protocadherin-gamma gene cluster. *Development*, *135*, 4153–4164.

Prasad, T., & Weiner, J. A. (2011). Direct and indirect regulation of spinal cord ia afferent terminal formation by the gamma-protocadherins. *Frontiers in Molecular Neuroscience*, *4*, 54.

Rubinstein, R., Thu, C. A., Goodman, K. M., Wolcott, H. N., Bahna, F., Mannepalli, S., et al. (2015). Molecular logic of neuronal self-recognition through protocadherin domain interactions. *Cell*, *163*, 629–642.

Schreiner, D., & Weiner, J. A. (2010). Combinatorial homophilic interaction between gamma-protocadherin multimers greatly expands the molecular diversity of cell adhesion. *Proceedings of the National Academy of Sciences of the United States of America*, *107*, 14893–14898.

Shao, Z., Noh, H., Bin Kim, W., Ni, P., Nguyen, C., Cote, S. E., et al. (2019). Dysregulated protocadherin-pathway activity as an intrinsic defect in induced pluripotent stem cell-derived cortical interneurons from subjects with schizophrenia. *Nature Neuroscience*, *22*, 229–242.

Styfhals, R., Seuntjens, E., Simakov, O., Sanges, R., & Fiorito, G. (2018). In silico identification and expression of protocadherin gene family in Octopus vulgaris. *Frontiers in Physiology*, *9*, 1905.

Su, H., Marcheva, B., Meng, S., Liang, F. A., Kohsaka, A., Kobayashi, Y., et al. (2010). Gamma-protocadherins regulate the functional integrity of hypothalamic feeding circuitry in mice. *Developmental Biology*, *339*, 38–50.

Sugino, H., Hamada, S., Yasuda, R., Tuji, A., Matsuda, Y., Fujita, M., et al. (2000). Genomic organization of the family of CNR cadherin genes in mice and humans. *Genomics*, *63*, 75–87.

Taguchi, Y., Koide, T., Shiroishi, T., & Yagi, T. (2005). Molecular evolution of cadherin-related neuronal receptor/protocadherin(alpha) (CNR/Pcdh(alpha)) gene cluster in Mus musculus subspecies. *Molecular Biology and Evolution*, *22*, 1433–1443.

Tasic, B., Nabholz, C. E., Baldwin, K. K., Kim, Y., Rueckert, E. H., Ribich, S. A., et al. (2002). Promoter choice determines splice site selection in protocadherin alpha and gamma pre-mRNA splicing. *Molecular Cell*, *10*, 21–33.

Thu, C. A., Chen, W. V., Rubinstein, R., Chevee, M., Wolcott, H. N., Felsovalyi, K. O., et al. (2014). Single-cell identity generated by combinatorial homophilic interactions between α, β, and γ protocadherins. *Cell*, *158*, 1045–1059.

Toyoda, S., Kawaguchi, M., Kobayashi, T., Tarusawa, E., Toyama, T., Okano, M., et al. (2014). Developmental epigenetic modification regulates stochastic expression of clustered protocadherin genes, generating single neuron diversity. *Neuron*, *82*, 94–108.

Vadodaria, K. C., Ji, Y., Skime, M., Paquola, A. C., Nelson, T., Hall-Flavin, D., et al. (2019). Altered serotonergic circuitry in SSRI-resistant major depressive disorder patient-derived neurons. *Molecular Psychiatry*, *24*, 808–818.

Wang, X., Weiner, J. A., Levi, S., Craig, A. M., Bradley, A., & Sanes, J. R. (2002). Gamma protocadherins are required for survival of spinal interneurons. *Neuron*, *36*, 843–854.

Weiner, J. A., Wang, X., Tapia, J. C., & Sanes, J. R. (2005). Gamma protocadherins are required for synaptic development in the spinal cord. *Proceedings of the National Academy of Sciences of the United States of America*, *102*, 8–14.

Wu, Q., & Maniatis, T. (1999). A striking organization of a large family of human neural cadherin-like cell adhesion genes. *Cell*, *97*, 779–790.

Wu, Q., & Maniatis, T. (2000). Large exons encoding multiple ectodomains are a characteristic feature of protocadherin genes. *Proceedings of the National Academy of Sciences of the United States of America*, *97*, 3124–3129.

Wu, S., Ying, G., Wu, Q., & Capecchi, M. R. (2007). Toward simpler and faster genome-wide mutagenesis in mice. *Nature Genetics*, *39*, 922–930.

Yokota, S., Hirayama, T., Hirano, K., Kaneko, R., Toyoda, S., Kawamura, Y., et al. (2011). Identification of the cluster control region for the protocadherin-beta genes located beyond the protocadherin-gamma cluster. *Journal of Biological Chemistry*, *286*, 31885–31895.

Yu, W. P., Rajasegaran, V., Yew, K., Loh, W. L., Tay, B. H., Amemiya, C. T., et al. (2008). Elephant shark sequence reveals unique insights into the evolutionary history of vertebrate genes: A comparative analysis of the protocadherin cluster. *Proceedings of the National Academy of Sciences of the United States of America*, *105*, 3819–3824.

Yu, W. P., Yew, K., Rajasegaran, V., & Venkatesh, B. (2007). Sequencing and comparative analysis of fugu protocadherin clusters reveal diversity of protocadherin genes among teleosts. *BMC Evolutionary Biology*, *7*, 49.

CHAPTER SIX

Wound healing and regeneration in spiny mice (*Acomys cahirinus*)

Daryl M. Okamura[a,b,e], Elizabeth D. Nguyen[a,b,e], David R. Beier[a,b,f], and Mark W. Majesky[a,b,c,d,*]
[a]Center for Developmental Biology & Regenerative Medicine, Seattle Children's Research Institute, Seattle, WA, United States
[b]Department of Pediatrics, University of Washington, Seattle, WA, United States
[c]Department of Laboratory Medicine & Pathology, University of Washington, Seattle, WA, United States
[d]Division of Pediatric Cardiology, Seattle Children's Hospital, Seattle, WA, United States
[e]Division of Pediatric Nephrology, Seattle Children's Hospital, Seattle, Washington, WA, United States
[f]Division of Genetic Medicine, Seattle Children's Hospital, University of Washington, Seattle, WA, United States
*Corresponding author: e-mail address: mwm84@uw.edu

Contents

Abstract

The winds of Patagonia are referred to by locals as "The Broom of God" because they sweep away the less fit species that cannot survive there. Fitness as an evolutionary trait has been considered as fundamental for many aspects of morphogenesis and behavior in metazoans. Yet, it has not received much attention in the area of wound healing, despite the obvious relevance of this polygenic trait to an organism's survival in nature. In this chapter, we review the evidence that the rodent species *Acomys cahirinus* is an emerging mammalian model system that has evolved a non-typical (for mammals) wound healing response that offers unique opportunities for the study of organ regeneration without fibrosis in an adult mammalian species.

Current Topics in Developmental Biology, Volume 148
ISSN 0070-2153
https://doi.org/10.1016/bs.ctdb.2022.03.001

Abbreviations

BrdU	5-bromo-2′-deoxyuridine
CKD	chronic kidney disease
CM	cardiac myocyte
DF	dermal fibroblast
En1	engrailed1 lineage
IRI	ischemia–reperfusion injury
MF	myofibroblast
MI	myocardial infarction
Mus^{CFW}	Swiss Webster outbred mice from Charles River
My	million years
PoMT	polyoma virus middle T antigen
PP2A	protein phosphatase type 2A
ROS	reactive oxygen species
TA	tibialis anterior muscle
TEAD	TEA (TEF-1 and abaA) domain
UUO	unilateral ureteral obstruction
VP	verteporfin
Yap	yes-associated protein-1

1. Introduction

A small number of established metazoan model organisms are used by most investigators for their research. These established models are familiar and include mice, rats, zebrafish, *Xenopus*, *C. elegans*, and *Drosophila*. For many questions this is entirely appropriate as these models are genetically tractable with a stable of well-characterized tools for their analysis. However, for some questions the best model may be a little-studied organism that exhibits a unique phenotype that is ideal for the question being addressed (Goldstein & King, 2016). Some recent examples include the blind mole rat for cancer resistance (Gorbunova et al., 2012), Greenland sharks for longevity (Neilson et al., 2016), pencil squid and other cephalods for camouflage (Crookes et al., 2004), tardigrades for survival in extreme environments (Boothby et al., 2017), tetrahymena for cilia and basal bodies (Bayless, Navarro, & Winey, 2019), and bats for echolocation (Danilovich & Yovel, 2019).

These emerging model organisms offer investigators a different genetic landscape for interrogation, inspiration and ideas that enable the pursuit of novel and exciting aspects of biology. For example, as an emerging model *Acomys* species (spiny mice) (Fig. 1) present a number of unique

Fig. 1 *Acomys cahirinus.* Adult spiny mice have large ears, protruding black eyes, a narrow snout with long whiskers, and spiny hairs on their dorsum.

opportunities. Bellofiore et al. reported that spiny mice are the first rodent species to exhibit menstruation, which therefore offers a new approach to understanding human menstrual disorders and pregnancy-associated diseases (Bellofiore et al., 2017; Bellofiore, George, Vollenhoven, & Temple-Smith, 2021). Nevo and colleagues studied *Acomys* as an evolutionary model of sympatric speciation in a desert habitat (Hadid et al., 2014). Seifert et al. reported that *Acomys* species develop skin that is structurally less able to withstand tensile forces before tearing than *Mus musculus* (Seifert et al., 2012). This property underpins the ability of these animals to shed their skin as a means of escape when attacked by predators. Remarkably, *Acomys* can then regenerate entirely normal skin without overgrowth or scar formation even as adults (Gawriluk et al., 2016; Seifert et al., 2012). These data raise the intriguing possibility that during evolution of spiny mice the mammalian response to tissue injury may have been redirected from inflammation and fibrosis which can result in loss of organ function to one of restoration of organ function by regeneration without fibrosis. If so, spiny mice may provide an evolved genetic template to characterize mechanisms of complex tissue regeneration in an adult mammal (Maden, 2018; Maden & Varholick, 2020).

2. Natural history and evolution of *Acomys* species

The genus *Acomys* is native to Africa and the Middle East. Aghová et al. analyzed 700 genotyped individuals from 282 different locations within *Acomys* home range with four genetic markers (Aghova et al., 2019). They date the origin of the genus *Acomys* to the late Miocene period ~8.7 My ago in the savannas of East Africa. Today, *Acomys* species are distributed mainly in arid and open ecologic niches in Africa and the Middle East. This habitat includes the African grasslands, scrublands, and wooded savannas that contain one of the most diverse collection of mammalian species on Earth (Timbuka & Kabigumila, 2006). Aghová et al. hypothesize that the genus *Acomys* arose by colonizing dry, open habitats that were empty niches at the time as progressively arid changes in climate led to a regression of woodland habitats (Aghova et al., 2019). These authors described 26 genetic lineages as potential species of *Acomys* that likely arose by sequential divisions of a parent lineage due to changes in geographical or climatic features over time. The exact number of *Acomys* species is still debated. Ellerman reported 25 species and 17 subspecies (Ellerman, 1941) while Musser and Carleton (Musser & Carleton, 2005) argued for 18 species and Monadjem et al. found 15 species (Mondajem, Taylor, Denys, & Cotterill, 2015). The most recent and comprehensive Handbook of Mammals of the World indicates 21 species and 12 subspecies of the genus *Acomys* (Denys, Taylor, & Aplin, 2017). Volobouev et al. conclude that *A. dimidiates*, long considered a subspecies of *A. cahirinus* despite similar diploid chromosome numbers (2n = 36 *vs* 38), are actually two different species of *Acomys* based on chromosome banding analysis (Volobouev, Gautun, Sicard, & Tranier, 1996).

For colonies established in research institutions throughout the world, two species of *Acomys* are most often found. *Acomys cahirinus* is usually employed while for some comparative studies of adaptation to desert ecology and physiology *Acomys russatus* (golden spiny mouse) is favored (Katz, Dayan, & Kronfeld-Schor, 2018; Kronfeld-Schor et al., 2000). For those using *Acomys cahirinus*, the adults are about twice the size of laboratory mice of the C57BL/6J strain (Mus^{B6}), the litters are smaller (~2–4 pups/litter), gestation times longer (~40–45 days) and the newborn pups are covered with hair, eyes open, ears unfurled, and are precocious compared to *Mus* (Haughton, Gawriluk, & Seifert, 2016; Keller & Burns, 1989; Pinheiro, Prata, Araújo, & Tiscornia, 2018) (Fig. 1). Consistent with the longer gestation times, organogenesis in spiny mice is largely complete at the time of birth (Brunjes, 1989; Dickinson, Walker, Cullen-McEwen, Wintour, & Moritz, 2005; Oosterhuis, Mooren, Charles, & Lamers, 1984). Spiny mice develop long, stiff, and sharp-tipped hairs on the dorsum that are modified awl hairs and are thought to serve a protective or display function (Montandon, Tzika, Martins, Chopard, & Milinkovitch, 2014). They characteristically exhibit large ears, a narrow snout, and large protruding black eyes (Fig. 1). Finally, at least three species of *Acomys* have been shown to possess weak skin that tears readily at low tensile forces in response to handling or predation, a phenotype presumed to allow escape from predators *via* autotomy in their native habitat (Seifert et al., 2012).

While the common-use name for *Acomys* species is "spiny mice," these animals are only distantly related to African *Murinae* (Lecompte et al., 2008) in general and *Mus musculus* in particular. Using DNA:DNA hybridization data, Chevret et al. reported that *Acomys* together with the "brush-furred mice" of the rodent genera *Uranomys* and *Lophuromys* are members of a monophyletic clade that is more closely related to *Gerbillinae* than to *Murinae* (Chevret, Denys, Jaeger, Michaux, & Catzeflis, 1993). However, spiny mice and gerbils are still considerably different both morphologically and genetically (Gustavsen, Kvicerova, Dickinson, & Heller, 2009). For comparative studies it is important to realize that *Acomys* species have been evolving separately from *Mus* for at least 8 My (Lecompte et al., 2008). This phylogenetic distance poses particular challenges to interpretation of differentially expressed transcriptomes between *Acomys* and *Mus* for any particular phenotype (Gerstein et al., 2014; Saxena et al., 2022; Scheenstra et al., 2020).

3. *Acomys* and the evolution of mammalian wound healing

In contrast to the dominant roles of inflammation and fibrosis in mammalian wound healing, those vertebrate species that are capable of organ regeneration as adults will heal wounds without fibrotic scars (Brockes & Kumar, 2008; Maden, 2018; Poss, 2010). However, most of those vertebrates are not mammals and do not have a committed terrestrial life style. Therefore, identification of a terrestrial mammal that can repair external and internal injuries with minimal or undetectable scarring is of great interest to the biomedical community. Currently, there are over 65,000 species of vertebrates including ~6500 species of mammals on the planet (Burgin, Colella, Kahn, & Upham, 2018; Schipper et al., 2008). The incredible wealth of biological diversity this represents raises the distinct possibility that overreliance on mouse models is limiting our insights into potential mechanisms for anti-fibrotic, regenerative therapies (Goldstein & King, 2016; Libby, 2015).

Given the damaging effect of tissue fibrosis on organ function (Friedman, Sheppard, Duffield, & Violette, 2013; Gurtner, Werner, Brandon, & Longaker, 2008; Humphreys, 2018), a useful adaptation for mammalian species would be to evolve mechanisms of wound healing that minimize fibrosis and promote organ regeneration. One current example might be bottlenose dolphins (*Tursiops truncates)* that are commonly subjected to shark bites in their native habitat. Several groups have reported studies of wound healing in dolphins in captivity or in the wild (Bruce-Allen & Geraci, 1985; Corkeron, Morris, & Bryden, 1987; Zasloff, 2011). They report that following the initial trauma of wounding, there is little bleeding, the individuals seem indifferent to pain of the injury, there is little or no infection at the wound site, and the missing tissue is replaced in about 4 weeks by reconstitution of the original tissue with minimal or no scar formation (Bloom & Yeager, 1994; Bruce-Allen & Geraci, 1985; Corkeron et al., 1987). This mode of wound healing produces a repaired skin surface for the dolphin that moves through the water with minimal disturbance of the hydrodynamics of normal dolphin skin (Bandyopadhyay & Hellum, 2014; Pavlov, 2006). This pattern of wound healing is, in many aspects, strikingly similar to what has been described for spiny mice (Brant, Yoon, Polvadore, Barbazuk, & Maden, 2016; Matias Santos et al., 2016; Seifert et al., 2012).

This may also speak to the possibility that tissue regeneration is a more broadly distributed property in mammalian genomes than is generally thought. It seems increasingly clear that many mammalian species can heal injuries by regeneration without fibrotic scars as embryos (Colwell, Longaker, & Lorenz, 2003; Drenckhahn et al., 2008; Pratsinis, Mavrogonatou, & Kletsas, 2019). Yet, the ability of spiny mice to heal external and internal injuries without fibrosis as adults appears unique among mammalian experimental models. Rabbits are reported to close ear punch injuries without scars (Gawronska-Kozak, Grabowska, Kopcewicz, & Kur, 2014), as can MRL/MpJ mice (Gourevitch et al., 2003) and p21-deficient mice (Bedelbaeva et al., 2010). Yet, ear repair in these models ultimately failed to extend to internal organ wound healing with little, if any, regenerative repair demonstrated in MRL/MpJ heart (Robey & Murry, 2008) or kidney (Shiozuru et al., 2016). Neonatal mouse myocardium exhibits regeneration potential after ventricular apex resection, albeit limited to the first few days after birth (Porrello et al., 2011). The same apex injuries heal by fibrotic scar from 7 days postnatal and thereafter. Mouse digit tip regenerates after amputation (Lehoczky, Robert, & Tabin, 2011; Rinkevich, Lindau, Ueno, Longaker, & Weissman, 2011) but it is only the extreme tip of the digit that regenerates, limiting the general usefulness of this model for other tissues and organs. Taken together, these results suggest that the information for tissue regeneration may exist in many, if not most, mammalian genomes but the expression of this information is generally silenced after birth. This may suggest the possibility that spiny mice have avoided silencing certain developmental pathways that when reexpressed in adults, enable tissue regeneration without fibrosis. Although the "regeneration recapitulates development" hypothesis may appear to be an attractive one, this idea has not been rigorously tested in spiny mice and is unlikely to fully explain adult tissue regeneration in vertebrates more broadly (Lepper, Conway, & Fan, 2009; Nacu & Tanaka, 2011; Wang & Conboy, 2010).

4. Repair of skin wounds in *Acomys*

Much of the increased interest in *Acomys* species over the last decade (Fig. 2) can be attributed to the report by Seifert, et al. that spiny mice can shed their dorsal skin as a deterrent against predation and then fully regenerate the lost skin without fibrosis or scarring (Seifert et al., 2012). These authors showed that the mean tensile strength of *Acomys* skin is at least 20 times weaker than Swiss-Webster mice (Mus^{CFW}) and that skin shedding in spiny mice occurs by tearing without a preexisting fracture plane. Of particular interest, was the finding that the extracellular matrix of regenerating

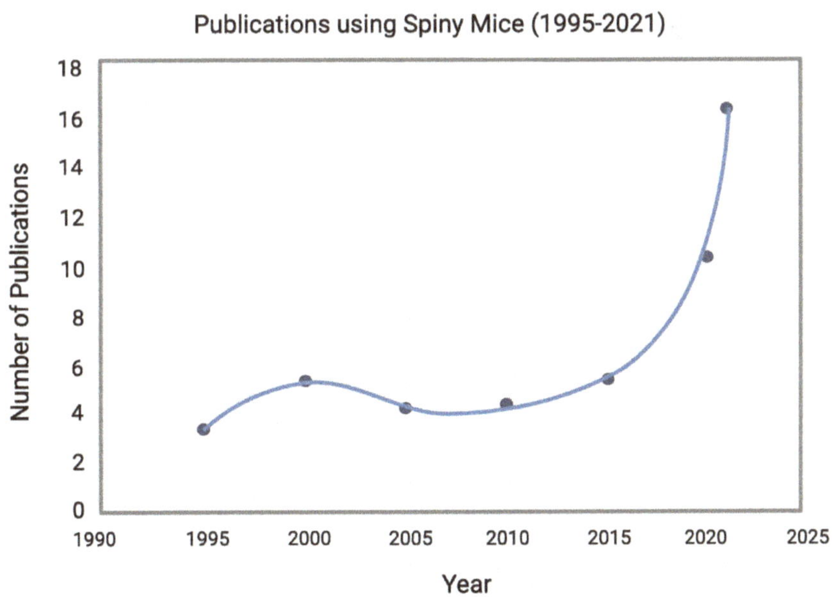

Fig. 2 Publications using spiny mice. Publications from 1995 to 2021 identified by search of the PubMed database using the search terms <acomys> or <spiny mice>. Interest in spiny mice as an experimental model organism has increased exponentially in the last decade, and particularly so in the last 5 years.

skin tissue in *Acomys* was enriched in type III collagen compared to type I collagen while the reverse was found for *Mus*^{CFW} wounds progressing to fibrotic scars (Seifert et al., 2012). Expression of keratin 17, LEF1, and Sox2 was examined during hair follicle neogenesis and the results suggested that reexpression of developmental pathways underlies, at least in part, *Acomys* regeneration of missing skin tissue (Seifert et al., 2012).

Tissue injury is generally accompanied by an influx of inflammatory cells to initiate a wound healing response (Eming, Wynn, & Martin, 2017). A comprehensive analysis of leukocyte infiltration kinetics during regeneration of ear hole injuries was reported by Simkin et al. (Simkin, Gawriluk, Gensel, & Seifert, 2017). After first ascertaining that no significant differences in circulating leukocytes (monocytes, neutrophils, lymphocytes, eosinophils) preexisted in uninjured *Acomys* compared to *Mus*^{CFW}, the authors investigated a time course of changes in leukocyte content in ear pinnae tissue after ear hole injury. They found similar changes in CD11b-positive cells (neutrophils, macrophages) over 15 days after ear injury in *Mus*^{CFW} and *Acomys* although higher numbers of neutrophils were found at day 1 for *Mus*^{CFW}. Using a pan-macrophage marker, IBA-1, no significant differences were observed in the changes in content of IBA-1-positive cells over 15 days after injury when *Mus*^{CFW} was compared to *Acomys*. However, when reactive oxygen species (ROS) production was assessed *in vivo* using lucigenin chemiluminescence, a much higher ROS production was found for *Acomys* than *Mus*^{CFW} at day 1 that was maintained up to day 5 after wounding (Simkin et al., 2017). This is an interesting finding since a role for sustained ROS production in tissue regeneration was also reported for zebrafish heart (Han et al., 2014) and Xenopus tadpole tail (Love et al., 2013) (reviewed in Helston & Amaya, 2021).

A series of experiments was also performed to deplete phagocytes (mainly macrophages) using clodronate liposomes (Van Rooijen & Hendrikx, 2010) injected at the base of the ear immediately prior to injury and again at days 2 and 5 after injury (Simkin et al., 2017). These experiments resulted in delays in blastema formation, reepithelization, cartilage histolysis, and ear hole closure in clodronate-treated *Acomys*. They provide evidence for an important role for one or more macrophage subtypes in regeneration of ear skin in *Acomys* similar to roles proposed for these cells in neonatal mouse heart (Aurora et al., 2014) and salamander limb (Godwin, Pinto, & Rosenthal, 2013) regeneration.

Myofibroblasts (MFs) are the major scar-forming cell type in skin and most other organs (Hinz et al., 2012). Injury-induced MFs restore structural integrity of disrupted tissue but limit regenerative potential in most mammalian models (Darby, Skalli, & Gabbiani, 1990; Tomasek, Gabbiani, Hinz, Chaponnier, & Brown, 2002). It has become clear in recent years that the term dermal fibroblast (DF) actually includes a number of distinct subtypes with different embryonic origins, subtype-specific transcriptomes, potentials for fibrosis or regeneration, and anatomical positions within skin (Abbasi et al., 2020; Driskell et al., 2013; Rinkevich et al., 2015) (reviewed in Plikus et al., 2021). It has been reported that DFs in the mouse arise from at least two independent origins that can be distinguished by whether they derive from progenitor cells that express *engrailed-1* (En1) or not (Rinkevich et al., 2015). The En1-positive lineage produces DFs that are responsible the bulk of connective tissue formation in development and the fibrotic response to injury (Mascharak et al., 2021; Rinkevich et al., 2015). In addition, tissue mechanics interacts with lineage-specific DF potentials to influence the balance between fibrosis and regeneration (Harn et al., 2021; Mascharak et al., 2021). Interestingly, while single-cell transcriptome studies provide evidence for distinct DF lineages (Driskell et al., 2013; Guerrero-Juarez et al., 2019; Philippeos et al., 2018), a recent single-cell ATAC-sequencing (scATAC-seq) study revealed accessible chromatin for myofibroblast markers in all DF populations in neonatal skin regardless of their transcriptomes (Thompson, Phan, Winuthayanon, Driskell, & Driskell, 2022). These results point to a significant degree of plasticity among fibroblast subtypes (Abbasi et al., 2020; Plikus et al., 2021) that may have allowed evolutionary selection of DF states that mediate wound healing without scars and regeneration of normal structure and function in *Acomys* species.

For example, in a time-course study following ear punch injury, Brewer et al. reported that MFs appeared rapidly after injury in both *Acomys* and outbred CD1 mice (Mus^{CD1}), but they persisted at the wound site only in Mus^{CD1} (Brewer et al., 2021). In response to injury, *Acomys* MFs exhibit different phenotypes and fates compared to Mus^{CD1} MFs *in vivo*. The timing of MF loss in *Acomys* correlated with wound closure, blastema differentiation, and nuclear localization of the Hippo pathway effector protein Yap that functions in many contexts as a pro-regenerative transcriptional coactivator (Heallen et al., 2011; Leach et al., 2017; Moya & Halder, 2019). Isolated dermal fibroblasts (DFs) from *Acomys* exhibited select differences in gene expression responses to TGFβ1, a well-studied fibrogenic cytokine, when compared to Mus^{CD1} DFs, including failure to induce expression of Col(I)α1, a major form of fibrillar collagen in scar tissue (Brewer et al., 2021). Of particular interest, was the finding of accelerated protein phosphatase 2A (PP2A)-dependent Yap phosphatase activity, which maintained biologically active nuclear Yap localization in *Acomys* DFs *in vitro*. Treatment of spiny mice *in vivo* with the nuclear Yap-TEAD inhibitor verteporfin (VP) (Chen et al., 2018; Liu-Chittenden et al., 2012) prolonged MF persistence and inhibited tissue regeneration (Brewer et al., 2021). VP-treated spiny mice exhibited loss of correct patterning of differentiated cell types in skin, and ectopic hair follicle formation in the ventral compartment, which correlated with loss of *de novo* cartilage formation. These findings suggest that *Acomys* may have evolved modifications of Yap localization, Yap-TEAD transcriptional activity, and MF fate after wounding that may be important for scar-free regenerative wound healing *in vivo* (Brewer et al., 2021; Hou, Astrowski, & Plikus, 2021).

Protein phosphorylation is a major mechanism used in evolution to constrain protein function and make it contingent upon the activity of signaling pathways that are responsive to extracellular cues (Battacharyya, Reményi, Yeh, & Lim, 2006; Gerhart & Kirschner, 1997; Hunter, 2012). In addition to gene duplication and evolution of different families of protein kinases (McDonald, Trost, & Napper, 2018), Goldman et al. reported evidence that protein phosphatases and their substrate docking sites are also important evolutionary targets for diversification of signaling networks regulated by protein phosphorylation (Goldman et al., 2014). With respect to phospho-Yap itself, Brewer et al., note that theirs is not the first report that evolution has targeted PP2A-dependent Yap de-phosphorylation (Brewer et al., 2021). A particularly interesting example is the evolutionary capture of tumor forming activity by a mammalian polyomavirus. The polyomavirus middle T antigen (PoMT) is the primary polyoma gene product responsible for tumor formation (Rouleau et al., 2016). Residues R103 and D182 of PoMT can bind to the central WW domains of Yap leading to formation of a PoMT-pYap complex that recruits PP2A leading to accelerated rates of dephosphorylation of pSer127-Yap and generation of active Yap (Rouleau et al., 2016). In an example of parallel evolution, the human adenovirus E4orf4 protein binds to the B55 class of PP2A regulatory subunits and inhibits Yap phosphatase activity resulting in maintenance of inactive pYap and tumor cell killing (Mui et al., 2015). Finally, studies of the evolution of protein kinase signaling pathways have shown that naturally arising sequence variants in phosphoprotein docking sites for existing protein phosphatases play fundamental roles in the acquisition of novel phenotypes (Roy & Cyert, 2009).

5. Repair of skeletal muscle injury in *Acomys*

A single injection of the myonecrotic factor cardiotoxin leads to a rapid breakdown of skeletal muscle fibers that is then restored by proliferation and differentiation of resident muscle stem cells, called satellite cells (Garry, Antony, & Garry, 2016). Following one injection of cardiotoxin, Maden et al. reported that both C57BL/6 mice (Mus^{B6}) and spiny mice completely regenerated damaged tibialis anterior (TA) muscle by 14 days with spiny mice exhibiting consistently faster regeneration kinetics (Maden et al., 2018). These processes occurred with similar changes in gene expression for myogenic transcription factors (*Myf5*, *MyoD1*, *MyoG*), TGFβ1, although the maximal levels of these factors were higher in *Mus* (Maden et al., 2018). Macrophages expressing the M1 activation markers F4/80 or CD86 were present in abundance in damaged Mus^{B6} muscle at 4 days after cardiotoxin injection while they were undetectable in *Acomys* muscle at 4 days. By contrast, cells expressing the M2 marker CD206 (Simkin et al., 2017) were present in equal numbers at 4 days after injury.

However, striking differences in muscle regeneration between Mus^{B6} and *Acomys* were found after a series of five cardiotoxin injections each spaced 3 weeks apart. Using this experimental regimen, Mus^{B6} muscles regenerated very poorly after the fifth injection of cardiotoxin with large numbers of adipocytes appearing among the few regenerated muscle fibers that were present (Maden et al., 2018). In contrast, in spiny mice, however, skeletal muscle regeneration was robust with few, if any, adipocytes appearing in the regenerated muscle tissue. Therefore, in addition to regeneration of panniculus carnosus muscle in skin (Brant et al., 2016; Gawriluk et al., 2016; Matias Santos et al., 2016; Seifert et al., 2012), spiny mice are capable of complete regeneration of TA skeletal muscle tissue even after repeated necrotic injuries *in vivo* (Maden et al., 2018).

6. Repair of kidney injury and regeneration of organ function in *Acomys*

Among the many clinically significant examples of loss of organ function due to progressive tissue fibrosis (Gurtner et al., 2008; Rockey, Bell, & Hill, 2015; Wynn & Ramalingam, 2012), end-stage kidney disease is among the most pervasive (Duffield, 2014; Hill et al., 2016; Humphreys, 2018). Cortical interstitial fibrosis is a consistent feature of most, if not all, chronic kidney diseases (CKD) (Duffield, 2014). Fibrosis is a strong and reliable predictor of clinically significant renal functional decline (Coresh et al., 2007) and CKD is among the strongest risk factors for cardiovascular disease (Go, Chertow, Fan, McCulloch, & Hsu, 2004). There are currently very few treatment options for patients with end-stage kidney disease who will frequently progress to needing long-term dialysis or kidney transplantation to survive.

To test whether scarless wound healing observed in *Acomys* skin extended to internal vital organs such as the kidney, two aggressive models of acute kidney injury were utilized (Okamura et al., 2021). Unilateral ureteral obstruction (UUO) is a surgical procedure for complete ligation of the ureter leading to renal tubule expansion, hydronephrosis, and activation of interstitial fibroblasts and connective tissue cells (Fig. 3). This sequence of events leads to extensive interstitial fibrosis and rapid loss of organ function in $Mus^{B6\text{-}C3HF1}$ (Moriyama et al., 1998) and Mus^{B6} (Li et al., 2013; Tapmeier et al., 2008). A second injury model tested was an ischemia–reperfusion injury (IRI) produced by clamping the vascular pedicle for 40 min followed by release of the clamp for kidney reperfusion (Pennathur et al., 2015; Singbartl, Green, & Ley, 2000). In both cases, Mus^{B6} or Mus^{CD1} develop rapidly progressive renal fibrosis resulting in loss of kidney function and organ failure (Okamura et al., 2021). Yet spiny mice, while exhibiting similar acute kidney damage as either *Mus* strain, fail to progress to kidney fibrosis. Rather, injured *Acomys* kidneys completely regenerated renal function by 2 weeks after injury (Okamura et al., 2021) (Fig. 3). Transcriptome analysis at early times after obstructive injury (2 days and 5 days) provided evidence that the spiny mouse genome appears poised to initiate a regenerative repair response at the time of the initial injury event (Okamura et al., 2021). While there does not appear to be formation of a regeneration blastema in the kidney, as there is after ear hole injuries in skin (Brant et al., 2016; Brewer et al., 2021; Gawriluk et al., 2016), there is robust angiogenesis, and apparent proliferation of cell types, such as podocytes, that are thought to be post-mitotic in *Mus* and humans (Kaverina et al., 2019; Okamura et al., 2021).

Comparative transcriptome analysis identified 843 genes that fulfilled criteria for distinct species-specific changes in expression between uninjured and UUO-treated kidneys at day 2 after surgery (Okamura et al., 2021). These genes clustered into six groups that contained secreted factors, extracellular matrix proteins, and transcriptional regulators (Okamura et al., 2021).

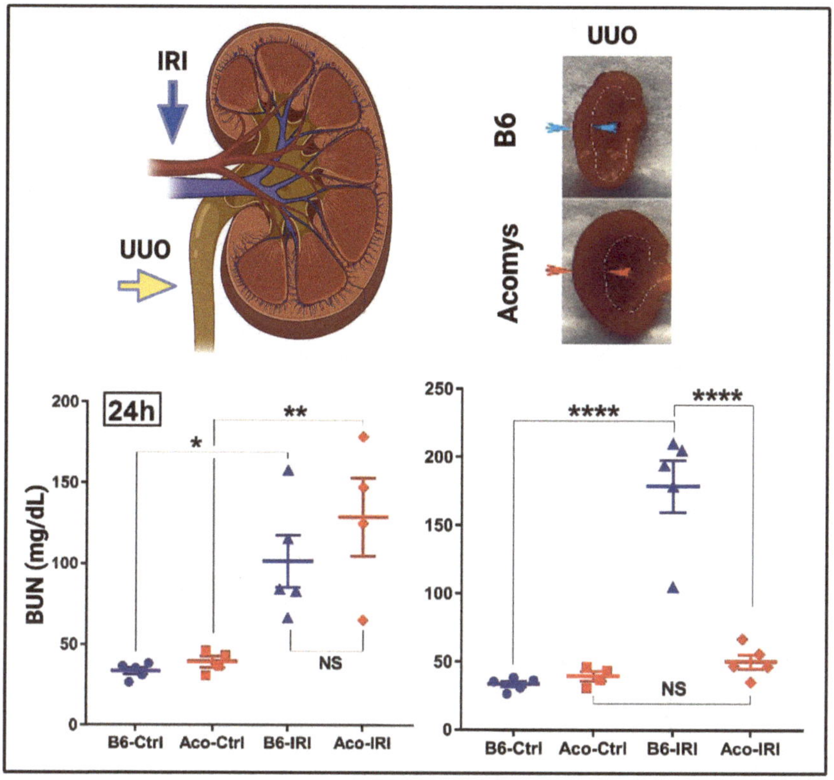

Fig. 3 Regeneration of kidney function after acute renal injury in *Acomys*. (*top left*) Kidneys were injured either by permanent obstruction of the ureter (UUO, yellow) or clamping of the renal artery for 40 min followed by release of the clamp allowing reperfusion of the ischemic kidney (IRI, blue). (*top right*) Gross view of kidneys 14 days after UUO surgery. Note preservation of the cortex (between the red arrowheads) in *Acomys* compared to the atrophic and fibrotic cortex (blue arrowheads) in Mus^{B6}. (*bottom left*) Kidney function measured by blood urea nitrogen (BUN) levels 24 h after IRI injury. Note that both *Acomys* (Aco, red) and Mus^{B6} (blue) exhibit similar levels of elevated BUN indicative of acute loss of kidney function. (*bottom right*) BUN levels (mg/dL) at 2 weeks after IRI injury. Note further declines in kidney function in Mus^{B6} (blue) progressing to organ failure. By contrast, BUN levels in *Acomys* (red) have returned to normal baseline values indicative of complete regeneration of organ function. ns – not significant. $^{*}P<0.05$, $^{**}P<0.01$, $^{***}P<0.001$, $^{****}P<0.0001$.

Given the evolutionary distance between *Acomys* and *Mus*, any comparison of transcriptomes will contain genes whose expression differs between species because of evolved non-coding sequence variation that may not have any direct relevance to scarless wound healing (Gerstein et al., 2014; Saxena et al., 2022; Scheenstra et al., 2020). However, it is likely that the clusters of differentially expressed genes described above reflect a genome that is poised to respond to tissue injury in such a way as to set into motion a regenerative response that, after 14 days, results in restoration of normal kidney structure and function without fibrosis (Okamura et al., 2021). The same argument could be made for the inflammatory cells that are recruited to sites of acute tissue injury in *Acomys* as their functions in tissue repair must interact with those of resident cells in such a way as to avoid activation of fibrotic pathways, minimize production of pro-inflammatory cytokines while contributing to tissue regeneration (Gawriluk et al., 2020; Simkin et al., 2017).

7. Repair of cardiac injury in *Acomys*

Three studies have recently reported on the response of *Acomys* hearts to myocardial ischemia (MI) due to coronary ligation (Koopmans et al., 2021; Peng et al., 2021; Qi et al., 2021). The three reports are highly concordant and the common findings are (1) that *Acomys* hearts exhibit increased resistance to ischemic damage, (2) a robust angiogenic response occurs after MI in spiny mouse hearts that is persistent over time, (3) a minimal infarct scar appears in *Acomys* hearts that is more porous and compliant than fibrotic infarct scars in *Mus*, and (4) *Acomys* hearts exhibit an improvement in cardiac function over time after MI while similarly injured *Mus* strains exhibit a progressive decline in cardiac function over the same time period.

Qi et al. reported that permanent occlusion of the left main anterior descending coronary artery (LAD) in adult spiny mice resulted in an acute drop in ejection fraction from about 87% to about 25% of uninjured controls at 1 day after MI (Qi et al., 2021). This value was very similar to ejection fractions observed Mus^{CD1} at 1 day after MI. Over the next 4 weeks, spiny mice recovered ejection fraction values to about 65% while Mus^{CD1} exhibited no improvement. At 13 weeks after MI, Mus^{CD1} hearts showed a 37% decrease in thickness of the left ventricular free wall while *Acomys* hearts had no decline in wall thickness. They also reported that *Acomys* hearts exhibited a smaller infarct zone with greatly reduced area of fibrotic scar than did Mus^{CD1} hearts. Using micro-CT analysis, *Acomys* hearts were shown to exhibit a much more robust vascular network within the infarct zone with large diameter conduit-like arteries distributed throughout the fibrotic scar tissue. By contrast, CD1 hearts had only a sparse distribution of small diameter blood vessels in the infarct zone with large areas appearing avascular. Using daily 5-bromo-2′-deoxyuridine (BrdU) injections starting at day 3 after MI and continuing until 14 days after injury, Qi et al. reported that up to 20% of the BrdU-labeled cells in the heart were also labeled by sarcomeric α-actinin and thus provisionally identified as proliferating cardiac myocytes (Qi et al., 2021).

Koopmans et al. likewise reported that the "scar" tissue that forms in the infarct zone in *Acomys* hearts is considerably different than scar tissue in similarly injured *Mus*^{B6} (Koopmans et al., 2021). The collagen fibers exhibit a porous "basket weave" structure with higher compliance and ventricular electrical conduction in *Acomys* hearts, consistent with a recovery of ejection fraction in spiny mice hearts (Koopmans et al., 2021). By contrast, collagen fibers in *Mus*^{B6} hearts are more densely packed and parallel in alignment, an organization that would resist compression during ventricular contraction and lead to progressive loss of cardiac function as the scar tissue expands over time. Their analysis of cardiac transcriptomes included a comparison of *Acomys* and *Mus*^{B6} neonatal hearts, since newborn mice are known to possess limited regeneration potential (Porrello et al., 2011). Of particular interest was their principal component analysis showing that adult *Acomys* hearts exhibit very little difference in gene expression when compared to neonatal *Acomys* hearts suggesting adult spiny mouse cardiac myocytes exist in an immature phenotype. On the other hand, there were many differences in gene expression in adult compared to neonatal *Mus* hearts corresponding to the maturation steps known to take place in the transition from newborn to adult cardiac physiology (Guo et al., 2021). These results suggest that adult *Acomys* hearts exist in a more immature state than adult *Mus* hearts (Koopmans et al., 2021) and offer a possible explanation for the differences in response to acute MI injury between the two species.

Finally, Peng et al. reported a higher frequency of mononucleated cardiomyocytes in normal adult *Acomys* hearts (26% *vs* 6% in *Mus*^{B6} hearts). This resembles the predominant mononucleated cardiomyocyte population in zebrafish hearts that are known to proliferate during cardiac regeneration (Kikuchi et al., 2010; Peng et al., 2021). Like the other studies cited above, Peng et al. also found a resistance to ischemic injury, a significant reduction in adverse left ventricular remodeling, and a more porous and less compressed collagen fiber orientation in infarct scars in *Acomys* hearts (Peng et al., 2021). They also reported finding an increased vascular density within *Acomys* scar tissue, with an elevated rate of endothelial cell proliferation and a strikingly greater distribution of Acta2-positive mature arteries in the center of the infarct scar than found in either *Mus*^{B6} or *Mus*^{CFW} strains. Therefore, all three studies demonstrated that adult *Acomys* hearts were more resistant to ischemic injury, exhibited a smaller infarct scar with a different structure and organization of collagen fibers suggesting more porous and compliant properties, and a robust angiogenic response that maintained vascular perfusion of injured myocardium thus leading to preservation of cardiac function.

8. Common features of *Acomys* regeneration studies

From the studies reviewed above, a number of common features begin to emerge in the way that *Acomys* cells respond to injury in different tissues and organs *in vivo* (Fig. 4). One is a general resistance to oxidative injury as seen experiments involving induction of tissue ischemia with or without reperfusion. For example, *Acomys* kidney, skin, and heart tissues resist oxidative damage and preserve organ function with minimal or no fibrosis

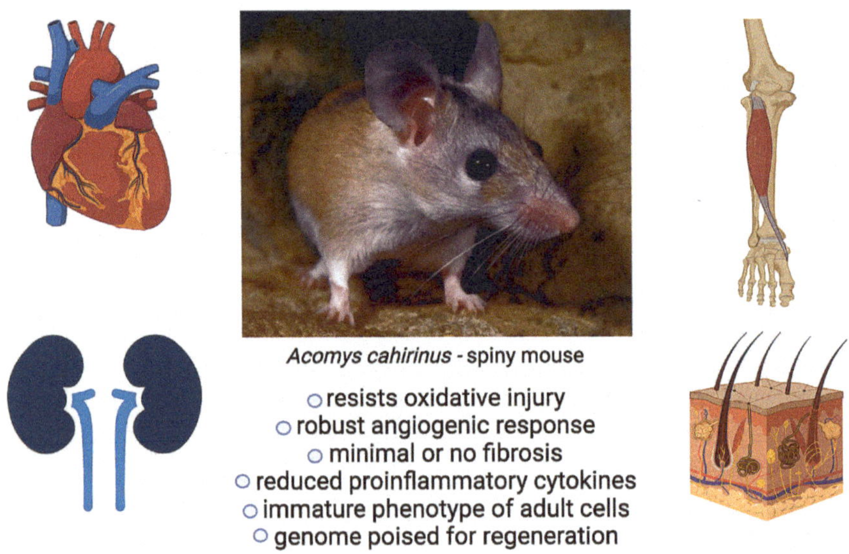

Fig. 4 *Acomys cahirinus* – an emerging model for mammalian regenerative wound healing. Comparative studies in the spiny mouse kidney (Okamura et al., 2021), heart (Koopmans et al., 2021; Peng et al., 2021; Qi et al., 2021), skeletal muscle (Maden et al., 2018), and skin (Brant et al., 2016; Brewer et al., 2021; Gawriluk et al., 2016; Gawriluk et al., 2020; Harn et al., 2021; Matias Santos et al., 2016; Seifert et al., 2012) provide evidence that *Acomys* species have evolved a different wound healing response to tissue injury than *Mus* or humans. Studies published to date indicate that some of the common features of wound repair among the organs indicated are those listed in the figure. Additional significant findings are a requirement for macrophage function in skin regeneration and a stronger and more prolonged ROS production at wound sites in *Acomys* (Saxena, Vekaria, Sullivan, & Seifert, 2019; Simkin et al., 2017). Figure created in BioRender.

when compared to various strains of *Mus* (Koopmans et al., 2021; Okamura et al., 2021; Peng et al., 2021; Qi et al., 2021; Saxena et al., 2019). Second, is a robust angiogenic response that quickly restores blood flow to the site of injury thus serving to provoke or maintain cell proliferation, cell survival and tissue remodeling required for a regenerative response (Brewer et al., 2021; Koopmans et al., 2021; Peng et al., 2021; Qi et al., 2021). Third is the proliferation in *Acomys* of fully differentiated adult cell types that are often regarded as post-mitotic in *Mus* and humans such as podocytes (Okamura et al., 2021) and cardiac myocytes (Koopmans et al., 2021; Qi et al., 2021), and finally a different composition, porosity, and time-dependent remodeling of the extracellular matrix (Koopmans et al., 2021; Peng et al., 2021; Seifert et al., 2012). In addition, the cardiac injury studies revealed that *Acomys* cardiac myocytes (CMs) in the adult heart exist in a more immature state with respect to content of mononucleated CMs (Peng et al., 2021), electrophysiological properties (Koopmans et al., 2021; Peng et al., 2021), proliferation (Qi et al., 2021) and whole ventricle transcriptomes (Koopmans et al., 2021) than *Mus* CMs. Thus, adult *Acomys* CMs may bear a stronger resemblance to regeneration-competent zebrafish CMs (Kikuchi & Poss, 2012) than to adult *Mus* or human CMs, at least for the parameters directly measured in those reports.

9. Summary

Scarless wound healing with regeneration of organ function is a long-sought goal of regenerative medicine. The work reviewed above suggest that the spiny mouse (*Acomys cahirinus*) is an emerging model organism with potential to identify mechanisms of tissue and organ regenerative repair that operate in the context of mammalian physiology directed by a mammalian genome. Although these animals have distinct husbandry needs compared to laboratory mice (Haughton et al., 2016), and their evolutionary distance from *Mus* presents a computational challenge for comparative transcriptomics, the remarkable wound healing properties that have evolved in the genus *Acomys* promise unique insights into potential applications for mammalian tissue regeneration.

Acknowledgments

This work was supported by a grant from the W.M. Keck Foundation; the US National Institutes of Health grants 1R01DK114149 and 5R01HL121877, the Loie Power Robinson Stem Cell & Regenerative Medicine Fund, Seattle Children's Research Institute, and the

Seattle Children's Foundation. We thank our colleagues in the Center for Developmental Biology & Regenerative Medicine, Seattle Children's Research Institute, University of Washington for critical input and helpful discussions, and Teresa Majesky for editing the manuscript.

References

Abbasi, S., Sinha, S., Labit, E., Rosin, N. L., Yoon, G., Rahmani, W., et al. (2020). Distinct regulatory programs control the latent regenerative potential of dermal fibroblasts during wound healing. *Cell Stem Cell, 27*, 396–412. https://doi.org/10.1016/j.stem.2020.07.008. pmid: 32755548.

Aghova, T., Palupčiková, J., Sumbera, R., Frynta, D., Lavrenchenko, L. A., Meheretu, Y., et al. (2019). Multiple radiations of spiny mice (Rodentia: Acomys) in dry open habitats of afro-Arabia: Evidence from a multi-locus phylogeny. *BMC Evolutionary Biology, 19*, 69. https://doi.org/10.1186/s12862-019-1380-9. pmid: 30832573.

Aurora, A. B., Porrello, E. R., Tan, W., Mahmoud, A. I., Hill, J. A., Bassel-Duby, R., et al. (2014). Macrophages are required for neonatal heart regeneration. *The Journal of Clinical Investigation, 124*, 1382–1392. https://doi.org/10.1172/JCI72181. pmid: 24569380.

Bandyopadhyay, P. R., & Hellum, A. M. (2014). Modeling how shark and dolphin skin patterns control transitional wall-turbulence vorticity patterns using spatiotemporal phase reset mechanisms. *Scientific Reports, 4*, 6650. https://doi.org/10.1038/srep06650. pmid: 25338940.

Battacharyya, R. P., Reményi, A., Yeh, B. J., & Lim, W. A. (2006). Domains, motifs, and scaffolds: The role of modular interactions in the evolution and wiring of cell signaling circuits. *Annual Review of Biochemistry, 75*, 655–680. https://doi.org/10.1146/annurev.biochem.75.103004.142710. pmid: 16756506.

Bayless, B. A., Navarro, F. M., & Winey, M. (2019). Motile cilia: Innovation and insight from ciliate model organisms. *Frontiers in Cell and Development Biology, 7*, 265. https://doi.org/10.3389/fcell.2019.00265. pmid: 31737631.

Bedelbaeva, K., Snyder, A., Gourevitch, D., Clark, L., Zhang, X. M., Leferovic, J., et al. (2010). Lack of p21 expression links cell cycle control and appendage regeneration in mice. *Proceedings of the National Academy of Sciences of the United States of America, 107*, 5845–5850. https://doi.org/10.1073/pnas.1000830107. pmid: 20231440.

Bellofiore, N., Ellery, S. J., Mamrot, J., Walker, D. W., Temple-Smith, P., & Dickinson, H. (2017). First evidence of a menstruating rodent: The spiny mouse (Acomys cahirinus). *American Journal of Obstetrics and Gynecology, 216*, 40e1–40e11. https://doi.org/10.1016/j.ajog.2016.07.041. pmid: 27503621.

Bellofiore, N., George, E., Vollenhoven, B., & Temple-Smith, P. (2021). Reproductive aging and menopause-like transition in the menstruating spiny mouse (Acomys cahirinus). *Human Reproduction, 36*, 3083–3094. https://doi.org/10.1093/humrep/deab215. pmid: 34601586.

Bloom, P., & Yeager, M. (1994). The injury and subsequent healing of a serious propeller strike to a bottle nose dolphin resident in the cold waters of the Northumberland coast of England. *Aquatic Mammals, 20*, 59–64.

Boothby, T. C., Tapia, H., Brozena, A. H., Piszkiewicz, S., Smith, A. E., Giovannini, I., et al. (2017). Tardigrades use intrinsically disordered proteins to survive desiccation. *Molecular Cell, 65*, 975–984. https://doi.org/10.1016/j.molcel.2017.02.018. pmid: 28306513.

Brant, J. O., Yoon, J. H., Polvadore, T., Barbazuk, W. B., & Maden, M. (2016). Cellular events during scar-free skin regeneration in the spiny mouse. *Wound Repair and Regeneration, 24*, 75–88. https://doi.org/10.1111/wrr.12385. pmid: 26606280.

Brewer, C. M., Nelson, B. R., Wakenight, P., Collins, S. J., Okamura, D. M., Dong, X. R., et al. (2021). Adaptations in hippo-yap signaling and myofibroblast fate underlie scar-free ear appendage wound healing in spiny mice. *Developmental Cell*, *56*, 2722–2740. https://doi.org/10.1016/j.devcel.2021.09.008. pmid: 34610329.

Brockes, J. P., & Kumar, A. (2008). Comparative aspects of animal regeneration. *Annual Review of Cell and Developmental Biology*, *24*, 525–549. https://doi.org/10.1146/annurev.cellbio.24.110707.175336. pmid: 18598212.

Bruce-Allen, L. J., & Geraci, J. R. (1985). Wound healing in the bottlenose dolphin (*Tursiops truncates*). *Canadian Journal of Fisheries and Aquatic Sciences*, *42*, 216–228. https://doi.org/10.1139/f85-029.

Brunjes, P. C. (1989). A comparative study of prenatal development in the olfactory bulb, neocortex, and hippocampal region of the precocial mouse *Acomys cahirinus* and rat. *Brain Research. Developmental Brain Research*, *49*, 7–25. https://doi.org/10.1016/0165-3806(89)90055-2. pmid: 2791267.

Burgin, C. J., Colella, J. P., Kahn, P. L., & Upham, N. S. (2018). How many species of mammals are there? *Journal of Mammalogy*, *99*, 1–14. https://doi.org/10.1093/jmammal/gyx147.

Chen, J., You, H., Li, Y., Xu, Y., He, Q., & Harris, R. C. (2018). EGF receptor-dependent YAP activation is important for renal recovery from AKI. *Journal of the American Society of Nephrology*, *29*, 2372–2385. https://doi.org/10.1681/ASN.2017121272. pmid: 30072422.

Chevret, P., Denys, C., Jaeger, J. J., Michaux, J., & Catzeflis, F. M. (1993). Molecular evidence that the spiny mouse (*Acomys*) is more closely related to gerbils (Gerbillinae) than to true mice (Murinae). *Proceedings of the National Academy of Sciences of the United States of America*, *90*, 3433–3436. https://doi.org/10.1073/pnas.90.8.3433. pmid: 8475093.

Colwell, A. S., Longaker, M. T., & Lorenz, H. P. (2003). Fetal wound healing. *Frontiers in Bioscience*, *8*, s1240–s1248. https://doi.org/10.2741/1183. pmid: 12957846.

Coresh, J., Selvin, E., Stevens, L. A., Manzi, J., Kusek, J. W., Eggers, P., et al. (2007). Prevalence of chronic kidney disease in the United States. *JAMA*, *298*, 2038–2047. https://doi.org/10.1001/jama.298.17.2038. pmid: 17986697.

Corkeron, P. J., Morris, R. J., & Bryden, M. M. (1987). A note on healing of large wounds in bottlenose dolphins, Tursiops truncates. *Aquatic Mammals*, *13*, 96–98.

Crookes, W. J., Ding, L. L., Huang, Q. L., Kimbell, J. R., Horwitz, J., & McFall-Ngai, M. J. (2004). Reflectins: The unusual proteins of squid reflective tissues. *Science*, *303*, 235–238. https://doi.org/10.1126/science.1091288. pmid: 14716016.

Danilovich, S., & Yovel, Y. (2019). Integrating vision and echolocation for navigation and perception in bats. *Science Advances*, *5*. https://doi.org/10.1126/sciadv.aaw6503. eaaw6503. pmid: 31249874.

Darby, I., Skalli, O., & Gabbiani, G. (1990). Alpha-smooth muscle actin is transiently expressed by myofibroblasts during experimental wound healing. *Laboratory Investigation*, *63*, 21–29. pmid: 2197503.

Denys, C., Taylor, P. J., & Aplin, K. P. (2017). Family muridae. In D. E. Wilson, T. E. Lacher Jr.,, & R. A. Mittemeier (Eds.), *Handbook of the mammals of the world, volume 7: Rodents II*. Barcelona: Lynx Edicions.

Dickinson, H., Walker, D. W., Cullen-McEwen, L., Wintour, E. M., & Moritz, K. (2005). The spiny mouse (*Acomys cahirinus*) completes nephrogenesis before birth. *American Journal of Physiology. Renal Physiology*, *289*, F273–F279. https://doi.org/10.1152/ajprenal.00400.2004. pmid: 15741606.

Drenckhahn, J. D., Schwarz, Q. P., Gray, S., Laskowski, A., Kiriazis, H., Ming, Z., et al. (2008). Compensatory growth of healthy cardiac cells in the presence of diseased cells restores tissue homeostasis during heart development. *Developmental Cell*, *15*, 521–533. https://doi.org/10.1016/j.devcel.2008.09.005. pmid: 18854137.

Driskell, R. R., Lichtenberger, B. M., Hoste, E., Kretzschmar, K., Simons, B. D., Charalambous, M., et al. (2013). Distinct fibroblast lineages determine dermal architecture in skin development and repair. *Nature, 504*, 277–281. https://doi.org/10.1038/nature12783. pmid: 24336287.

Duffield, J. S. (2014). Cellular and molecular mechanisms in kidney fibrosis. *The Journal of Clinical Investigation, 124*, 2299–2306. https://doi.org/10.1172/JCI72267. pmid: 24892703.

Ellerman, J. R. (1941). The families and genera of living rodents. In; family Muridae, London: British museum of. *Natural History*.

Eming, S. A., Wynn, T. A., & Martin, P. (2017). Inflammation and metabolism in tissue repair and regeneration. *Science, 356*, 1026–1030. https://doi.org/10.1126/science.aam7928. pmid: 28596335.

Friedman, S. L., Sheppard, D., Duffield, J. S., & Violette, S. (2013). Therapy for fibrotic diesases: Nearing the starting line. *Science Translational Medicine, 5*, 167sr1. https://doi.org/10.1126/scitranslmed.3004700. pmid 23303606.

Garry, G. A., Antony, M. L., & Garry, D. J. (2016). Cardiotoxin induced injury and skeletal muscle regeneration. *Methods in Molecular Biology, 1460*, 61–71. https://doi.org/10.1007/978-1-4939-3810-0_6. pmid: 27492166.

Gawriluk, T. R., Simkin, J., Hacker, C. K., Kimani, J. M., Kiama, S. G., Ezenwa, V. O., et al. (2020). Complex tissue regeneration in mammals is associated with reduced inflammatory cytokines and an influx of T cells. *Frontiers in Immunology, 11*, 1695. https://doi.org/10.3389/fimmu.2020.01695. pmid: 32849592.

Gawriluk, T. R., Simkin, J., Thompson, K. L., Biswas, S. K., Clare-Salzler, Z., Kimani, J. M., et al. (2016). Comparative analysis of ear-hole closure identifies epimorphic regeneration as a discrete trait in mammals. *Nature Communications*, 7, 11164. https://doi.org/10.1038/ncomms11164. pmid: 27109826.

Gawronska-Kozak, B., Grabowska, A., Kopcewicz, M., & Kur, A. (2014). Animal models of skin regeneration. *Reproductive Biology, 14*, 61–67. https://doi.org/10.1016/j.repbio.2014.01.004. pmid: 24607256.

Gerhart, J., & Kirschner, M. (1997). *Cells, Embryos, and Evolution* (pp. 238–295). Malden, MA: Blackwell Science.

Gerstein, M. B., Rozowsky, J., Yan, K. K., Wang, D., Cheng, C., Brown, J. B., et al. (2014). Comparative analysis of the transcriptome across distant species. *Nature, 512*, 445–448. https://doi.org/10.1038/nature13423. pmid: 25164755.

Go, A. S., Chertow, G. M., Fan, D., McCulloch, C. E., & Hsu, C. Y. (2004). Chronic kidney disease and the risks of death, cardiovascular events, and hospitalization. *The New England Journal of Medicine, 351*, 1296–1305. https://doi.org/10.1056/NEJMoa041031. pmid: 15385656.

Godwin, J. W., Pinto, A. R., & Rosenthal, N. A. (2013). Macrophages are required for adult salamander limb regeneration. *Proceedings of the National Academy of Sciences of the United States of America, 110*, 9415–9420. https://doi.org/10.1073/pnas.1300290110. 23690624.

Goldman, A., Roy, J., Bodenmiller, B., Wanka, S., Landry, C. R., Aebersold, R., et al. (2014). The calcineurin signaling network evolves via conserved kinase-phosphatase modules that transcend substrate identity. *Molecular Cell, 55*, 422–435. https://doi.org/10.1016/j.molcel.2014.05.012. pmid: 24930733.

Goldstein, B., & King, N. (2016). The future of cell biology: Emerging model organisms. *Trends in Cell Biology, 26*, 818–824. https://doi.org/10.1016/j.tcb.2016.08.005. pmid: 27639630.

Gorbunova, V., Hine, C., Tian, X., Ablaeva, J., Gudkov, A. V., Nevo, E., et al. (2012). Cancer resistance in the blind mole rat is mediated by concerted necrotic cell death mechanism. *Proceedings of the National Academy of Sciences of the United States of America, 109*, 19392–19396. https://doi.org/10.1073/pnas.1217211109. pmid: 23129611.

Gourevitch, D., Clark, L., Chen, P., Seitz, A., Samulewicz, S. J., & Heber-Katz, E. (2003). Matrix metalloproteinase activity correlates with blastema formation in the regenerating MRL mouse ear hole model. *Developmental Dynamics*, *226*, 377–387. https://doi.org/10.1002/dvdy.10243. pmid: 12557216.

Guerrero-Juarez, C. F., Dedhia, P. H., Jin, S., Ruiz-Vega, R., Ma, D., Liu, Y., et al. (2019). Single-cell analysis reveals fibroblast heterogeneity and myeloid-derived adipocyte progenitors in murine skin wounds. *Nature Communications*, *10*, 650. https://doi.org/10.1038/s41467-018-08247-x. pmid: 30737373.

Guo, Y., Cao, Y., Jardin, B. D., Sethi, I., Ma, Q., Moghadaszadeh, B., et al. (2021). Sarcomeres regulate murine cardiomyocyte maturation through MRTF-SRF signaling. *Proceedings of the National Academy of Sciences of the United States of America*, *118*. https://doi.org/10.1073/pnas.2008861118, e2008861118. pmid: 33361330.

Gurtner, G. C., Werner, S., Brandon, Y., & Longaker, M. T. (2008). Wound repair and regeneration. *Nature*, *453*, 314–321. https://doi.org/10.1038/nature07039. pmid: 18480812.

Gustavsen, C. R., Kvicerova, J., Dickinson, H., & Heller, R. S. (2009). *Acomys*, the closest relatives to gerbils, do express Pdx-1 protein and have similar islet morphology to gerbils. *Islets*, *1*, 191–197. https://doi.org/10.4161/isl.1.3.9557. pmid: 21099272.

Hadid, Y., Pavlicek, T., Beiles, A., Ianovici, R., Raz, S., & Nevo, E. (2014). Sympatric incipient speciation of spiny mice *Acomys* at "Evolution Canyon," Israel. *Proceedings of the National Academy of Sciences of the United States of America*, *111*, 1043–1048. https://doi.org/10.1073/pnas.1322301111. pmid: 24402169.

Han, P., Zhou, X. H., Chang, N., Xiao, C. L., Yan, S., Ren, H., et al. (2014). Hydrogen peroxide primes heart regeneration with a derepression mechanism. *Cell Research*, *24*, 1091–1107. https://doi.org/10.1038/cr.2014.108. pmid: 25124925.

Harn, H. I., Wang, S. P., Lai, Y. C., Van Handel, B., Liang, Y. C., Tsai, S., et al. (2021). Symmertry breaking of tissue mechanics in wound induced hair follicle regeneration of laboratory and spiny mice. *Nature Communications*, *12*, 2595. https://doi.org/10.1038/s41467-021-22822-9. pmid: 33972536.

Haughton, C. L., Gawriluk, T. R., & Seifert, A. W. (2016). The biology and husbandry of the African spiny mouse (*Acomys cahirinus*) and the research uses of a laboratory colony. *Journal of the American Association for Laboratory Animal Science*, *55*, 9–17. pmid: 26817973.

Heallen, T., Zhang, M., Wang, J., Bonilla-Claudio, M., Klysik, E., Johnson, R. L., et al. (2011). Hippo pathway inhibits Wnt signaling to restrain cardiomyocyte proliferation and heart size. *Science*, *332*, 458–461. https://doi.org/10.1126/science.1199010. pmid: 21512031.

Helston, O., & Amaya, E. (2021). Reactive oxygen species during heart regeneration in zebrafish: Lessons for future clinical therapies. *Wound Repair and Regeneration*, *29*, 211–224. https://doi.org/10.1111/wrr.12892. pmid: 33471940.

Hill, N. R., Fatoba, S. T., Oke, J. L., Hirst, J. A., O'Callaghan, C. A., Lasserson, D. S., et al. (2016). Global prevalence of chronic kidney disease – A systematic review and meta-analysis. *PLoS One*, *11*, e0158765. https://doi.org/10.1371/journal.pone.0158765. pmid: 27383068.

Hinz, B., Phan, S. H., Thannikal, V. J., Prunotto, M., Desmoulière, A., Varga, J., et al. (2012). Recent developments in myofibroblast biology: Paradigms for connective tissue remodeling. *The American Journal of Pathology*, *180*, 1340–1355. https://doi.org/10.1016/j.ajpath.2012.02.004. pmid: 22387320.

Hou, R., Astrowski, A. A., & Plikus, M. V. (2021). Fibroblasts feel evolutionary pressure to regenerate. *Developmental Cell*, *56*, 2685–2687. https://doi.org/10.1016/j.devcel.2021.09.019. pmid: 34637705.

Humphreys, B. D. (2018). Mechanisms of renal fibrosis. *Annual Review of Physiology*, *80*, 309–326. https://doi.org/10.1146/annurev-physiol-022516-034227. pmid: 29068765.

Hunter, T. (2012). Why nature chose phosphate to modify proteins. *Philosophical Transactions of the Royal Society B, 367*, 2513–2516. https://doi.org/10.1098/rstb.2012.0013. pmid: 22889903.

Katz, N., Dayan, T., & Kronfeld-Schor, N. (2018). Fitness effects of interspecific competition between two species of desert rodents. *Zoology (Jena, Germany), 128*, 62–68. https://doi.org/10.1016/j.zool.2018.03.002. pmid: 29699824.

Kaverina, N. V., Eng, D. G., Freedman, B. S., Kutz, J. N., Chozinski, T. J., Vaughan, J. C., et al. (2019). Dual lineage tracing shows that glomerular parietal epithelial cells can transdifferentiate toward the adult podocyte fate. *Kidney International, 96*, 597–611. https://doi.org/10.1016/j.kint.2019.03.014. pmid: 31200942.

Keller, G. L., & Burns, K. A. (1989). Husbandry and hematology of captive spiny mice (*Acomys cahirinus*). *Laboratory Animal Science, 39*, 625–626. pmid: 2593647.

Kikuchi, K., Holdway, J. E., Werdich, A. A., Anderson, R. M., Fang, Y., Egnaczyk, G. F., et al. (2010). Primary contribution to zebrafish heart regeneration by gata4(+) cardiomyocytes. *Nature, 464*, 601–605. https://doi.org/10.1038/nature08804. pmid: 20336144.

Kikuchi, K., & Poss, K. D. (2012). Cardiac regenerative capacity and mechanisms. *Annual Review of Cell and Developmental Biology, 28*, 719–741. https://doi.org/10.1146/annurev-cellbio-101011-155739. pmid: 23057748.

Koopmans, T., van Beijnum, H., Roovers, E. F., Tomasso, A., Malhotra, D., Boeter, J., et al. (2021). Ischemic tolerance and cardiac repair in the spiny mouse (Acomys). *NPJ Regenerative Medicine, 6*, 78. https://doi.org/10.1038/s41536-021-00188-2. pmid: 34789755.

Kronfeld-Schor, N., Haim, S., Dayan, T., Zisapel, N., Klingenspor, M., & Heldmaier, G. (2000). Seasonal thermogenic acclimation of diurnally and nocturnally active desert spiny mice. *Physiological and Biochemical Zoology, 73*, 37–44. https://doi.org/10.1086/316718. 10685905.

Leach, J. P., Heallen, T., Zhang, M., Rahmani, M., Morikawa, Y., Hill, M. C., et al. (2017). Hippo pathway deficiency reverses systolic heart failure after infarction. *Nature, 550*, 260–264. https://doi.org/10.1038/nature24045. pmid: 28976966.

Lecompte, E., Aplin, K., Denys, C., Catzeflis, F., Chades, M., & Chevret, P. (2008). Phylogeny and biogeography of African Murinae based on mitochondrial and nuclear gene sequences, with a new tribal classification of the subfamily. *BMC Evolutionary Biology, 8*, 199. https://doi.org/10.1186/1471-2148-8-199. pmid: 18616808.

Lehoczky, J. A., Robert, B., & Tabin, C. J. (2011). Mouse digit tip regeneration is mediated by fate-restricted progenitor cells. *Proceedings of the National Academy of Sciences of the United States of America, 108*, 20609–20614. https://doi.org/10.1073/pnas.1118017108. pmid: 22143790.

Lepper, C., Conway, S. J., & Fan, C. M. (2009). Adult satellite cells and embryonic muscle progenitors have distinct genetic requirements. *Nature, 460*, 627–631. https://doi.org/10.1038/nature08209. pmid: 19554048.

Li, S., Mariappan, N., Megyesi, J., Shank, B., Kannan, K., Theus, S., et al. (2013). Proximal tubule PPARα attenuates renal fibrosis and inflammation caused by unilateral ureteral obstruction. *American Journal of Physiology. Renal Physiology, 305*, F618–F627. https://doi.org/10.1152/ajprenal.00309.2013. pmid: 23804447.

Libby, P. (2015). Murine "model" monotheism: An iconoclast at the altar of mouse. *Circulation Research, 117*, 921–925. https://doi.org/10.1161/CIRCRESAHA.115.307523. pmid: 26541681.

Liu-Chittenden, Y., Huang, B., Shim, J. S., Chen, Q., Lee, S. J., Anders, R. A., et al. (2012). Genetic and pharmacological disruption of the TEAD-YAP complex suppresses the oncogenic activity of YAP. *Genes & Development, 26*, 1300–1305. https://doi.org/10.1101/gad.192856.112. pmid: 22677547.

Love, N. R., Chen, Y., Ishibashi, S., Kritsiligkou, P., Lea, R., Koh, Y., et al. (2013). Amputation-induced reactive oxygen species are required for successful Xenopus tadpole tail regeneration. *Nature Cell Biology*, *15*, 222–228. https://doi.org/10.1038/ncb2659. pmid: 23314862.

Maden, M. (2018). The evolution of regeneration - where does that leave mammals? *The International Journal of Developmental Biology*, *62*, 369–372. https://doi.org/10.1387/ijdb.180031mm. pmid: 29938749.

Maden, M., Brant, J. O., Rubiano, A., Sandoval, A. G. W., Simmons, C., Mitchell, R., et al. (2018). Perfect chronic skeletal muscle regeneration in adult spiny mice, *Acomys cahirinus*. *Science Reports*, *8*, 8920. https://doi.org/10.1038/s41598-018-27178-7. pmid: 29892004.

Maden, M., & Varholick, J. A. (2020). Model systems for regeneration: The spiny mouse, *Acomys cahirinus*. *Development*, *147*. https://doi.org/10.1242/dev.167718, 167718. pmid: 32098790.

Mascharak, S., desJardins-Park, H. E., Davitt, M. F., Griffin, M., Borrelli, M. R., Moore, A. L., et al. (2021). Preventing *engrailed-1* activation in fibroblasts yields wound regeneration without scarring. *Science*, *372*. https://doi.org/10.1126/science.aba2374. eaba2374. pmid: 33888614.

Matias Santos, D., Rita, A. M., Casanellas, I., Brito Ova, A., Araujo, I. M., Power, D., et al. (2016). Ear wound regeneration in the African spiny mouse *Acomys cahirinus*. *Regeneration*, *3*, 52–61. https://doi.org/10.1002/reg2.50. pmid: 27499879.

McDonald, M., Trost, B., & Napper, S. (2018). Conservation of kinase-phosphorylation site pairings: Evidence for an evolutionarily dynamic phosphoproteome. *PLoS One*, *13*. https://doi.org/10.1371/journal.pone.0202036, e0202036. pmid: 30106995.

Mondajem, A., Taylor, A., Denys, C., & Cotterill, F. P. D. (2015). *Rodents of sub-Saharan Africa*. A biogeographic and taxonomic synthesis. Berlin: De Gruyter.

Montandon, S. A., Tzika, A. C., Martins, A. F., Chopard, B., & Milinkovitch, M. C. (2014). Two waves of anisotropic growth generate enlarged follicles in the spiny mouse. *EvoDevo*, *5*, 33. https://doi.org/10.1186/2041-9139-5-33. pmid: 25705371.

Moriyama, T., Kawada, N., Ando, A., Yamauchi, A., Horio, M., Nagata, K., et al. (1998). Up-regulation of HSP47 in the mouse kidney with unilateral ureteral obstruction. *Kidney International*, *54*, 110–119. https://doi.org/10.1046/j.1523-1755.1998.00964.x. pmid: 9648069.

Moya, I. M., & Halder, G. (2019). Hippo-yap/Taz signaling in organ regeneration and regenerative medicine. *Nature Reviews. Molecular Cell Biology*, *20*, 211–226. https://doi.org/10.1038/s41580-018-0086-y. pmid: 30546055.

Mui, M. Z., Zhou, Y., Blanchette, P., Chughtai, N., Knight, J. F., Gruosso, T., et al. (2015). The human adenovirus type 5 E4orf4 protein targets two phosphatase regulators of the hippo signaling pathway. *Journal of Virology*, *89*, 8855–8870. https://doi.org/10.1128/JVI.03710-14. pmid: 26085163.

Musser, G. G., & Carleton, M. D. (2005). Superfamily Muroidea. In D. E. Wilson, & R. Dam (Eds.), *Mammal species of the world. A taxonomic and geographic reference* (pp. 894–1531). The Johns Hopkins University Press.

Nacu, E., & Tanaka, E. M. (2011). Limb regeneration; a new development? *Annual Review of Cell and Developmental Biology*, *27*, 409–440. https://doi.org/10.1146/annurev-cellbio-092910-154115. pmid: 21801016.

Neilson, J., Hedeholm, R. B., Heinemeier, J., Bushnell, P. G., Christiansen, J. S., Olsen, J., et al. (2016). Eye lens radiocarbon reveals centuries of longevity in the Greenland shark (Somniosus microcephalus). *Science*, *353*, 702–704. https://doi.org/10.1126/science.aaf1703. pmid: 27516602.

Okamura, D. M., Brewer, C. M., Wakenight, P., Bahrami, N., Bernardi, K., Tran, A., et al. (2021). Spiny mice activate unique transcriptional programs after severe kidney injury regenerating organ function without fibrosis. *iScience*, *24*, 103269. https://doi.org/10.1016/j.isci.2021.103269. pmid: 34849462.

Oosterhuis, W. P., Mooren, P. G., Charles, R., & Lamers, W. H. (1984). Perinatal development of the lung in rat and spiny mouse: Its relation to altricial and precocial timing of birth. *Biology of the Neonate*, *45*, 236–243. https://doi.org/10.1159/000242011. pmid: 6326867.

Pavlov, V. V. (2006). Dolphin skin as a natural anisotropic compliant wall. *Bioinspiration & Biomimetics*, *1*, 31–40. https://doi.org/10.1088/1748-3182/1/2/001.

Peng, H., Shindo, K., Donahue, R. R., Gao, E., Ahern, B. M., Levitan, B. M., et al. (2021). Adult spiny mice (Acomys) exhibit endogenous cardiac recovery in response to myocardial infarction. *NPJ Regenerative Medicine*, *6*, 74. https://doi.org/10.1038/s41536-021-00186-4. pmid: 34789749.

Pennathur, S., Pasichnyk, K., Bahrami, N. M., Zeng, L., Febbraio, M., Yamaguchi, I., et al. (2015). The macrophage phagocytic receptor CD36 promotes fibrogenic pathways on removal of apoptotic cells during chronic kidney injury. *The American Journal of Pathology*, *185*, 2232–2245. https://doi.org/10.1016/j.ajpath.2015.04.016. pmid: 26092500.

Philippeos, C., Telerman, S. B., Oules, B., Pisco, A. O., Shaw, T. J., Elgueta, R., et al. (2018). Spatial and single-cell transcriptional profiling identifies functionally distinct human dermal fibroblast subpopulations. *The Journal of Investigative Dermatology*, *138*, 811–825. https://doi.org/10.1016/i.jid.2018.01.016. pmid: 29391249.

Pinheiro, G., Prata, D. F., Araújo, I. M., & Tiscornia, G. (2018). The African spiny mouse [Acomys spp.] as an emerging model for development and regeneration. *Laboratory Animals*, *52*, 565–576. https://doi.org/10.1177/0023677218769921. pmid: 29699452.

Plikus, M. V., Wang, X., Sinha, S., Forte, E., Thompson, S. M., Herzog, E. L., et al. (2021). Fibroblasts: Origins, definitions, and functions in health and disease. *Cell*, *184*, 3852–3872. https://doi.org/10.1016/j.cell.2021.06.024. pmid: 34297930.

Porrello, E. R., Mahmoud, A. I., Simpson, E., Hill, J. A., Olson, E. N., & Sadek, H. A. (2011). Transient regenerative potential of the neonatal mouse heart. *Science*, *331*, 1078–1080. https://doi.org/10.1126/science. pmid: 21350179.

Poss, K. D. (2010). Advances in understanding tissue regenerative capacity and mechanisms in animals. *Nature Reviews. Genetics*, *11*, 710–722. https://doi.org/10.1038/nrg2879. pmid: 20838411.

Pratsinis, H., Mavrogonatou, E., & Kletsas, D. (2019). Scarless wound healing: From development to senescence. *Advanced Drug Delivery Reviews*, *146*, 325–343. https://doi.org/10.1016/j.addr.2018.04.011. pmid: 29654790.

Qi, Y., Dasa, O., Maden, M., Vohra, R., Batra, A., Walter, G., et al. (2021). Functional heart recovery in an adult mammal, the spiny mouse. *International Journal of Cardiology*, *338*, 196–203. https://doi.org/10.1016/j.ijcard.2021.06.015. pmid: 34126132.

Rinkevich, Y., Lindau, P., Ueno, H., Longaker, M. T., & Weissman, I. L. (2011). Germ-layer and lineage-restricted stem/progenitors regenerate the mouse digit tip. *Nature*, *476*, 409–413. https://doi.org/10.1038/nature10346. pmid: 21866153.

Rinkevich, Y., Walmsley, G. G., Hu, M. S., Maan, Z. N., Newman, A. M., Drukker, M., et al. (2015). Identification and isolation of a dermal lineage with intrinsic fibrogenic potential. *Science*, *348*. https://doi.org/10.1126/science.aaa2151. aaa2151. pmid: 25883361.

Robey, T. E., & Murry, C. E. (2008). Absence of regeneration in the MRL/MpJ mouse heart following infarction or cryoinjury. *Cardiovascular Pathology*, *17*, 6–13. https://doi.org/10.1016/j.carpath.2007.01.005. pmid: 18160055.

Rockey, D. C., Bell, P. D., & Hill, J. A. (2015). Fibrosis: A common pathway to organ injury and failure. *The New England Journal of Medicine*, *372*, 1138–1149. https://doi.org/10.1056/NEJMra1300575. pmid: 25785971.

Rouleau, C., Pores Fernando, A. T., Hwang, J. H., Faure, N., Jiang, T., White, E. A., et al. (2016). Transformation by polyomavirus middle T antigen involves a unique bimodal interaction with the hippo effector yap. *Journal of Virology*, *90*, 7032–7045. https://doi.org/10.1128/JVI.0041-16. pmid: 27194756.

Roy, J., & Cyert, M. S. (2009). Cracking the phosphatase code: docking interactions determine substrate specificity. *Science Signaling*, *2*. https://doi.org/10.1126/scisignal.2100re9. :re9. pmid: 19996458.

Saxena, A., Sharma, V., Muthuirulan, P., Nerfeld, S. J., Tran, M. P., Guiterrez, H. L., et al. (2022). Interspecies transcriptomics identify genes that underlie disproportionate foot growth in jerboas. *Current Biology*, *32*, 1–15. https://doi.org/10.1016/j.cub.2021.10.063. pmid: 34793695.

Saxena, S., Vekaria, H., Sullivan, P. G., & Seifert, A. W. (2019). Connective tissue fibroblasts from highly regenerative mammals are refractory to ROS-induced cellular senescence. *Nature Communications*, *10*, 4400. https://doi.org/10.1038/s41467-019-12398-w. pmid: 31562333.

Scheenstra, M. R., Martinez-Botia, P., Acebes-Huerta, A., Brouwer, R. W. W., Caballero-Snachez, N., Gillemans, N., et al. (2020). Comparison of the PU.1 transcriptional regulome and interactome in human and mouse inflammatory dendritic cells. *Journal of Leukocyte Biology*. https://doi.org/10.1002/JLB.6A1219-711RRR. pmid: 33289106.

Schipper, J., Chanson, J. S., Chiozza, F., Cox, N. A., Hoffmann, M., Katariya, V., et al. (2008). The status of the world's land and marine animals: Diversity, threat, and knowledge. *Science*, *322*, 225–230. https://doi.org/10.1126/science.1165115. pmid: 18845749.

Seifert, A. W., Kiama, S. G., Seifert, M. G., Goheen, J. R., Palmer, T. M., & Maden, M. (2012). Skin shedding and tissue regeneration in African spiny mice (Acomys). *Nature*, *489*, 561–565. https://doi.org/10.1038/nature11499. pmid: 23018966.

Shiozuru, D., Ichii, O., Kimura, J., Nakamura, T., Ali Elewa, Y. H., Otsuka-Kanazawa, S., et al. (2016). MRL/MpJ mice show unique pathological features after experimental kidney injury. *Histology and Histopathology*, *31*, 189–204. pmid: 26345626.

Simkin, J., Gawriluk, T. R., Gensel, J. C., & Seifert, A. W. (2017). Macrophages are necessary for epimorphic regeneration in African spiny mice. *eLife*, *6*. https://doi.org/10.7554/eLife.24623, e24623. pmid: 28508748.

Singbartl, K., Green, S. A., & Ley, K. (2000). Blocking P-selectin protects from ischemia/reperfusion-induced acute renal failure. *The FASEB Journal*, *14*, 48–54. https://doi.org/10.1096/fasebj.14.1.48. pmid: 10627279.

Tapmeier, T. T., Brown, K. L., Tang, Z., Sacks, S. H., Sheerin, N. S., & Wong, W. (2008). Reimplantation of the ureter after unilateral ureteral obstruction provides a model that allows functional evaluation. *Kidney International*, *73*, 885–889. https://doi.org/10.1038/sj.ki.5002797. pmid: 18200000.

Thompson, S. M., Phan, Q. M., Winuthayanon, S., Driskell, I. M., & Driskell, R. R. (2022). Parallel single cell multi-omics analysis of neonatal skin reveals transitional fibroblast states that restricts differentiation into distinct fates. *The Journal of Investigative Dermatology*. https://doi.org/10.1016/j/jid/2021.11.032. 34922949.

Timbuka, C. D., & Kabigumila, J. D. (2006). Diversity and abundance of small mammals in the Serengeti kopjes, Tanzania. *Tanzania Journal of Science*, *32*, 1–12. https://doi.org/10.4314/tjs.v32i1.18424.

Tomasek, J. J., Gabbiani, G., Hinz, B., Chaponnier, C., & Brown, R. A. (2002). Myofibroblasts and mechano-regulation of connective tissue remodeling. *Nature Reviews. Molecular Cell Biology*, *3*, 349–363. https://doi.org/10.1038/nrm809. pmid: 11988769.

Van Rooijen, N., & Hendrikx, E. (2010). Liposomes for specific depletion of macrophages from organs and tissues. *Methods in Molecular Biology*, *605*, 189–203. https://doi.org/10.1007/978-1-60327-360-2_13. pmid: 20072882.
Volobouev, V., Gautun, J. C., Sicard, B., & Tranier, M. (1996). The chromosome complement of Acomys spp. (Rodentia, Muridae) from Oursi, Burkina Faso – The ancestral karyotype of the cahirinus-dimidiatus group? *Chromosome Research*, *4*, 526–530. https://doi.org/10.1007/BF02261780. pmid: 8939364.
Wang, J., & Conboy, I. (2010). Embryonic vs adult myogenesis: Challenging the 'regeneration recapitulates development' paradigm. *Journal of Molecular Cell Biology*, *2*, 1–4. https://doi.org/10.1093/jmcb/mjp027. pmid: 19797316.
Wynn, T. A., & Ramalingam, T. R. (2012). Mechanisms of fibrosis: therapeutic translation for fibrotic disease. *Nature Medicine*, *18*, 1028–1040. https://doi.org/10.1038/nm.2807. pmid: 22772564.
Zasloff, M. (2011). Observations on the remarkable (and mysterious) wound-healing process of the bottlenose dolphin. *The Journal of Investigative Dermatology*, *131*, 2503–2505. https://doi.org/10.1038/jid.2011.220. pmid: 21776005.

CHAPTER SEVEN

Plumbing our organs: Lessons from vascular development to instruct lab generated tissues

Anne R. Ryan and Ondine Cleaver*
Department of Molecular Biology, Center for Regenerative Science and Medicine (CRSM), University of Texas Southwestern Medical Center, Dallas, TX, United States
*Corresponding author: e-mail address: ondine.cleaver@utsouthwestern.edu

Contents

Current Topics in Developmental Biology, Volume 148
ISSN 0070-2153
https://doi.org/10.1016/bs.ctdb.2022.02.013

Abstract

The formation, growth and maintenance of our organs, such as our kidneys or pancreas, requires their coordinated growth alongside the intricate vasculature that pervades them. Blood vessels course through nearly every tissue in our bodies, facilitating the essential exchange of gases, nutrition and wastes, as well as the rapid circulation of hormones and other signaling molecules. Endothelial cells (ECs) that line all of our blood vessels are therefore the gatekeepers for communication between the circulation and organ-specific cell types. We and many others have sought to understand: (1) how endothelial cell progenitors initially assemble to form blood vessels in the embryo, and (2) how the embryonic vascular tree expands to perfuse growing organs. Here, we review what we have learned from embryonic blood vessels and how this knowledge instructs our approaches to vascularize laboratory generated tissues, such as organoids. We will assess our general understanding of blood vessel formation, and discuss recent studies of the growing kidney vasculature. Furthermore, we will assess the challenges and limitations faced by organoid technologies, including the difficulties in achieving the patterned vascular network that is essential to organ function. Lastly, we will then review recent studies of kidney organoid blood vessels and propose approaches that improve vascularization. Understanding the ontogeny of organ-specific vasculatures will help propel regenerative therapeutic approaches.

1. Introduction

As organs emerge in the embryo and grow in size both pre- and postnatally, so does their associated vasculature. For the most part, studies of organ development have largely overlooked the mechanisms at play in building local vessels. Where do vessel progenitor cells come from? How do they assemble and expand in tissues that that can increase in size up to 100 fold? What signals vessels to remain stable within a tissue? How do blood vessels become adapted to local tissue needs? We have only begun to answer these questions for normal organ blood vessels, and little to nothing is known about the vessels of laboratory engineered tissues, including organoids.

Why are we interested in generating organs in the lab? Recent efforts have been motivated by the stark demand by patients needing transplants. In the US, over 100K people are currently on the national transplant waiting list. Hence, the demand far outpaces available transplantable organs. Unfortunately efforts to bioengineer organs have achieved minimal success, as recapitulating complex tissues present obvious difficulties. Understanding and recapitulating the full anatomical and physiological characteristics of organs, with their full panoply of cell types, is a momentous task. In addition, and importantly for the discussion herein, proper vascularization of replacement tissues and organs presents considerable obstacles.

We argue here that part of the challenge is that we currently have little understanding of vascular formation and diversification during organogenesis, and hence no tools to assess faithfulness of laboratory-generated blood vessels. This chapter will review what we know about the coordinated growth of blood vessels and tissues during embryonic development, as well as the patterning, heterogeneity, and remodeling of the endothelium. Basic knowledge of the growing endothelium will pave the way to developing fully vascularized and therapeutically viable tissues for transplantation. In addition, we will discuss our current understanding of the vasculature of the organoid, a three-dimensional culture system widely used to model organ development *ex vivo*.

2. Vascular development: Basic step-wise processes

Blood vessels are essential for viability of multicellular organisms whose tissues are large and complex enough that they cannot experience gas and nutrient exchange through simple diffusion. Blood vessels develop by two recognized mechanisms: vasculogenesis and angiogenesis (Risau, 1997), which are discussed below. Mammalian embryos, such as those of mice and humans, develop progressively following gastrulation and initially lack any organs. However, the first functional organ to arise is the cardiovascular system, which courses through the embryonic mesoderm. As organs begin to bud and expand, the initial vascular plexus expands coordinately with those organs (Fig. 1). This growth continues throughout embryogenesis, into post-fetal life, until the organism achieves its full adult size.

2.1 Vasculogenesis: De novo formation of vessels from angioblasts

The formation of the first embryonic vessels occurs *via* vasculogenesis. During this process, endothelial progenitors, or angioblasts, coalesce and form blood vessels *de novo*. Vasculogenesis occurs early in embryogenesis when angioblasts emerge from the mesoderm (Fig. 2A), aligning as two linear aggregates or endothelial "cords" down the length of the embryo (Fig. 2B) (Drake & Fleming, 2000). These first cords will transform into the paired dorsal aortae. During this transformation, the cords will undergo tubulogenesis, opening up a central canal to form lumenized blood vessels and allowing blood flow (Fig. 2C). In the mouse this is a relatively rapid process occurring in only a matter of hours. During embryogenesis, many blood vessels, including the yolk sac vessels, the aortae and the cardinal veins, are formed through vasculogenesis. Once vasculogenesis gives rise to the

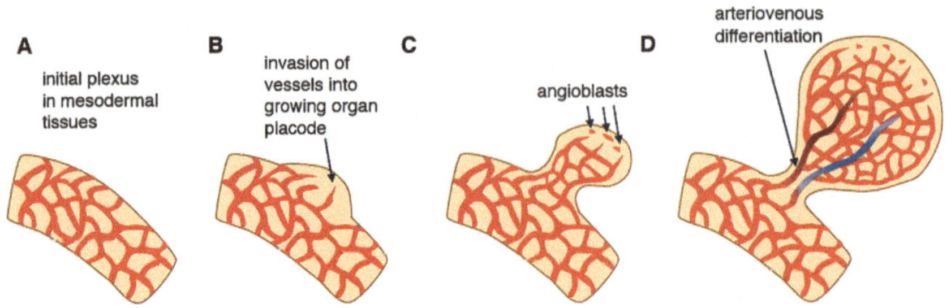

Fig. 1 Coordinated growth of organs and vascular beds. (A–D) Sequential outgrowth of a theoretical organ into which the vasculature expands. (A) Prior to organ growth, a primitive vascular plexus already is present within the embryonic mesoderm. (B) As the organ grows, vessels must expand into the newly growing tissues. In a number of cases, this has been shown to occur *via* sprouting angiogenesis. (C) In organs such as the kidney and lung, newly sprouting vessels have been shown to connect with existing pools of angioblasts in the organ periphery. (D) As the organ becomes much larger, blood vessels remodel. As some vessels preferentially acquire increased flow, they will expand and differentiate into arteries and veins. The number of vessels at the proximal connection of the organ are often reduced in number during remodeling and arteriovenous differentiation.

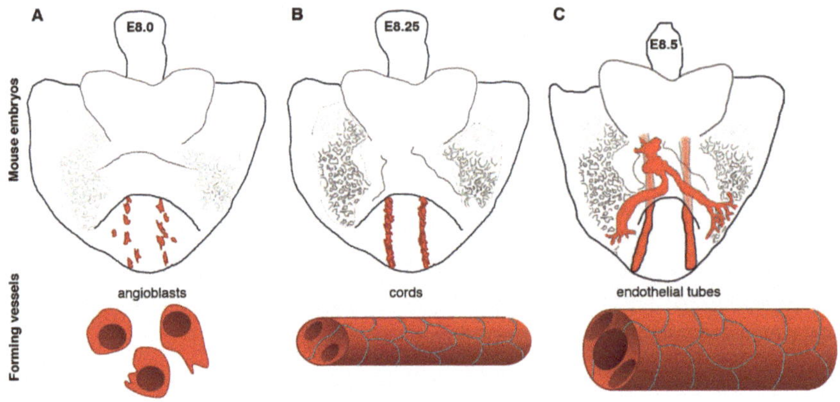

Fig. 2 Formation of the first mammalian vessels occurs *via* vasculogenesis. (A–C) Early mouse embryos during formation of the first embryonic vessels (in red). Embryos are shown in top row. Embryonic stages indicated at top. All embryos are a stage prior to gut tub closure and embryonic turning. Structures indicated: allantois (al); cardiogenic plate (cp); open gut tube (g); neural folds (nf); yolk sac (ys). Schematic close up view of angioblasts and endothelial cells during vasculogenesis at bottom. (A) At embryonic day 8.0 (E8.0) angioblasts differentiate within the mesoderm, migrate and begin to adhere to each other. (B) At 8.25, the angioblasts have aligned in two parallel cords, presaging the formation of the paired dorsal aortae. Cords are lumenless aggregates of angioblasts. (C) Cords transform into tubes that allow blood flow when endothelial cells change shape (become thin) and they clear adhesions from their apical membranes.

primary vascular plexus (or primitive cardiovascular system), the primary method of new vessel formation and vascular growth switches to angiogenesis.

2.2 Angiogenesis: Elaboration of new vessels from pre-existing ones

Angiogenesis is the creation of new blood vessels from existing, pre-formed and lumen bearing vessels. Angiogenesis can be subdivided into sprouting angiogenesis and remodeling angiogenesis (Wilting & Christ, 1996). Vessels that develop by the former mechanism sprout from currently existing vessels, while blood vessels that develop *via* the latter mechanism arise by either splitting an existing vessel into two (intussusception) or by fusion with adjacent vessels (anastomosis). During sprouting angiogenesis, specialized distal endothelial cells (ECs) called "tip cells" become activated and extend out of a pre-existing vessel to form a new vascular branch (Fig. 3A). These new sprouts are then shaped by the influence of cues in the microenvironment, both positive such as vascular endothelial growth factor (VEGF) and negative such semaphorins, flt-1 (Chappell, Taylor, Ferrara, & Bautch, 2009) and a wide array of other cues (Meadows et al., 2012). This sculpting of the vasculature is referred to as "patterning."

After blood vessels develop, they further differentiate into different types of vessels including arteries, veins, capillaries, and lymphatic vessels This is primarily associated with remodeling angiogenesis whereby vessels either enlarge in diameter and give rise to arteries and veins. In addition, smaller vessels can either regress *via* pruning or split *via* intussusception (Fig. 3B). Although the cardiovascular system is arguably one of the most important organ systems in our body, we are still discovering fundamental aspects of how cells come together to form these tubular networks. Only recently have we come to appreciate how diverse ECs can be and how they are able to form complex, individualized and organotypic vascular beds.

2.3 Necessity of flow for blood vessel differentiation and maintenance

The requirement of blood flow for vessel differentiation and maintenance has been known for decades. The vasculature is highly dynamic, especially during initial formation, and it is sensitive to both microenvironmental changes, to signaling factors and to biomechanical cues. In fact, as the vasculature expands, blood vessels remodel by selective sprouting, expansion of lumens, or regression, all partly responding to rerouting of blood flow as

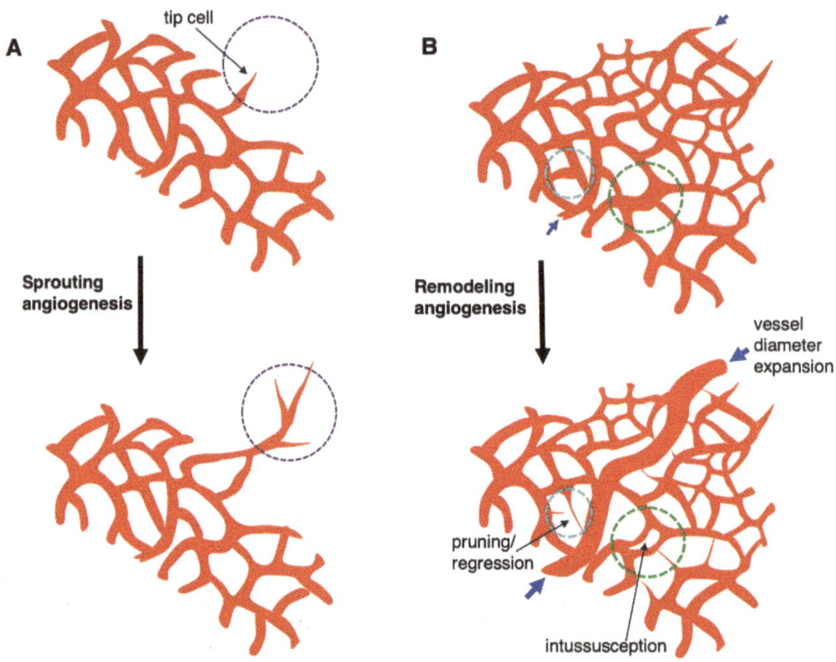

Fig. 3 Sprouting and remodeling angiogenesis. When new vessels emerge from preexisting vessels, the process is called angiogenesis. (A) Sprouting angiogenesis occurs when an endothelial cell within the wall of a preexisting vessel becomes activated, degrades any existing basement membrane, and extends filopodial protrusions. This pioneering cell is called a tip cell. The sprout then grows, and endothelial cells in the growing stalk, behind the tip cell, undergo proliferation and extend the existing lumen into the new sprout. (B) Remodeling angiogenesis refers to a battery of endothelial behaviors. These include pruning or regression of vessels (often due to rerouted or reduced blood flow), expansion of existing vessels (snort arrows), or splitting of larger vessels into smaller ones (intussusception).

vascular networks develop. Together these changes allow the vasculature to adapt to local needs, contributing to a more hemodynamically effective network.

Blood flow in developing vessels is particularly critical to blood vessel remodeling and stability. Disruption of flow during development, or the inability of ECs to respond to flow, results in vascular remodeling defects and retention of an immature plexus (Lucitti et al., 2007; Udan, Vadakkan, & Dickinson, 2013). This remodeling in response to flow is necessary for arteriovenous specification and formation of the vascular hierarchy. Indeed, arteries, veins, and capillaries experience different types and velocities of flow, and flow has been shown to influence their developmental trajectories (le Noble et al., 2004; Udan et al., 2013).

Pruning is a key process catalyzed by flow cessation. Indeed withdrawal of flow results in rapid destabilization and regression of vessels, as ECs detach from surrounding extracellular matrix (ECM) (Lucitti et al., 2007). Vascular regression is both a normal part of development, such as in the remodeling of the postnatal retinal vascular plexus, as well as a variety of pathological responses (Korn & Augustin, 2015). Whether pathological or developmental, regression is preempted by apoptosis of the downstream ECs following cessation or blockage of flow (Meeson, Palmer, Calfon, & Lang, 1996). Therefore, not only is flow necessary for proper vascular development, it is also required for blood vessel maintenance.

The impact of flow on endothelium is evident at both the cellular and molecular levels. Shear stress produced by blood flow induces many changes in ECs, influencing EC polarity, migration, proliferation, gene expression and apoptosis (Udan et al., 2013). Three well-characterized endothelial membrane proteins, platelet endothelial cell adhesion molecule-1 (PECAM-1), vascular endothelial cadherin (VE-CAD), and vascular endothelial growth factor receptor 2 (VEGFR2) form a mechanosensory complex that is responsible for EC response to flow, with PECAM-1 acting as the direct sensor (Tzima et al., 2005). The response of these proteins to shear stress results in the activation of multiple signaling cascades and pathways including ERK, SRC, AKT, NFkB, and PI3K (Tzima et al., 2005). Loss of PECAM-1, which directly transmits mechanical forces across the cell, causes defective endothelial response to flow, in that ECs no longer activate NFκB and downstream inflammatory genes in regions of disturbed flow. Many other pathways and molecules are known to act as mechanosensors (Lim, Jang, & Kim, 2018) and deciphering their roles in blood vessel differentiation and maintenance will be of great interest in coming years.

2.4 Endothelial heterogeneity

One striking feature of vascular ECs is that they are widely heterogeneous and participate in forming all kinds of different blood vessels (Aird, 2003; Aird, 2012). One obvious set of differences can be seen at the anatomical level. Blood vessels display a remarkable variety of large and small vessels, including arteries designed to withstand high blood pressure from the heart, and veins that experience low pressure and contain valves designed to help return de-oxygenated blood to the lungs. The larger arteries and veins taper off and join at capillary beds, where most gas and nutrient exchange happens.

Beyond arteriovenous differentiation, ECs are also heterogeneous both structurally and transcriptionally, between and within different organs and tissues. These structural and molecular differences play an important role in regionally distinct vascular functions. For example, sinusoidal ECs in the liver regulate the immune response through their fenestrations and scavenger receptors, while ECs in the brain have tight cell–cell junctions and low permeability to create the protective blood–brain barrier (Aird, 2012). In the kidney, fenestrated ECs with a unique basement membrane allow blood filtration into the nephron through the glomerulus, while ECs of the vasa recta form a unique architecture that facilitates slow blood flow, maintaining the concentration gradient of the medulla (Guyton & Hall, 2006). All these examples demonstrate the critical importance of regional specialization of ECs across and within organs. As technologies have advanced and methods such as single cell RNA-sequencing (scRNA-seq) studies have become commonplace, there has been an increased appreciation for the full range of transcriptional heterogeneity present in all tissues, and ECs are no exception (Marcu et al., 2018). The challenge now is to make sense of the information, validate it, and use the it to develop a toolbox of regionally specific transcriptional signatures that can be used to recognize organotypic and specialized vasculatures.

2.5 Angiocrine signaling

An interesting aspect of the endothelium to be considered is its ability to act as a distributed signaling network with the capacity to actively communicate with surrounding tissues. Given its heterogeneity, it suggests ECs could act to signal in regionally specific ways. How can ECs signal to neighboring cells? There is growing evidence that ECs produce what have been termed "angiocrine signals," which comprise diverse factors capable of influencing neighboring cells, ranging from promoting differentiation of organs to stem cell maintenance (Cleaver, 2004; Leibing et al., 2018; Rafii et al., 1995; Rafii, Butler, & Ding, 2016; Singhal & Augustin, 2020).

Indeed, ECs express a wide range of regionalized cell-surface and secreted signaling molecules that have been slowly recognized by the field. For example, Rspondin3 expression in the central vein is necessary to both initially establish and maintain liver zonation (Rocha et al., 2015). In the developing brain, crosstalk between neural progenitor cells and ECs through the angiopoietin-Tie2 pathway induces TGFβ1 production by the ECs, which signals to the neural progenitors, promoting oligodendrocyte differentiation (Paredes et al., 2021). These studies suggest that blood vessels may be necessary to produce properly differentiated and patterned tissues *in vitro*. Additionally, crosstalk with ECs is also important to tissue responses to stress or injury (Dumas et al., 2020). In the lung, matrix metalloproteinase 14 produced by ECs promotes alveolarization after lung injury (Ding et al., 2011). Angiocrine signaling has therefore been investigated in a number of organs, although interestingly few studies have been carried out in the kidney. Further studies will be necessary to determine whether the vasculature influences cells of the developing or mature kidney in an angiocrine manner, and if so, what signals are involved.

3. Vascularization of the kidney

Development of the kidney, which includes its intimately associated and complex vasculature, has been the subject of much interest. Indeed, classical studies have assessed both EC progenitor assembly and formation of new vessels *via* angiogenesis (Abrahamson, Robert, Hyink, St John, & Daniel, 1998; Robert, St John, & Abrahamson, 1998; Sequeira-Lopez et al., 2015). Kidney development in mice begins at E10.5, when the Wolffian ducts bud into the metanephric mesenchyme. At this early stage, capillaries closely associate with the ureteric bud forming a plexus that surrounds, but does not invade, the metanephric mesenchyme (Daniel et al., 2018; Munro, Hohenstein, & Davies, 2017). Indeed, the most peripheral cortical capillaries do not penetrate the caps of nephron progenitor cells, but rather surround them. As the kidney continues to develop, so does the vasculature. Capillaries wrap around the maturing renal vesicle and S-shaped body in a predictable manner as the nephron takes shape. By E13.5, vascular diversification can first be detected by the differentiation of large arteries and veins. Arteriogenesis occurs around this time, as some vessels remodel and enlarge to form the first distinguishable arteries in the kidney. They also begin to express arterial genes, such as Connexin40. As development proceeds, these initial renal arteries branch under the cortex, at the periphery of the developing kidney (Daniel et al., 2018).

While we have gained some insights into the events that occur during renal vascularization and development of the vascular architecture, we have little resolution on endothelial diversification during this process or on the precise step-wise association of distinct vessels with various kidney structures. We cannot at this point distinguish what type of ECs develop alongside the proximal tubules *versus* the descending or ascending loop of Henle, for instance. Studying the relationship between renal blood vessels and the developing renal epithelial structures, as well as the process of renal vascular maturation, will help us to develop the ability to recognize properly vascularized organoids.

4. Organoid technology as a step toward organ replacement

The use of organoids to model organogenesis and to generate potentially transplantable replacement tissues is an exciting new area within the realms of both developmental biology and personalized medicine. Organoids have proven to be excellent manipulable models to study organ development, to model disease, and to test novel drugs (Kim, Koo, & Knoblich, 2020; Takebe & Wells, 2019). Organoids are 3-dimensional, self-organizing structures that differentiate into the main cell types of a particular organ following tailored, directed differentiation. Unlike traditional cell culture platforms, organoids allow the generation of complex tissues *in vitro*. In the search for transplantable tissue grown in the laboratory, many have looked to organoids as the solution. Organoids can be created by directed differentiation of embryonic stem cells (ESCs), induced pluripotent stem cells (iPSCs) or from the progenitors isolated from tissues. Many organoid models currently exist, including liver, brain, intestine, and kidneys.

Enticingly, organoids can be grown from human induced pluripotent stem cells (hiPSCs), meaning that tissue can be derived directly from patient somatic cells. This allows any tissue grown from these cells to be a perfect immunologic match, avoiding the immune response and freeing the patient from a lifetime of immunosuppressants. If the patient has a genetic disease, the cells can be studied, used to test pharmacological approaches or even edited at the DNA level using CRISPR/Cas9 to correct genetic errors to generate healthy cells. Fig. 4 demonstrates the goal of personalized medicine in kidney disease patients: iPSCs derived from patients can be used to generate an immune-matched kidney transplant. (See Fig. 5.)

Despite the promising potential of organoids, the field is likely at least 20 years away from developing transplantable tissue. In order to turn organoids into transplantable organs, three key components not yet achieved are still needed: (1) tissue patterning and organization, including both

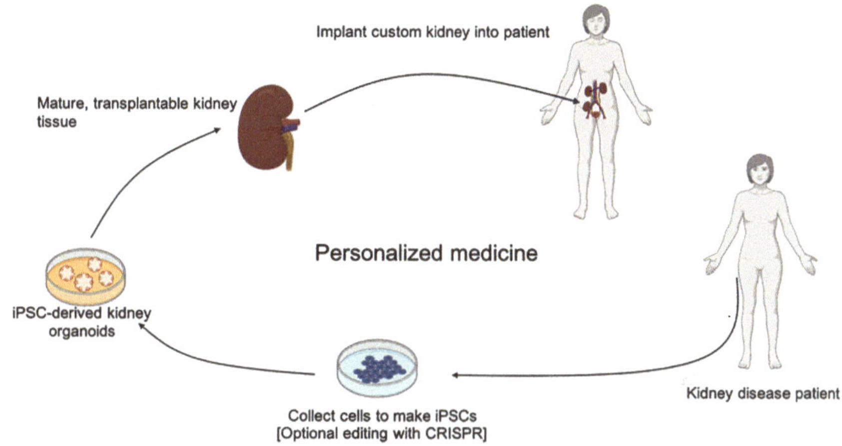

Fig. 4 Process of developing personalized, transplantable kidneys for patients with kidney disease. Cells from patients can be used to generate induced pluripotent stem cells. These cells can be used to study drug approaches, or they can be genetically edited such as using CRISPR approaches. In addition, these naïve cells can be used to generate organoids. While transplantable organs have not yet been generated from organoids, that lofty goal is the ultimate hope. This will require solving both proper vascularization and functional integration of engineered tissues.

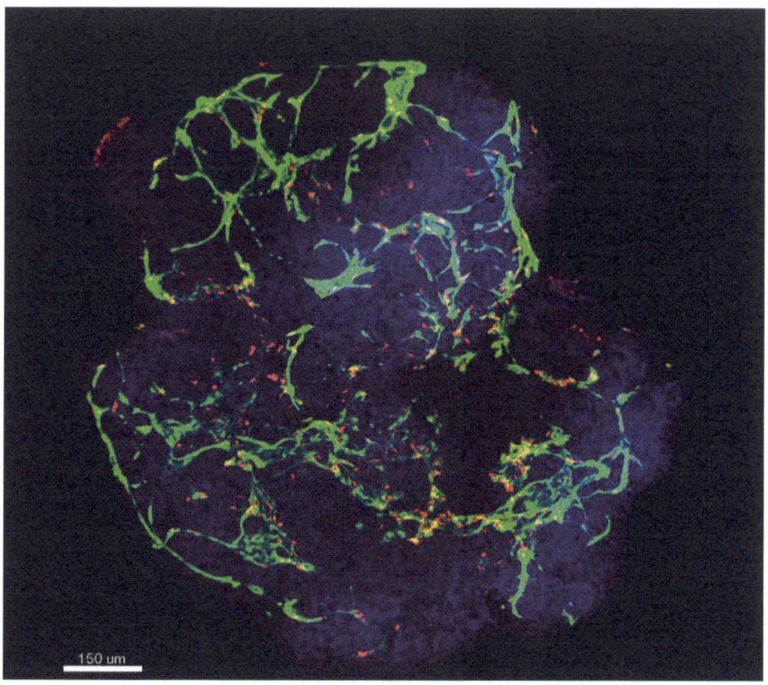

Fig. 5 Vasculature of a day 12 kidney organoid. Immunofluorescent staining for Pecam in green; Sox17 in red; DAPI stained nuclei in blue. Note the sparsity of the organoid vessels and the lack of overall patterning.

vascular and epithelial plumbing; (2) epithelial specification and maturity; and (3) scalability. The most significant reason organoids are not yet transplantable is that while organoids can be coaxed to form the major structures of an organ, they still remain disorganized, immature, small, and they have yet to be shown to integrate properly in host tissues. In addition, the vasculature has been observed to be abnormal and largely unpatterned (Ryan et al., 2021).

Using kidney organoids as an example of the potential and challenges of this promising platform, their small size and tissue disorganization have presented significant obstacles to their therapeutic potential. In addition, a major requirement to organoid viability is proper integration into host tissues. Conceptually, this requires functional "plumbing" of the tissue, including both a proper inlet, for blood, and an outlet, for urine excretion. Once implanted, kidney organoids would quickly die from a lack of oxygen and nutrients, if not rapidly connected to a perfused cardiovascular system. Similarly, their filtering abilities would be useless if they were unable to shuttle away toxins and wastes *via* an appropriate conduit. The requirement for functional vascular connection to the kidneys is underscored by a condition termed "renal artery stenosis," whereby one or more of the arteries carrying blood to the kidneys narrows, which is a major cause of kidney transplant loss and patient death (Bruno, Remuzzi, & Ruggenenti, 2004). Hence proper connectivity of kidney organoids to the host renal apparatus is likely a fundamental requirement for their use as replacement tissue.

4.1 Can organoids be properly vascularized?

One critical challenge to generating transplantable and functional organoids is that of properly patterning the organoid vasculature. Normal renal vasculature in living kidneys is highly patterned, with a wide variety of different vessels types including arcuate, interlobar and interlobular arteries and veins, and not surprisingly this vascular network is strikingly functionally heterogeneous. For the kidney to function, the vasculature must form fenestrated capillary tufts within the glomeruli, where afferent arterioles enter the nephron for filtration. Efferent arterioles then exit the glomerulus to form the peritubular capillaries, which surround the cortical nephron, and the vasa recta, which surround the loop of Henle. Peritubular capillaries are necessary for draining the nephron, absorbing secreted water and solutes. Can this be recapitulated in laboratory generated organoids? That is the challenge. Can these vessels be recognized once generated? That is another challenge.

Recent studies have shown evidence that ingrowth of host vasculature into organoids implanted can occur following implantation under the kidney capsule (Kumar Gupta et al., 2020). The hope was that functional connection of the host and organoid vasculature would "plumb" the implanted tissue allowing the blood flow and subsequent vascular remodeling needed. However, organoid vessels following implantation are observed to be mispatterned. In addition, a lack of maturity in kidney organoids has been encountered in most recent studies and this indicates the organoids are not functional. Kidney organoids analyzed by one group exhibited a transcriptomic profile similar to that of a second trimester embryonic kidney (Takasato et al., 2015). There was only evidence of rudimentary function at a cellular level in the proximal tubule epithelium, but no functional resorption by the vasculature. These findings demonstrate that kidney organoids achieved to date, while promising, fail to progress in their development to functional maturation.

Lastly, an additional challenge of using organoids as a platform to grow patient tissues is the scalability problem. Organoids are approximately 1–2 mm in diameter and contain less than one thousand nephrons, whereas an adult kidney contains about one million nephrons and is the size of a fist (Guyton & Hall, 2006; Takasato, Er, Chiu, & Little, 2016). This size differential encountered with kidney organoids exemplifies the challenges that will be faced by all organoid models: in order to turn organoids into transplantable tissue, they will need to be much larger, or to be aggregated, and properly vascularized. Case in point, the primary function of the kidney is to filter blood and size is proportional to their function. Twenty five percent of the heart's output is pumped into the kidneys from the renal arteries, which are direct offshoots of the dorsal aorta which experiences the highest flow of the entire circulatory system. Kidney organoid vessels must thus be able to withstand normal pressure and volume of blood. Therefore, in order to better establish organoids as developmental models and to create transplantable organs, we must further characterize and understand vascular development in normal renal vessels in the context of their function and apply this knowledge to *in vitro* organ culture.

5. Organoids and vascularization

In 2015, a landmark study detailing the generation of kidney organoids was the first to show their generation from human iPSCs (Takasato et al., 2015). This study was the first to show that hiPSC-derived kidney organoids developed expected cell types, and also exhibited rudimentary proximal tubule function, as shown by proximal tubule dextran endocytosis. Since then, multiple published methods have described kidney organoid generation, both from mouse embryonic kidney progenitors and pluripotent stem cells. These studies include generation of kidney organoids from mouse ESCs (mESCs) (Taguchi & Nishinakamura, 2015), from mouse embryonic kidney (Brown, Muthukrishnan, & Oxburgh, 2015), from hiPSCs (Takasato et al., 2015), from both human ESCs (hESCs) and iPSCs (Freedman et al., 2015; Kumar Gupta et al., 2020; Morizane et al., 2015; Przepiorski et al., 2018; Taguchi & Nishinakamura, 2015) and from mouse embryonic kidney (Ryan et al., 2021; Zeng et al., 2021).

Directed differentiation of iPSCs to kidney organoids recapitulates much of the step-wise process of development observed during embryonic kidney induction. This approach begins with activating Wnt signaling in iPSCs to induce mesoderm specification. Cells further differentiate into metanephric mesenchyme, splitting off into both stromal and nephron progenitor cell fates. Wnt activation is followed by addition of FGF9 to induce mesenchymal-to-epithelial transition (MET), mimicking formation of pretubular aggregates. From here, the cells self-organize into nephron-like structures including podocytes, proximal and distal tubules, and loop of Henle (Takasato et al., 2015). Originally, it was thought that through the organoid differentiation process, both nephrons and collecting ducts developed. However, this is likely due to the misattribution of collecting duct to distal tubule (Howden et al., 2021). New methods have since been introduced to specifically develop collecting duct organoids, which can then be recombined with nephron organoids (Zeng et al., 2021).

In addition to epithelial tubules (which includes nephrons and collecting ducts), kidney organoids contain ECs and stroma. The origin of organoid angioblasts is unclear. However, as the directed differentiation process mimics induction of the intermediate mesoderm, it is possible an off-target effect is to generate some amount of lateral plate mesoderm as well, which contains ECs. In the developing embryo, these different types of mesoderm are in close proximity and angioblasts first arise from the lateral plate mesoderm.

We can glean some insights about the vasculature of laboratory-generated tissues from the intersection of developmental biology and earlier work carried out using embryoid bodies (EBs). EBs are aggregates of either ESCs or PSCs which are allowed to spontaneously differentiate by removal of stem-maintaining factors (Kurosawa, 2007). EBs form when stem cells differentiate into the three embryonic germ layers: mesoderm, endoderm, and ectoderm. Intriguingly, EBs develop blood vessels spontaneously without the addition of growth factors (Gerecht-Nir, Ziskind, Cohen, & Itskovitz-Eldor, 2003; Wang, Clark, & Bautch, 1992). These models of early embryonic development have been used as tools to study the origin of ECs, vasculogenesis, and tissue-specification of ECs (Kim et al., 2008).

Similar to the step-wise processes that occur during embryogenesis, EB and organoid ECs mature over time in culture. EB derived ECs appear on day 3 of culture and they display sequential expression of endothelial specification and maturation genes, starting with VEGFR2, followed by CD31, Tie2, and then VE-Cadherin, and finally LVAP and CD34 (Vittet et al., 1996). This may also be assessed by cellular morphology, as ECs develop into cord and vessel-like stages later during culture, around day 11. This progressive differentiation of ECs closely resembles the processes in the embryo (Fig. 2). However, it is debated whether EB and organoid vessels reach a fully mature state, even following addition of angiogenesis-promoting growth factors. Partial maturation of vessels as assessed by gene expression has also been observed in kidney organoids and is improved by interstitial flow (Homan et al., 2019). Maturation of engineered tissue vessels is clearly the next frontier which will require increased resolution and additional tools to allow evaluation.

The importance of vasculature in laboratory-generated tissues has largely gone unexplored, including in organoids. It is not uncommon for pioneer organoid studies to only glancingly evaluate ECs. Additionally, some organoid systems such as cerebral organoids, yield little to no vasculature at all (Lancaster et al., 2013). While the differentiation of specific types of organoids, such as kidney organoids, mimics specific developmental pathways, it is clear conditions used to do not uniformly promote EC fate in parallel. We argue that knowledge of vascular development is essential to promote improvements in establishment and functionality of organoid vasculature.

6. The promise of vascular transcriptomes

As we learn more about blood vessel developmental mechanisms and diversification of ECs in developing and adult organs, we can use this information to guide the creation of better, more tissue-specific vascularized organoids. The transcriptional signature of organoid ECs can indicate what type of vasculature organoids are able to produce—and what they cannot. In the kidney, one study utilized scRNA-seq to identify blood vessel types in the mouse kidney both during embryogenesis and in the adult (Barry et al., 2019). The transcriptional signatures identified in this study can provide a road map for what vessels in organoids should express. However, despite laudable progress, there is still so much unknown about both EC heterogeneity, and its developmental and functional impact.

In order to create properly vascularized organoids, it is necessary to know what the vasculature should look like. Blood vessel anatomy in the adult kidney is well established, but we still need higher resolution of the transcriptional identity of renal blood vessels during development. We previously assessed heterogeneity of ECs in developing mouse kidneys by examining a variety of known EC genes, which revealed early emergence of EC heterogeneity during embryonic kidney vascularization (Daniel et al., 2018). Other critical studies have contributed to our understanding of renal EC heterogeneity, especially with the advancement of scRNA-seq (Aird, 2007; Barry et al., 2019; Dumas et al., 2020; Molema & Aird, 2012). However, despite the growing number of published scRNA-seq studies, there has been a frustrating lack of substantial *in vivo* validation of gene expression.

We propose that a deeper dive into available scRNA-seq data is necessary to enhance our understanding of blood vessel development and heterogeneity. Only then will we be able to properly recognize divergent vasculatures of *in vitro* tissues. It is imperative to connect these powerful *in silico* analyses to the biological questions at hand and determine what these RNA sequencing datasets are telling us. Can scRNA-seq analysis help us answer questions such as how many different types of ECs are present in the kidney? Will this data allow us to discover new renal specific EC genes or will it further help characterize spatial heterogeneity? Can we use this data to identify angiocrine signaling pathways involved in either development or regeneration? The mounting data available in the public (published) domain provides much optimism that these questions can be answered.

7. Vascularization of organoids increases epithelial maturity

A properly formed and mature vascular network is clearly essential to the function and survival of tissues, such as the kidney. One interesting concept that has emerged is that vascularization may also increase tissue maturation, thereby improving function. This idea was demonstrated by one study where the addition of ECs to cerebral organoids resulted in improved neuronal action potentials (Cakir et al., 2019). Similarly, enhanced vascularization of kidney organoids improved podocyte maturity, as well as proximal tubule cell polarity and ciliation (Homan et al., 2019; van den Berg et al., 2018). The importance of vasculature in fostering maturation of laboratory-generated tissues has thus been recognized, and efforts are currently being made to hone existing protocols to further improve vascularized tissues. Below, we discuss current methods for creating and maintaining organoid vasculature, and we evaluate their strengths and weaknesses.

8. Methods for vascularization of organoids

8.1 Vascularization through addition of ECs

Cerebral organoids generated to date have largely been avascular (Lancaster et al., 2013). This may be due to the differentiation protocol not fostering EC production, or it may be the result of regression of blood vessels following to the long culture time required for generation of cerebral organoids. In order to overcome the lack of blood vessels, some groups have added ECs to their differentiating neural cells and seen beneficial effects. Various forms of ECs have been added, including cell lines such as HUVECs or ECs differentiated from hESCs (Cakir et al., 2019; Pham et al., 2018; Shi et al., 2020). Given that ECs in normal organs and tissues are by nature highly heterogeneous and tissue-specific, the addition HUVECs or other transformed EC lines to create vasculature in organoids is not likely ideal. The endogenous vasculature of organoids has been shown to be transcriptionally similar to ECs *in vivo* (Combes, Zappia, Er, Oshlack, & Little, 2019), suggesting cell lines would be poor substitutes for ECs grown and matured in association with renal cell types. Additionally, ECs lose their unique properties once cultured (Israely et al., 2014; Lacorre et al., 2004).

One promising approach has harnessed the power of an endothelial "master regulator." It has recently been shown that expression of the endothelial transcription factor ETV-2 in EC lines gives them the plasticity to adapt to a new environment (Palikuqi et al., 2020). ETV-2 expression, even transient, appears to revert ECs into an angiogenic state that allows them to form a network of stable vessels. In fact, overexpression of

ETV-2 in hESCs appears sufficient to differentiate into ECs (Cakir et al., 2019; Morita et al., 2015). Stimulation of endothelial progenitor cell fate in this manner might help provide a wider source of adaptable ECs to be used in vascularization efforts.

8.2 In vitro flow studies

Regression of blood vessels has been noted in multiple tissue culture and organoid systems where culture is carried out under static conditions (Holloway et al., 2020; Ryan et al., 2021; Takasato et al., 2015). As discussed above, blood flow contributes two important functions to the endothelium that carries it: (1) Flow provides ECs and surrounding tissues with essential oxygen and nutrients, while removing metabolic wastes, and (2) Flow exerts shear stress on the ECs, providing biomechanical cues that result in production of survival and differentiation signals. Hence, it is rather unsurprising that organoid vasculature regresses due to the lack of hemodynamic flow in standard culture conditions. In order to develop organoids with a stable vasculature, hemodynamic flow must be provided. There are two ways of doing this: through *in vitro* fluidics systems and through implantation into a host animal.

8.3 Fluidics devices and other approaches to providing ECs with flow

Adding hemodynamic flow to *in vitro* tissues or organoids can be done using microfluidics, engineered devices that produce the flow of liquid, colloquially known as "tissue on a chip." There are many variations of these tools, with their own specific objectives and advantages. A simple way to introduce flow into organoids is using an orbital shaker or a spinning flask bioreactor. This creates flow in a circular motion by moving all the fluid. While most studies focus on the hemodynamic flow within a vessel, even interstitial flow has been shown to induce the sprouting of ECs (Song & Munn, 2011). Orbital shakers are a standard part of the Lancaster et al., 2013 protocol to prevent necrosis in cerebral organoids, although vasculature is not present. Przepiorski and colleagues uses a rotating bioreactor to generate hiPSC-derived kidney organoids, but in this study there does not appear to be a significant improvement in vasculature over other organoid culture systems (Przepiorski, Crunk, Espiritu, Hukriede, & Davidson, 2020).

A "macrofluidics" device applies shear stress to as many as 100 organoids at once (Homan et al., 2019). Most of the flow occurs around and over the organoids, although some fluid moves through the vessels as well. This method has only been used with kidney organoids, but it can likely be used with multiple other types of organoids. The mechanisms at play remain uncertain, as there is not yet clear evidence whether the improvement of vasculature from this type of flow is due to an increase in oxygenation and nutrients or due to the physical forces effected by flow applied to the organoid ECs.

Bioreactors and microfluidics chambers may stabilize organoid vasculature, but it is clear that in order to create a functional kidney tissue, the ideal culture system will require a vasculature with an inlet and outlet. One such microfluidics chip was generated to create a *de novo* capillary bed with unidirectional flow through lumenized blood vessels (Moya, Hsu, Lee, Hughes, & George, 2013; Wang et al., 2016). Its utility in creating the ideal plumbing for organoids has yet to be tested.

8.4 Implantation of organoids to support and sustain organoid vasculature

Beyond man-made engineering devices, there are also biological platforms for stimulating tissues such as organoids. Many researchers have achieved organoid vascularization by implanting them into immunocompromised mice. This approach allows for host blood vessel invasion into the organoid. The most widely used method for implantation is under the kidney capsule of an adult mouse, due to the relative accessibility of the area and the capsule holding the implanted tissue in place. In addition, the outer cortex of the adult kidney contains a rich vascular bed. Within weeks, vasculature from the host invades into the organoid tissue implantation, resulting in significant growth of the organoids and maturation of the tissue (Finkbeiner et al., 2015; van den Berg et al., 2018). Organoids implanted at earlier time points in culture display improved growth (Bantounas et al., 2018). Additionally, the presence of pre-implantation vasculature may increase and accelerate the extent of host invasion (Cakir et al., 2019). Interestingly, VEGF deficient mESCs implanted into nude mice do not form teratomas as readily as wildtype mESCs and have less vascularization, demonstrating the importance of endogenous VEGF in blood vessel recruitment after implantation (Ferrara & Bunting, 1996).

The vasculature that grows following tissue implantation has been shown to exhibit two possible behavioral outcomes: Either the host vasculature anastomoses with the existing organoid vasculature present, creating hybrid vessels, or the host vasculature entirely takes over the organoid, with the graft vasculature disappearing over time (Francipane et al., 2019; van den Berg et al., 2018). However, these phenomena are not necessarily mutually exclusive, and more studies on the dynamics between the host and graft ECs are needed to shed light on this process. One possibility is that give the normal tendency of ECs to migrate against the direction of flow (Sprague, Luo, & Palmaz, 1997), it is possible they are simply diluted by the host ECs as they migrate. Alternatively, deficiencies in organoid ECs, which have not yet been identified, may result in them being out competed by host vessels if anastomosis, and therefore blood flow, is not quickly established.

Drawbacks of implantation include the expense and difficulty of the technique. Immunocompromised mice are expensive to obtain and house. This method also diminishes the *in vitro* advantages that organoids provide over *in vivo* models. Additionally, the surgery can result in an overgrowth of scar tissue, stroma, and other unwanted cell types such as cartilage (Kumar Gupta et al., 2020; Przepiorski, et al., 2020). Solving these issues is a clear challenge for the future of replacement therapeutic approaches.

8.5 Alternative pathways to promoting blood vessel maturation and maintenance

It is clear from both embryonic blood vessels and numerous *in vitro* studies showing that growth and stabilization of vasculature requires hemodynamic flow. However, microfluidics and implantation experiments are technically challenging. Therefore, pharmacological interventions are become attractive alternative approaches for maintaining vasculature. Piezo-1 is mechanosensory ion channel expressed on ECs. It is necessary for the endothelial response to flow and particularly important for vascular development and remodeling. In fact, the genetic deletion of Piezo1 in ECs is embryonic lethal (Ranade et al., 2014). Yoda-1, a Piezo1 agonist, has been shown to activate Icam1 and Vcam1 expression in a concentration-dependent manner, reaching similar levels as 5 dyn/cm^2 shear stress at 2 μM Yoda-1 (Davies et al., 2019). At the highest dose of Yoda-1 used, organoids displayed significantly more vasculature by PECAM$^+$ percent area. Further experiments will be necessary in order to determine if Yoda-1 treatment expands the longevity of vasculature and how it affects the epithelium.

While considerable attention has been placed on the necessity of flow for vascular maintenance and maturation, the vasculature must be stabilized before flow can be established. Here we discuss developmental signaling pathways known to be important in embryonic blood vessels that may be activated for this purpose. The addition of pro-angiogenic and pro-maintenance (anti-regression) factors are attractive potential tools for increasing and maintaining vasculature before flow can be established. The addition of VEGF increases EC number in EBs, with a significant improvement using a growth factor (GF) cocktail that increased the number of both ECs and hematopoietic cells. These GFs include VEGF, FGF2, IL-6, and EPO. There has been success in both increasing and maintaining the vasculature in intestinal organoids through the addition of VEGF, BMP4, and FGF2 (Holloway et al., 2020). VEGF signaling is necessary for vascular development, as VEGF deficient mESCs fail to form either ECs or vascular-like structures in embryoid bodies (Ferrara et al., 1996).

Another interesting approach is modulating the levels of oxygen in organoid cultures. Hypoxia, or low oxygen, is a well-known inducer of angiogenesis. Low oxygen levels induce hypoxia inducible factors (HIFs) which turn on a number of genes including angiogenic factors such as VEGF (Krock, Skuli, & Simon, 2011; Liu, Cox, Morita, & Kourembanas, 1995). This results in blood vessels sprouting into the hypoxic region in an effort to provide oxygen to the ischemic tissue (Shweiki, Itin, Soffer, & Keshet, 1992). Hypoxia-induced angiogenesis is integral to both physiological processes such as wound healing as well as pathological phenomena like tumorigenesis. *In utero*, the embryo experiences an environment that is around 3% oxygen. However, cell culture is performed at ambient oxygen levels of 20%. In fact, standard cell culture should be referred to as "hyperoxia" and 3–9% O_2 as "normoxia" or "physoxia" since it is closer to physiological levels (Ivanovic, 2009). Manipulating oxygen levels or HIF-controlled signaling pathways has already demonstrated some success in stimulating vascularization of organoids and holds interesting promise (Coyle et al., 2021).

In vivo, many signals that are important for vessel stability and maintenance come from supportive cell types, pericytes and macrophages, as well as extracellular matrix (ECM). Pericytes are the cells that architecturally and environmentally support blood vessels. Not only do pericytes provide vessels with physical support and ECM, but they also create biochemical signals that stabilize vessels, and their recruitment is necessary for vessel maturation. EC-pericyte communicate in multiple ways including *via* the Angiopoietin-Tie, VEGF-VEGR2, and TGFβ signaling pathways

(Armulik, Abramsson, & Betsholtz, 2005; Chatterjee & Naik, 2012; Dickson et al., 1995; Wakui et al., 2006). Early vascular plexuses are initially comprised of "naked" cords of angioblasts that require VEGF for maintenance (Alon et al., 1995). Once vascular lumens open, flow begins, and remodeling is completed, the vessels then recruit pericytes, which stimulate VEGF-independent stability and maturation. Disruption of the EC-pericyte relationship results in aberrant remodeling of the capillary plexus, including regression and breaking of the vessels (Benjamin, Hemo, & Keshet, 1998). Pericytes vary depending on vessel type, from the smooth muscle actin (SMA) tightly covering arteries to more the more sparsely scattered pericytes of the capillaries and veins. Pericytes are heterogeneous like ECs; notably, pericytes have disputed origins, and may differentiate from various sources, including mesenchyme and macrophages in a tissue-specific manner (Muller et al., 2008; Yamamoto et al., 2017; Yamazaki & Mukouyama, 2018). In organoids, pericytes appear to develop along with ECs, as they are prominent in blood vessel organoids. Embryoid bodies contain a small amount of two types of pericytes, Desmin$^+$ and α-SMA$^+$, pericytes. However, they are not closely associated with the CD31$^+$ ECs. VEGF treatment increases the number of ECs and pericytes present while also increasing pericyte-EC association (Hagedorn et al., 2004).

Macrophages are also essential for regulating angiogenesis, vascular regression, and anastomosis. In order to recreate the diverse population of cells in the brain, organoids have been made by combining precursors of the various cell types present, including microglial/macrophage precursors and ECs in cerebral organoids, or Kupffer cells—specialized macrophages—and ECs in liver organoids (Rennert et al., 2015; Schwartz et al., 2015). Embryoid bodies produce CD34$^+$CD35$^+$ hematopoietic cells which can be induced to form mature macrophages through treatment with cytokines IL3, SCF and M-CSF (Subramanian et al., 2009; Wiles & Keller, 1991). Might these cell types play a role in support of developing organoid vasculatures? These questions are wide open and in need of exploration.

9. The challenges of organoid culture

Despite immense progress in less than a decade, many challenges still exist across the organoid and tissue engineering fields. In fact, we are probably discovering new challenges as soon as we solve some. As outlined in this chapter, there are numerous stark defects within organoid vasculatures, from their patterning, to their maintenance and maturity. With many methods constantly being innovated to mimic developmental processes and attempt to generate functional tissue organoids, it is still unclear what makes the "best" organoid. Additionally, there is significant variability between organoids, particularly for different batches of ESCs or iPSCs, but even the same batch of cells can drift over time (Phipson et al., 2019). Compounding this problem, there is no certified list of attributes that are commonly agreed upon by the field to decisively determine if an organoid is a "success." Currently, the best estimation available is measured by gene expression of hallmark genes through antibody staining, RNA sequencing, or qPCR. Continued, in-depth and often "incremental" developmental biology studies, like investigations into renal endothelial heterogeneity (Barry et al., 2019; Daniel et al., 2018), are integral to establishing landmarks, toolsets and goals for ultimately achieving implantable renal tissue.

10. Perspectives for the future of vascularized engineered tissues

Blood vessel development and maintenance is essential to viability of tissues and exceedingly complicated. Forming and maintaining healthy vessels require both physical signals from shear stress and biochemical signals provided by the surrounding pericytes, stroma, and parenchyma. At this point, the field is still working to understand even the most fundamental aspects of vascular biology. Therefore, recreating a healthy vasculature *in vitro* is a tremendous challenge. The possibility that a single specific factor or condition will be enough to create and sustain a functional vasculature in organoids is likely zero. Through trial and error, we can incorporate our growing knowledge of blood vessel development into the organoid system, as we continue to learn more about vascular dynamics both *in vivo* and *in vitro*.

The future of tissue engineering and generation of replacement organs will require either a laboratory-generated or organic fluidic system to properly plumb tissue blood vessels and provide engineered organs with an inlet and outlet to their associated vasculature. Not only is this necessary for growth and function of the cultured tissues, such as kidneys, but flow is required for maintenance of a healthy vascular system. Once blood flow is properly established, this will allow the necessary scaling-up required for successful organoid culture systems. Only then will we be able to achieve an artificial kidney that is large enough to carry out the filtration requirements for an adult. We argue that these challenges can only be overcome by better understanding the ontogeny of the renal vasculature, in normal kidneys.

Acknowledgments

We are grateful to the entire ReBuilding the Kidney (RBK) consortium for discussion on every aspect of this work. Specifically, we want to thank Dr. Melissa Little, Dr. Jessica Vanslambrouck, and Dr. Sara Howden for their invaluable advice regarding kidney organoids, as well as Drs. Jennifer Lewis and Kimberly Homan for discussions regarding hemodynamic flow and organoids. We are grateful to the Cleaver and Carroll labs for discussions over the course of Annie Ryan's PhD work, which resulted in this chapter.

Funding

This work was supported by National Institute of Diabetes and Digestive and Kidney Diseases DK106743, DK079862 to O.C.; National Institute of Heart, Lung and Blood HL113498 to O.C.; a Center for Regenerative Science and Medicine (CRSM) predoctoral fellowship to A.R.; an ASN Pre-Doctoral Fellowship Award to A.R and an F31 DK124046.

References

Abrahamson, D. R., Robert, B., Hyink, D. P., St John, P. L., & Daniel, T. O. (1998). Origins and formation of microvasculature in the developing kidney. *Kidney International. Supplement*, *67*, S7–11. https://doi.org/10.1046/j.1523-1755.1998.06702.x.

Aird, W. C. (2003). Endothelial cell heterogeneity. *Critical Care Medicine*, *31*(Suppl. 4), S221–S230. https://doi.org/10.1097/01.CCM.0000057847.32590.C1.

Aird, W. C. (2007). Phenotypic heterogeneity of the endothelium: I. Structure, function, and mechanisms. *Circulation Research*, *100*(2), 158–173. https://doi.org/10.1161/01.RES.0000255691.76142.4a.

Aird, W. C. (2012). Endothelial cell heterogeneity. *Cold Spring Harbor Perspectives in Medicine*, *2*(1), a006429. https://doi.org/10.1101/cshperspect.a006429.

Alon, T., Hemo, I., Itin, A., Pe'er, J., Stone, J., & Keshet, E. (1995). Vascular endothelial growth factor acts as a survival factor for newly formed retinal vessels and has implications for retinopathy of prematurity. *Nature Medicine*, *1*(10), 1024–1028. https://doi.org/10.1038/nm1095-1024.

Armulik, A., Abramsson, A., & Betsholtz, C. (2005). Endothelial/pericyte interactions. *Circulation Research*, *97*(6), 512–523. https://doi.org/10.1161/01.RES.0000182903.16652.d7.

Bantounas, I., Ranjzad, P., Tengku, F., Silajdzic, E., Forster, D., Asselin, M. C., et al. (2018). Generation of functioning nephrons by implanting human pluripotent stem cell-derived kidney progenitors. *Stem Cell Reports*, *10*(3), 766–779. https://doi.org/10.1016/j.stemcr.2018.01.008.

Barry, D. M., McMillan, E. A., Kunar, B., Lis, R., Zhang, T., Lu, T., et al. (2019). Molecular determinants of nephron vascular specialization in the kidney. *Nature Communications*, *10*(1), 5705. https://doi.org/10.1038/s41467-019-12872-5.

Benjamin, L. E., Hemo, I., & Keshet, E. (1998). A plasticity window for blood vessel remodelling is defined by pericyte coverage of the preformed endothelial network and is regulated by PDGF-B and VEGF. *Development, 125*(9), 1591–1598. https://www.ncbi.nlm.nih.gov/pubmed/9521897.

Brown, A. C., Muthukrishnan, S. D., & Oxburgh, L. (2015). A synthetic niche for nephron progenitor cells. *Developmental Cell, 34*(2), 229–241. https://doi.org/10.1016/j.devcel.2015.06.021.

Bruno, S., Remuzzi, G., & Ruggenenti, P. (2004). Transplant renal artery stenosis. *Journal of the American Society of Nephrology, 15*(1), 134–141. https://doi.org/10.1097/01.asn.0000099379.61001.f8.

Cakir, B., Xiang, Y., Tanaka, Y., Kural, M. H., Parent, M., Kang, Y. J., et al. (2019). Engineering of human brain organoids with a functional vascular-like system. *Nature Methods, 16*(11), 1169–1175. https://doi.org/10.1038/s41592-019-0586-5.

Chappell, J. C., Taylor, S. M., Ferrara, N., & Bautch, V. L. (2009). Local guidance of emerging vessel sprouts requires soluble Flt-1. *Developmental Cell, 17*(3), 377–386. https://doi.org/10.1016/j.devcel.2009.07.011.

Chatterjee, S., & Naik, U. P. (2012). Pericyte-endothelial cell interaction: A survival mechanism for the tumor vasculature. *Cell Adhesion & Migration, 6*(3), 157–159. https://doi.org/10.4161/cam.20252.

Cleaver, O. (2004). Blood vessel signals during development and beyond. *Current Topics in Developmental Biology, 62*, 1–36. https://doi.org/10.1016/S0070-2153(04)62001-1.

Combes, A. N., Zappia, L., Er, P. X., Oshlack, A., & Little, M. H. (2019). Single-cell analysis reveals congruence between kidney organoids and human fetal kidney. *Genome Medicine, 11*(1), 3. https://doi.org/10.1186/s13073-019-0615-0.

Coyle, R. C., Barrs, R. W., Richards, D. J., Ladd, E. P., Menick, D. R., & Mei, Y. (2021). Targeting HIF-alpha for robust prevascularization of human cardiac organoids. *Journal of Tissue Engineering and Regenerative Medicine, 15*(2), 189–202. https://doi.org/10.1002/term.3165.

Daniel, E., Azizoglu, D. B., Ryan, A. R., Walji, T. A., Chaney, C. P., Sutton, G. I., et al. (2018). Spatiotemporal heterogeneity and patterning of developing renal blood vessels. *Angiogenesis, 21*(3), 617–634. https://doi.org/10.1007/s10456-018-9612-y.

Davies, J. E., Lopresto, D., Apta, B. H. R., Lin, Z., Ma, W., & Harper, M. T. (2019). Using Yoda-1 to mimic laminar flow in vitro: A tool to simplify drug testing. *Biochemical Pharmacology, 168*, 473–480. https://doi.org/10.1016/j.bcp.2019.08.013.

Dickson, M. C., Martin, J. S., Cousins, F. M., Kulkarni, A. B., Karlsson, S., & Akhurst, R. J. (1995). Defective haematopoiesis and vasculogenesis in transforming growth factor-beta 1 knock out mice. *Development, 121*(6), 1845–1854. https://www.ncbi.nlm.nih.gov/pubmed/7600998.

Ding, B. S., Nolan, D. J., Guo, P., Babazadeh, A. O., Cao, Z., Rosenwaks, Z., et al. (2011). Endothelial-derived angiocrine signals induce and sustain regenerative lung alveolarization. *Cell, 147*(3), 539–553. https://doi.org/10.1016/j.cell.2011.10.003.

Drake, C. J., & Fleming, P. A. (2000). Vasculogenesis in the day 6.5 to 9.5 mouse embryo. *Blood, 95*(5), 1671–1679.

Dumas, S. J., Meta, E., Borri, M., Goveia, J., Rohlenova, K., Conchinha, N. V., et al. (2020). Single-cell RNA sequencing reveals renal endothelium heterogeneity and metabolic adaptation to water deprivation. *Journal of the American Society of Nephrology, 31*(1), 118–138. https://doi.org/10.1681/ASN.2019080832.

Ferrara, N., & Bunting, S. (1996). Vascular endothelial growth factor, a specific regulator of angiogenesis. *Current Opinion in Nephrology and Hypertension, 5*(1), 35–44. https://doi.org/10.1097/00041552-199601000-00008.

Ferrara, N., Carver-Moore, K., Chen, H., Dowd, M., Lu, L., O'Shea, K. S., et al. (1996). Heterozygous embryonic lethality induced by targeted inactivation of the VEGF gene. *Nature*, *380*(6573), 439–442. http://www.ncbi.nlm.nih.gov/entrez/query.fcgi?cmd=Retrieve&db=PubMed&dopt=Citation&list_uids=8602242.

Finkbeiner, S. R., Freeman, J. J., Wieck, M. M., El-Nachef, W., Altheim, C. H., Tsai, Y. H., et al. (2015). Generation of tissue-engineered small intestine using embryonic stem cell-derived human intestinal organoids. *Biology Open*, *4*(11), 1462–1472. https://doi.org/10.1242/bio.013235.

Francipane, M. G., Han, B., Oxburgh, L., Sims-Lucas, S., Li, Z., & Lagasse, E. (2019). Kidney-in-a-lymph node: A novel organogenesis assay to model human renal development and test nephron progenitor cell fates. *Journal of Tissue Engineering and Regenerative Medicine*, *13*(9), 1724–1731. https://doi.org/10.1002/term.2924.

Freedman, B. S., Brooks, C. R., Lam, A. Q., Fu, H., Morizane, R., Agrawal, V., et al. (2015). Modelling kidney disease with CRISPR-mutant kidney organoids derived from human pluripotent epiblast spheroids. *Nature Communications*, *6*, 8715. https://doi.org/10.1038/ncomms9715.

Gerecht-Nir, S., Ziskind, A., Cohen, S., & Itskovitz-Eldor, J. (2003). Human embryonic stem cells as an in vitro model for human vascular development and the induction of vascular differentiation. *Laboratory Investigation*, *83*(12), 1811–1820. https://doi.org/10.1097/01.lab.0000106502.41391.f0.

Guyton, A. C., & Hall, J. E. (2006). In E. W. Schmitt, & R. Gruliow (Eds.), *Textbook of medical physiology* Elsevier.

Hagedorn, M., Balke, M., Schmidt, A., Bloch, W., Kurz, H., Javerzat, S., et al. (2004). VEGF coordinates interaction of pericytes and endothelial cells during vasculogenesis and experimental angiogenesis. *Developmental Dynamics*, *230*(1), 23–33. https://doi.org/10.1002/dvdy.20020.

Holloway, E. M., Wu, J. H., Czerwinski, M., Sweet, C. W., Wu, A., Tsai, Y. H., et al. (2020). Differentiation of human intestinal organoids with endogenous vascular endothelial cells. *Developmental Cell*, *54*(4), 516–528.e517. https://doi.org/10.1016/j.devcel.2020.07.023.

Homan, K. A., Gupta, N., Kroll, K. T., Kolesky, D. B., Skylar-Scott, M., Miyoshi, T., et al. (2019). Flow-enhanced vascularization and maturation of kidney organoids in vitro. *Nature Methods*, *16*(3), 255–262. https://doi.org/10.1038/s41592-019-0325-y.

Howden, S. E., Wilson, S. B., Groenewegen, E., Starks, L., Forbes, T. A., Tan, K. S., et al. (2021). Plasticity of distal nephron epithelia from human kidney organoids enables the induction of ureteric tip and stalk. *Cell Stem Cell*, *28*(4), 671–684 e6. https://doi.org/10.1016/j.stem.2020.12.001.

Israely, E., Ginsberg, M., Nolan, D., Ding, B. S., James, D., Elemento, O., et al. (2014). Akt suppression of TGFbeta signaling contributes to the maintenance of vascular identity in embryonic stem cell-derived endothelial cells. *Stem Cells*, *32*(1), 177–190. https://doi.org/10.1002/stem.1521.

Ivanovic, Z. (2009). Hypoxia or in situ normoxia: The stem cell paradigm. *Journal of Cellular Physiology*, *219*(2), 271–275. https://doi.org/10.1002/jcp.21690.

Kim, G. D., Kim, G. J., Seok, J. H., Chung, H. M., Chee, K. M., & Rhee, G. S. (2008). Differentiation of endothelial cells derived from mouse embryoid bodies: A possible in vitro vasculogenesis model. *Toxicology Letters*, *180*(3), 166–173. https://doi.org/10.1016/j.toxlet.2008.05.023.

Kim, J., Koo, B. K., & Knoblich, J. A. (2020). Human organoids: Model systems for human biology and medicine. *Nature Reviews. Molecular Cell Biology*, *21*(10), 571–584. https://doi.org/10.1038/s41580-020-0259-3.

Korn, C., & Augustin, H. G. (2015). Mechanisms of vessel pruning and regression. *Developmental Cell*, *34*(1), 5–17. https://doi.org/10.1016/j.devcel.2015.06.004.

Krock, B. L., Skuli, N., & Simon, M. C. (2011). Hypoxia-induced angiogenesis: Good and evil. *Genes & Cancer*, *2*(12), 1117–1133. https://doi.org/10.1177/1947601911423654.

Kumar Gupta, A., Sarkar, P., Wertheim, J. A., Pan, X., Carroll, T. J., & Oxburgh, L. (2020). Asynchronous mixing of kidney progenitor cells potentiates nephrogenesis in organoids. *Communications Biology*, *3*(1), 231. https://doi.org/10.1038/s42003-020-0948-7.

Kurosawa, H. (2007). Methods for inducing embryoid body formation: In vitro differentiation system of embryonic stem cells. *Journal of Bioscience and Bioengineering*, *103*(5), 389–398. https://doi.org/10.1263/jbb.103.389.

Lacorre, D. A., Baekkevold, E. S., Garrido, I., Brandtzaeg, P., Haraldsen, G., Amalric, F., et al. (2004). Plasticity of endothelial cells: Rapid dedifferentiation of freshly isolated high endothelial venule endothelial cells outside the lymphoid tissue microenvironment. *Blood*, *103*(11), 4164–4172. https://doi.org/10.1182/blood-2003-10-3537.

Lancaster, M. A., Renner, M., Martin, C. A., Wenzel, D., Bicknell, L. S., Hurles, M. E., et al. (2013). Cerebral organoids model human brain development and microcephaly. *Nature*, *501*(7467), 373–379. https://doi.org/10.1038/nature12517.

le Noble, F., Moyon, D., Pardanaud, L., Yuan, L., Djonov, V., Matthijsen, R., et al. (2004). Flow regulates arterial-venous differentiation in the chick embryo yolk sac. *Development*, *131*(2), 361–375. https://doi.org/10.1242/dev.00929.

Leibing, T., Geraud, C., Augustin, I., Boutros, M., Augustin, H. G., Okun, J. G., et al. (2018). Angiocrine Wnt signaling controls liver growth and metabolic maturation in mice. *Hepatology*, *68*(2), 707–722. https://doi.org/10.1002/hep.29613.

Lim, C. G., Jang, J., & Kim, C. (2018). Cellular machinery for sensing mechanical force. *BMB Reports*, *51*(12), 623–629. https://www.ncbi.nlm.nih.gov/pubmed/30293551.

Liu, Y., Cox, S. R., Morita, T., & Kourembanas, S. (1995). Hypoxia regulates vascular endothelial growth factor gene expression in endothelial cells. Identification of a 5′ enhancer. *Circulation Research*, 77(3), 638–643. https://doi.org/10.1161/01.res.77.3.638.

Lucitti, J. L., Jones, E. A., Huang, C., Chen, J., Fraser, S. E., & Dickinson, M. E. (2007). Vascular remodeling of the mouse yolk sac requires hemodynamic force. *Development*, *134*(18), 3317–3326. https://doi.org/10.1242/dev.02883.

Marcu, R., Choi, Y. J., Xue, J., Fortin, C. L., Wang, Y., Nagao, R. J., et al. (2018). Human organ-specific endothelial cell heterogeneity. *iScience*, *4*, 20–35. https://doi.org/10.1016/j.isci.2018.05.003.

Meadows, S. M., Fletcher, P. J., Moran, C., Xu, K., Neufeld, G., Chauvet, S., et al. (2012). Integration of repulsive guidance cues generates avascular zones that shape mammalian blood vessels. *Circulation Research*, *110*(1), 34–46. https://doi.org/10.1161/CIRCRESAHA.111.249847.

Meeson, A., Palmer, M., Calfon, M., & Lang, R. (1996). A relationship between apoptosis and flow during programmed capillary regression is revealed by vital analysis. *Development*, *122*(12), 3929–3938. https://www.ncbi.nlm.nih.gov/pubmed/9012513.

Molema, G., & Aird, W. C. (2012). Vascular heterogeneity in the kidney. *Seminars in Nephrology*, *32*(2), 145–155. https://doi.org/10.1016/j.semnephrol.2012.02.001.

Morita, R., Suzuki, M., Kasahara, H., Shimizu, N., Shichita, T., Sekiya, T., et al. (2015). ETS transcription factor ETV2 directly converts human fibroblasts into functional endothelial cells. *Proceedings of the National Academy of Sciences of the United States of America*, *112*(1), 160–165. https://doi.org/10.1073/pnas.1413234112.

Morizane, R., Lam, A. Q., Freedman, B. S., Kishi, S., Valerius, M. T., & Bonventre, J. V. (2015). Nephron organoids derived from human pluripotent stem cells model kidney development and injury. *Nature Biotechnology*, *33*(11), 1193–1200. https://doi.org/10.1038/nbt.3392.

Moya, M. L., Hsu, Y. H., Lee, A. P., Hughes, C. C., & George, S. C. (2013). In vitro perfused human capillary networks. *Tissue Engineering. Part C, Methods*, *19*(9), 730–737. https://doi.org/10.1089/ten.TEC.2012.0430.

Muller, S. M., Stolt, C. C., Terszowski, G., Blum, C., Amagai, T., Kessaris, N., et al. (2008). Neural crest origin of perivascular mesenchyme in the adult thymus. *Journal of Immunology*, *180*(8), 5344–5351. https://doi.org/10.4049/jimmunol.180.8.5344.

Munro, D. A. D., Hohenstein, P., & Davies, J. A. (2017). Cycles of vascular plexus formation within the nephrogenic zone of the developing mouse kidney. *Scientific Reports*, 7(1), 3273. https://doi.org/10.1038/s41598-017-03808-4.

Palikuqi, B., Nguyen, D. T., Li, G., Schreiner, R., Pellegata, A. F., Liu, Y., et al. (2020). Adaptable haemodynamic endothelial cells for organogenesis and tumorigenesis. *Nature*, *585*(7825), 426–432. https://doi.org/10.1038/s41586-020-2712-z.

Paredes, I., Vieira, J. R., Shah, B., Ramunno, C. F., Dyckow, J., Adler, H., et al. (2021). Oligodendrocyte precursor cell specification is regulated by bidirectional neural progenitor-endothelial cell crosstalk. *Nature Neuroscience*, *24*(4), 478–488. https://doi.org/10.1038/s41593-020-00788-z.

Pham, M. T., Pollock, K. M., Rose, M. D., Cary, W. A., Stewart, H. R., Zhou, P., et al. (2018). Generation of human vascularized brain organoids. *Neuroreport*, *29*(7), 588–593. https://doi.org/10.1097/WNR.0000000000001014.

Phipson, B., Er, P. X., Combes, A. N., Forbes, T. A., Howden, S. E., Zappia, L., et al. (2019). Evaluation of variability in human kidney organoids. *Nature Methods*, *16*(1), 79–87. https://doi.org/10.1038/s41592-018-0253-2.

Przepiorski, A., Crunk, A. E., Espiritu, E. B., Hukriede, N. A., & Davidson, A. J. (2020). The utility of human kidney organoids in modeling kidney disease. *Seminars in Nephrology*, *40*(2), 188–198. https://doi.org/10.1016/j.semnephrol.2020.01.009.

Przepiorski, A., Sander, V., Tran, T., Hollywood, J. A., Sorrenson, B., Shih, J. H., et al. (2018). A simple bioreactor-based method to generate kidney organoids from pluripotent stem cells. *Stem Cell Reports*, *11*(2), 470–484. https://doi.org/10.1016/j.stemcr.2018.06.018.

Rafii, S., Butler, J. M., & Ding, B. S. (2016). Angiocrine functions of organ-specific endothelial cells. *Nature*, *529*(7586), 316–325. https://doi.org/10.1038/nature17040.

Rafii, S., Shapiro, F., Pettengell, R., Ferris, B., Nachman, R. L., Moore, M. A., et al. (1995). Human bone marrow microvascular endothelial cells support long-term proliferation and differentiation of myeloid and megakaryocytic progenitors. *Blood*, *86*(9), 3353–3363. http://www.ncbi.nlm.nih.gov/entrez/query.fcgi?cmd=Retrieve&db=PubMed&dopt=Citation&list_uids=7579438.

Ranade, S. S., Qiu, Z., Woo, S. H., Hur, S. S., Murthy, S. E., Cahalan, S. M., et al. (2014). Piezo1, a mechanically activated ion channel, is required for vascular development in mice. *Proceedings of the National Academy of Sciences of the United States of America*, *111*(28), 10347–10352. https://doi.org/10.1073/pnas.1409233111.

Rennert, K., Steinborn, S., Groger, M., Ungerbock, B., Jank, A. M., Ehgartner, J., et al. (2015). A microfluidically perfused three dimensional human liver model. *Biomaterials*, *71*, 119–131. https://doi.org/10.1016/j.biomaterials.2015.08.043.

Risau, W. (1997). Mechanisms of angiogenesis. *Nature*, *386*(6626), 671–674. http://www.ncbi.nlm.nih.gov/entrez/query.fcgi?cmd=Retrieve&db=PubMed&dopt=Citation&list_uids=9109485.

Robert, B., St John, P. L., & Abrahamson, D. R. (1998). Direct visualization of renal vascular morphogenesis in Flk1 heterozygous mutant mice. *The American Journal of Physiology*, *275*(1), F164–F172. https://doi.org/10.1152/ajprenal.1998.275.1.F164.

Rocha, A. S., Vidal, V., Mertz, M., Kendall, T. J., Charlet, A., Okamoto, H., et al. (2015). The angiocrine factor Rspondin3 is a key determinant of liver zonation. *Cell Reports*, *13*(9), 1757–1764. https://doi.org/10.1016/j.celrep.2015.10.049.

Ryan, A. R., England, A. R., Chaney, C. P., Cowdin, M. A., Hiltabidle, M., Daniel, E., et al. (2021). Vascular deficiencies in renal organoids and ex vivo kidney organogenesis. *Developmental Biology*, *477*, 98–116.

Schwartz, M. P., Hou, Z., Propson, N. E., Zhang, J., Engstrom, C. J., Santos Costa, V., et al. (2015). Human pluripotent stem cell-derived neural constructs for predicting neural toxicity. *Proceedings of the National Academy of Sciences of the United States of America, 112*(40), 12516–12521. https://doi.org/10.1073/pnas.1516645112.
Sequeira-Lopez, M. L., Lin, E. E., Li, M., Hu, Y., Sigmund, C. D., & Gomez, R. A. (2015). The earliest metanephric arteriolar progenitors and their role in kidney vascular development. *American Journal of Physiology. Regulatory, Integrative and Comparative Physiology, 308*(2), R138–R149. https://doi.org/10.1152/ajpregu.00428.2014.
Shi, Y., Sun, L., Wang, M., Liu, J., Zhong, S., Li, R., et al. (2020). Vascularized human cortical organoids (vOrganoids) model cortical development in vivo. *PLoS Biology, 18*(5), e3000705. https://doi.org/10.1371/journal.pbio.3000705.
Shweiki, D., Itin, A., Soffer, D., & Keshet, E. (1992). Vascular endothelial growth factor induced by hypoxia may mediate hypoxia-initiated angiogenesis. *Nature, 359*(6398), 843–845. https://doi.org/10.1038/359843a0.
Singhal, M., & Augustin, H. G. (2020). Beyond angiogenesis: Exploiting angiocrine factors to restrict tumor progression and metastasis. *Cancer Research, 80*(4), 659–662. https://doi.org/10.1158/0008-5472.CAN-19-3351.
Song, J. W., & Munn, L. L. (2011). Fluid forces control endothelial sprouting. *Proceedings of the National Academy of Sciences of the United States of America, 108*(37), 15342–15347. https://doi.org/10.1073/pnas.1105316108.
Sprague, E. A., Luo, J., & Palmaz, J. C. (1997). Human aortic endothelial cell migration onto stent surfaces under static and flow conditions. *Journal of Vascular and Interventional Radiology, 8*(1 Pt 1), 83–92. https://doi.org/10.1016/s1051-0443(97)70521-9.
Subramanian, A., Guo, B., Marsden, M. D., Galic, Z., Kitchen, S., Kacena, A., et al. (2009). Macrophage differentiation from embryoid bodies derived from human embryonic stem cells. *Journal of Stem Cells, 4*(1), 29–45. doi:jsc.2009.4.1.29.
Taguchi, A., & Nishinakamura, R. (2015). Nephron reconstitution from pluripotent stem cells. *Kidney International, 87*(5), 894–900. https://doi.org/10.1038/ki.2014.358.
Takasato, M., Er, P. X., Chiu, H. S., & Little, M. H. (2016). Generation of kidney organoids from human pluripotent stem cells. *Nature Protocols, 11*(9), 1681–1692. https://doi.org/10.1038/nprot.2016.098.
Takasato, M., Er, P. X., Chiu, H. S., Maier, B., Baillie, G. J., Ferguson, C., et al. (2015). Kidney organoids from human iPS cells contain multiple lineages and model human nephrogenesis. *Nature, 526*(7574), 564–568. https://doi.org/10.1038/nature15695.
Takebe, T., & Wells, J. M. (2019). Organoids by design. *Science, 364*(6444), 956–959. https://doi.org/10.1126/science.aaw7567.
Tzima, E., Irani-Tehrani, M., Kiosses, W. B., Dejana, E., Schultz, D. A., Engelhardt, B., et al. (2005). A mechanosensory complex that mediates the endothelial cell response to fluid shear stress. *Nature, 437*(7057), 426–431. https://doi.org/10.1038/nature03952.
Udan, R. S., Vadakkan, T. J., & Dickinson, M. E. (2013). Dynamic responses of endothelial cells to changes in blood flow during vascular remodeling of the mouse yolk sac. *Development, 140*(19), 4041–4050. https://doi.org/10.1242/dev.096255.
van den Berg, C. W., Ritsma, L., Avramut, M. C., Wiersma, L. E., van den Berg, B. M., Leuning, D. G., et al. (2018). Renal subcapsular transplantation of PSC-derived kidney organoids induces neo-vasculogenesis and significant glomerular and tubular maturation in vivo. *Stem Cell Reports, 10*(3), 751–765. https://doi.org/10.1016/j.stemcr.2018.01.041.
Vittet, D., Prandini, M. H., Berthier, R., Schweitzer, A., Martin-Sisteron, H., Uzan, G., et al. (1996). Embryonic stem cells differentiate in vitro to endothelial cells through successive maturation steps. *Blood, 88*(9), 3424–3431. https://www.ncbi.nlm.nih.gov/pubmed/8896407.

Wakui, S., Yokoo, K., Muto, T., Suzuki, Y., Takahashi, H., Furusato, M., et al. (2006). Localization of Ang-1, -2, Tie-2, and VEGF expression at endothelial-pericyte interdigitation in rat angiogenesis. *Laboratory Investigation*, *86*(11), 1172–1184. https://doi.org/10.1038/labinvest.3700476.

Wang, R., Clark, R., & Bautch, V. L. (1992). Embryonic stem cell-derived cystic embryoid bodies form vascular channels: An in vitro model of blood vessel development. *Development*, *114*(2), 303–316. https://www.ncbi.nlm.nih.gov/pubmed/1591994.

Wang, L., Luo, J. Y., Li, B., Tian, X. Y., Chen, L. J., Huang, Y., et al. (2016). Integrin-YAP/TAZ-JNK cascade mediates atheroprotective effect of unidirectional shear flow. *Nature*, *540*(7634), 579–582. https://doi.org/10.1038/nature20602.

Wiles, M. V., & Keller, G. (1991). Multiple hematopoietic lineages develop from embryonic stem (ES) cells in culture. *Development*, *111*(2), 259–267. https://www.ncbi.nlm.nih.gov/pubmed/1893864.

Wilting, J., & Christ, B. (1996). Embryonic angiogenesis: A review. *Naturwissenschaften*, *83*(4), 153–164.

Yamamoto, S., Muramatsu, M., Azuma, E., Ikutani, M., Nagai, Y., Sagara, H., et al. (2017). A subset of cerebrovascular pericytes originates from mature macrophages in the very early phase of vascular development in CNS. *Scientific Reports*, 7(1), 3855. https://doi.org/10.1038/s41598-017-03994-1.

Yamazaki, T., & Mukouyama, Y. S. (2018). Tissue specific origin, development, and pathological perspectives of pericytes. *Frontiers in Cardiovascular Medicine*, *5*, 78. https://doi.org/10.3389/fcvm.2018.00078.

Zeng, Z., Huang, B., Parvez, R. K., Li, Y., Chen, J., Vonk, A. C., et al. (2021). Generation of patterned kidney organoids that recapitulate the adult kidney collecting duct system from expandable ureteric bud progenitors. *Nature Communications*, *12*(1), 3641. https://doi.org/10.1038/s41467-021-23911-5.

CHAPTER EIGHT

Innervation in organogenesis

Samuel E. Honeycutt, Pierre-Emmanuel Y. N'Guetta, and Lori L. O'Brien*
Department of Cell Biology and Physiology, University of North Carolina at Chapel Hill, Chapel Hill, NC, United States
*Corresponding author: e-mail address: lori_obrien@med.unc.edu

Contents

Abstract

Proper innervation of peripheral organs helps to maintain physiological homeostasis and elicit responses to external stimuli. Disruptions to normal function can result in pathophysiological consequences. The establishment of connections and communication between the central nervous system and the peripheral organs is accomplished through the peripheral nervous system. Neuronal connections with target tissues arise from ganglia partitioned throughout the body. Organ innervation is initiated during development with stimuli being conducted through several types of neurons including sympathetic, parasympathetic, and sensory. While the physiological modulation of mature organs by these nerves is largely understood, their role in mammalian development is only beginning to be uncovered. Interactions with cells in target tissues can

Current Topics in Developmental Biology, Volume 148
ISSN 0070-2153
https://doi.org/10.1016/bs.ctdb.2022.02.004

affect the development and eventual function of several organs, highlighting their significance. This chapter will cover the origin of peripheral neurons, factors mediating organ innervation, and the composition and function of organ-specific nerves during development. This emerging field aims to identify the functional contribution of innervation to development which will inform future investigations of normal and abnormal mammalian organogenesis, as well as contribute to regenerative and organ replacement efforts where nerve-derived signals may have significant implications for the advancement of such studies.

1. Introduction

The formation of physiologically competent organs is subject to strict molecular and cellular control. It is often dependent on the precise specification and differentiation of numerous cell types in addition to contributions from the blood, vasculature, lymphatic system, immune cells, and neuroendocrine cells. However, in recent years the role of neurons of the peripheral nervous system has gained attention in the understanding of organogenesis and the implications for physiological function. Comprised of billions of neurons, the nervous system consists of two major parts: (1) the central nervous system (CNS) which includes the brain and the spinal cord, and (2) the peripheral nervous system (PNS) which lies outside the CNS and incorporates cranial nerves and spinal nerves that innervate peripheral tissues.

The PNS is responsible for the relay of information to and from the CNS in order to coordinate molecular and physiological changes in various parts of the body. The PNS is partitioned into sensory (afferent) and motor (efferent) divisions. The sensory neurons transmit information from peripheral organs to the CNS. Motor neurons transmit impulses from the CNS to the target tissue thereby eliciting a response. The motor division of the PNS is comprised of the somatic nervous system and the autonomic nervous system. Somatic neurons transmit signals to skeletal muscles. Autonomic neurons, which are classified as either sympathetic or parasympathetic, supply impulses to visceral tissues. PNS functions rely on proper development of a robust neuronal network which innervates the target tissue. The crosstalk between neurons and the innervated tissue is mediated by connections with nerve axons. Axons are nerve fibers which project from the neuronal body to form synapses with cells in the target organ. Collections of neuronal cell bodies, or soma, lie in ganglia interspersed throughout the body that exist externally from the target organ or in some cases within the tissue itself. During development, axons will elongate as they respond to various neurotrophic factors and guidance cues until they establish precise junctions with their end target (Huang & Reichardt, 2001; O'Donnell, Chance, & Bashaw, 2009).

Axon extensions innervate organs during development, growing in tandem with the tissue. Perturbations to peripheral tissue innervation can lead to significant physiological defects and disease (Chrestian, McMillan, Poulin, Campbell, & Vajsar, 2015; Dietrich & Dragatsis, 2016; Indo et al., 1996; Saporta & Shy, 2013). A growing body of work interrogating the role of the PNS has uncovered novel roles for these neurons in organ development. Various methodologies, including genetic or chemical ablation of PNS-associated neurons, have enabled interrogation of these neurons and their crosstalk with the target tissue. These studies are broadening our understanding of developmental biology and challenging long-held views that these neuronal connections only modulate mature physiology. This chapter will cover three main concepts of peripheral innervation: 1) the developmental origin of neurons, 2) how neurons are attracted to, and innervate peripheral tissues during development, and 3) the role of innervation in organogenesis. A recent in-depth review covers these aspects of sympathetic innervation (Scott-Solomon, Boehm, & Kuruvilla, 2021). In this chapter we will cover sympathetic, parasympathetic, sensory, and enteric innervation and their associated roles in development. While the interrogation of these nerves has led to significant insights into both organogenesis and disease, much remains to be learned about these networks and their significance beyond basic physiology.

2. Developmental origins of neurons

2.1 Neural crest cells

Condensing from the ectodermal germ layer, the neural plate will give rise to a transient population of multipotent progenitors known as neural crest cells (NCCs). These cells migrate long distances from their origins to produce the peripheral and enteric ganglia and associated neurons (Fig. 1). NCCs were initially discovered in the chick embryo in 1868 by Swiss anatomist Wilhelm His Sr. who described them as *Zwischenstrang* or "the cord in between", due to their location between the neural plate and non-neural ectoderm (reviewed by Horstadius, 1950). Unique to vertebrates, NCCs ultimately differentiate to produce cell types and structures that are defining features of the subphylum *Vertebrata*. NCCs can adopt a wide range of fates, and their contribution to the many systems of the developing embryo have been

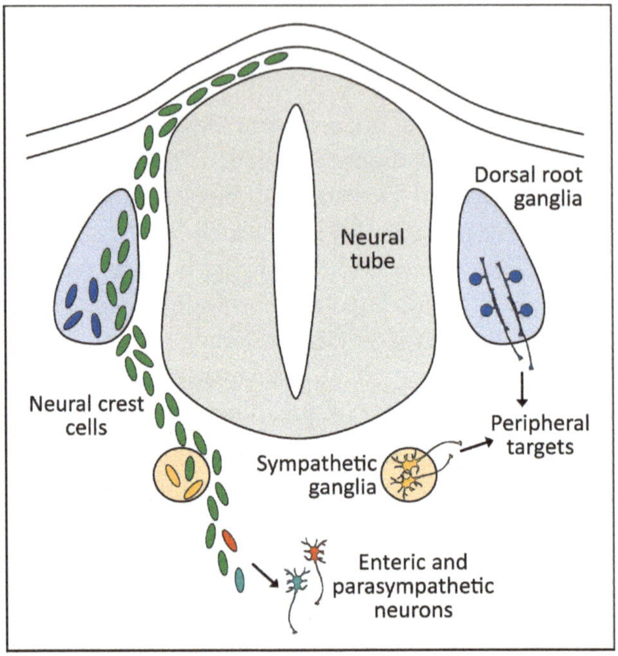

Fig. 1 During development neural crest cells migrate to sites such as ganglia, where they differentiate into neurons of the peripheral nervous system. Neural crest cells can directly give rise to sensory, sympathetic, and enteric neurons or indirectly to parasympathetic neurons.

thoroughly reviewed elsewhere (LaBonne & Bronner-Fraser, 1999; Le Douarin & Kalcheim, 1999; Takahashi, Sipp, & Enomoto, 2013). For the purposes of this chapter, the focus will be on NCCs role in the formation of the vertebrate peripheral and enteric nervous systems that form the basis of innervation during organogenesis.

After formation of the three germ layers and the notochord, secretion of signaling molecules including members of the Wnt, Bmp, and Fgf families drive the creation of a border between the non-neural ectoderm and the neural plate (Meulemans & Bronner-Fraser, 2004). As the paraxial mesoderm pushes in, the neural plate folds bilaterally to produce the neural crests. These crests then fuse along the dorsal midline of the developing neural tube (reviewed by Huang & Saint-Jeannet, 2004; Jacobson, 1991). Expression of *Foxd3*, *Sox9/10*, *c-Myc*, *AP-2*, and *Slug/Snail* along the length of the dorsal surface of the neural tube induces the formation of NCCs from the neuroepithelium (del Barrio & Nieto, 2002; Meulemans & Bronner-Fraser, 2004). Migration of NCCs begins after they undergo epithelial-to-mesenchymal transition and delaminate from the neural tube. NCCs from specific regions of the neural tube will differentiate into various neuronal populations once they reach their terminal destination. NCCs from the cranial region will become cranial ganglia. The enteric ganglia and enteric nervous system are derived from vagal and sacral NCCs, while NCCs from the trunk region of the neural tube become sympathetic and dorsal root ganglia, as well as sensory neurons and glia.

2.2 Sympathetic, sensory, and parasympathetic neuron formation from NCCs

Sympathetic neurons are derived from sympathoadrenal (SA) progenitors that are themselves derived from neural crest trunk cells (reviewed by Huber, 2006). Trunk NCCs first migrate ventrally along defined regions, often including the vasculature, to the dorsal aorta where they aggregate forming the primary sympathetic ganglia. BMPs expressed by the wall of the dorsal aorta drive NCCs to a sympathoadrenal commitment (Reissmann et al., 1996; Schneider, Wicht, Enderich, Wegner, & Rohrer, 1999). From the dorsal aorta SA cells undergo a second migration to either the secondary sympathetic ganglia or the adrenal gland. Dependent upon the extrinsic cues of the target site, the SA cells differentiate into either tyrosine hydroxylase (TH)-expressing sympathetic neurons or become chromaffin cells of the adrenal medulla, which act as modified sympathetic ganglia (Anderson & Axel, 1986; Anderson, Carnahan, Michelsohn, & Patterson, 1991; Seidl & Unsicker, 1989a, 1989b). From these final locations the terminally differentiated sympathetic neurons extend axonal projections to the developing organs of the body.

Originating from trunk NCCs, the sensory neurons of the peripheral nervous system are located in trigeminal and dorsal root ganglia (DRG) (reviewed by Pavan & Raible, 2012). Several rounds of neurogenesis produce the sensory neurons of the DRG. Mechanoreceptive and proprioceptive neurons are added in the first round, followed by thermoreceptive, nociceptive, and more mechanoreceptive neurons in the second round. The third round consists of nociceptive neurons from boundary cap glial cells which sit at the borders between the CNS and PNS (George, Chaverra, Todd, Lansford, & Lefcort, 2007; Hjerling-Leffler et al., 2005). Sustained β-catenin expression from Wnt signaling drives differentiation of sensory neurons from NCCs, at the same time blocking fate choice to a different lineage (Lee et al., 2004). In addition to β-catenin, two transcription factors, neurogenin1 (*Ngn1*) and neurogenin2 (*Ngn2*) are critical to DRG and sensory neuron development. Knockout of both *Ngn1* and *Ngn2* results in a complete absence of DRG neurons (Ma, Fode, Guillemot, & Anderson, 1999). Post-natal maturation of sensory neurons produces three subclasses, delineated by expression of Trk receptors. Nociceptive and thermoceptive neurons primarily express TrkA, subpopulations of mechanoreceptive neurons express TrkB, while TrkC is found on proprioceptive neurons (Bibel & Barde, 2000; Davey & Davies, 1998; Huang & Reichardt, 2003; Kramer, Sigrist, Taniuchi, Jessell, & Arber, 2006).

By contrast, parasympathetic neurons are not derived directly from NCCs, but from ASCL1-positive Schwann cell precursors. Cranial and trunk nerves are a source of glial progenitors which are recruited to serve as the precursor cells. These cells migrate along preganglionic nerve fibers into or near their target organs prior to specification and differentiation (Bibel & Barde, 2000). The expression of *Phox2b* in these "bi-potent" precursors during migration can determine whether these progenitors will give rise to Schwann cells or continue on to form parasympathetic ganglia (Espinosa-Medina et al., 2014).

2.3 Enteric NCCs

The enteric NCCs (ENCCs) that will eventually form the enteric nervous system (ENS) undergo the longest migration of the NCC populations. This migration takes place through roughly a quarter of gestation in mice, and the neural network they will produce is one of the most intricate systems in the vertebrate body. Capable of completely autonomous function with no input from the CNS, the ENS is a "complex, integrative brain in its own right", often referred to as the second brain (Gershon, 1999). Here it will be briefly covered with relevance to this chapter; several reviews provide a more comprehensive overview of development of the ENS (Anderson, Newgreen, & Young, 2006; Nagy & Goldstein, 2017; Gershon, 1999).

The ENS is derived from vagal and sacral NCCs (VNCCs and SNCCs, respectively) that begin migration from the neural tube toward the gut at E8.5 in the mouse and follow two distinct tracts to reach their target locations. VNCCs first enter the proximal foregut at E9.5 in the mouse (known as ENCCs after this stage), then traverse to the stomach, cecal region and hindgut between E10.5 and E14.5 (Anderson, Stewart, & Young, 2006; Kapur, Yost, & Palmiter, 1992; Young et al., 2004). Expression of the tyrosine kinase receptor, Ret, is critical for commitment of NCCs to an enteric fate. Induction of Ret occurs during migration through the somites, triggered by retinoic acid (RA) secreted by the paraxial mesoderm (Simkin, Zhang, Rollo, & Newgreen, 2013). Studies by Niederreither et al. (2003) demonstrated that mice that are unable to synthesize RA, through deletion of *Raldh2*, exhibit downregulation of *Ret* in VNCCs, leading to improper migration and ultimately agenesis of the enteric ganglia, and thus the ENS. Interestingly, this produces a condition similar to Hirschsprung's disease in humans. Hirschsprung's disease is a developmental disorder that results from an absence of enteric ganglion cells and innervation of the distal colon. These patients experience intestinal obstructions and difficulties passing stool (Butler, Tjaden, & Trainor, 2013).

NCCs from the sacral region of the neural tube also contribute to formation of the ENS, mainly in the hindgut. Similar to VNCCs, SNCCs migrate from the neural tube at E9.5 and accumulate at the hindgut, forming pelvic ganglia around E11.5. These cells will eventually enter the distal hindgut at E13.5, only after the establishment of the VNCCs. Populations of SNCCs continue to the proximal hindgut by following fibers from the newly established pelvic ganglia. Once there, they associate with VNCCs and contribute to the generation of ENS neurons and glia, eventually forming the pelvic plexus (Burns & Le Douarin, 1998; Wang, Chan, Sham, Burns, & Chan, 2011). SNCCs express four-fold lower levels of Ret than VNCCs. Overexpression or increases in Ret push SNCCs toward a more vagal phenotype, resulting in increased migration and invasive potential (Delalande et al., 2008).

3. Neurotrophins and axon guidance cues

Peripheral innervation during organogenesis occurs through the interactions of a number of secreted polypeptides, known as neurotrophic factors, and their target receptors, consisting of the tropomyosin receptor kinase (Trk) family. These interactions influence growth, differentiation and survival of neurons innervating the developing organs (Fig. 2). The Trk receptors are tyrosine kinase receptors that were initially discovered in colon carcinoma (Martin-Zanca, Hughes, & Barbacid, 1986). The Trk family consists of three members, TrkA, TrkB, and TrkC, which are genetically encoded by the neurotrophic tyrosine kinase receptor genes *Ntrk1*, *Ntrk2*, and *Ntrk3*, respectively (Chao, 2003; Huang & Reichardt, 2003; Lee, Kermani, Teng, & Hempstead, 2001; Lee, Kim, Khursigara, & Chao, 2001).

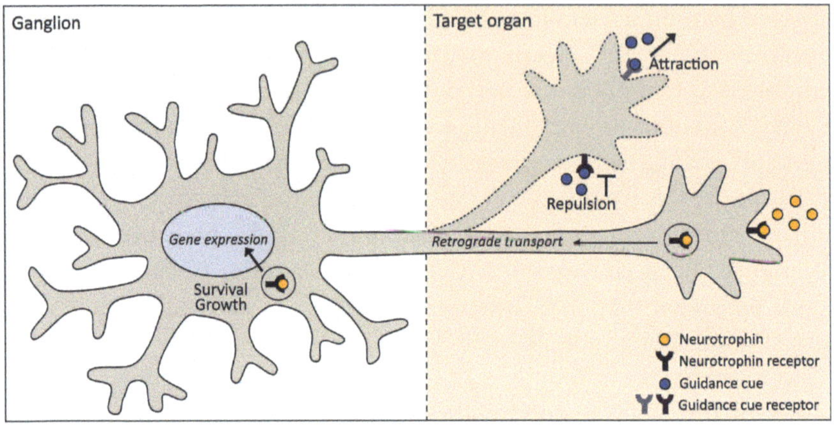

Fig. 2 Postganglionic neurons, such as those of the sympathetic nervous system, innervate target organs. Neurotrophins produced by the target tissue bind their specific neuronal receptors to promote growth and survival of the neuron. Additionally, guidance cues secreted by the target tissue interact with their respective neuronal receptors to guide axons within the tissue, inducing attractive or repulsive growth responses.

These catalytic transmembrane receptors are highly expressed by the peripheral and central nervous systems and have a diversity of functions. The ligands for these receptors consist of nerve growth factor (NGF) and the structurally similar proteins BDNF (brain-derived neurotrophic factor), NT-3 (neurotrophin-3), and NT-4 (neurotrophin-4) (Bibel & Barde, 2000; Huang & Reichardt, 2001; Patapoutian & Reichardt, 2001). NGF and NT-3 interact with TrkA. TrkB is activated by BDNF as well as NT-3/4 (Proenca, Song, & Lee, 2016; Rydén & Ibáñez, 1996). TrkC highly and specifically binds NT-3 (Chao, 2003; Ip et al., 1993; Squinto et al., 1991). In addition to the Trk receptors, a fourth receptor known as low-affinity nerve growth factor receptor (LNGFR/p75NTR) also interacts with all four of the known secreted neurotrophins, and is thought to act as a sink. LNGFR is often found in conjunction with Trk receptors and functions in balancing the amount of ligand available to the receptors. In cells that express high levels of LNGFR, but not Trk receptors, apoptosis is triggered. This is in opposition to Trk signaling which prevents cell death and mediates growth and differentiation (Chao, 2003; Kaplan & Miller, 2000; Lee, Kermani, et al., 2001; Lee, Kim, et al., 2001). Interestingly, neurotrophins are primarily responsible for regulating survival of sensory and sympathetic neurons during innervation in peripheral organ development (Bibel & Barde, 2000; Davies, 1994). In contrast, the central nervous system does not require neurotrophins for neuronal survival (Oppenheim, Prevette, Qin-Wei, Collins, & Mac Donald, 1991).

3.1 NGF, TrkA, and LNGFR/p75NTR

NGF was discovered by Nobel laureates Levi-Montalcini and Cohen, 1956. It was the first known growth factor, responsible for regulating survival, proliferation, and maintenance of its target cells. NGF affects a wide array of processes including pancreatic beta-cell survival (Pierucci et al., 2001), immunoregulation (Lambiase et al., 1997), ovulation (Ratto et al., 2012), and even the early rush of falling in love, known as limerence (Emanuele et al., 2006). However, NGF is most well-known for its role in regulating neuronal proliferation and the survival of cholinergic, sensory, and sympathetic neurons during embryogenesis (Bradshaw, Blundell, Lapatto, McDonald, & Murray-Rust, 1993; Dreyfus, 1989). The absence of NGF results in apoptosis of sensory and sympathetic neurons (Freeman et al., 2004). NGF binding to TrkA results in the positive growth and maintenance of the neuron based on signaling cascades through Bcl-2 and TRAF6/NF-kB (Sanes, Reh, & Harris, 2011). Alternatively, signaling that occurs through interactions with the receptors LNGFR/p75NTR and TrkA results in cell death (Lee, Kermani, et al., 2001; Lee, Kim, et al., 2001). As such, a fine balance is required for the careful maintenance of proper neuronal growth.

NGF's most important function is regulating the survival of sympathetic and sensory neurons during peripheral nervous system development. Studies in which neutralizing antibodies to NGF were administered in multiple newborn mammal species resulted in the irreversible death of nearly all sympathetic neurons (Levi-Montalcini & Booker, 1960). Several studies in which *Ngf* or *Ntrk1* were deleted found that mice homozygous for the mutations did not respond to painful stimulation and had a loss of up to 80% of sensory and sympathetic ganglia (Crowley et al., 1994; Silos-Santiago et al., 1995; Smeyne et al., 1994). These results also indicate that no compensatory mechanism exists for NGF via other neurotrophins. NGF's critical role in survival obscures the study of its impact on nerve growth. To investigate this Patel, Jackman, Rice, Kucera, and Snider (2000) created a genetic mouse model in which the pro-apoptotic factor *Bax* was deleted in combination with *Ngf* or *Ntrk1*. This method removed the role of NGF in preventing apoptosis of these cells, thus allowing for the study of its proliferative effects. The authors found that sympathetic neurons required NGF in order to proliferate, however another study (Glebova & Ginty, 2004) found that its necessity in various developing organs was highly variable, with sympathetic innervation in some organs completely unaffected by the loss of NGF.

Dysregulation of NGF signaling through mutations in *NTRK1* and *NGFR* (LNGFR/p75NTR) result in neurodevelopmental and neurodegenerative disorders. A prominent example occurs from mutations in *NTRK1* which result in a recessive disorder known as congenital insensitivity to pain with anhidrosis (CIPA). These patients cannot feel pain or temperature and do not have the ability to sweat (anhidrosis), thus they are unable to thermoregulate (Indo, 1993; Piña-Garza, 2013). Roughly 20% of CIPA patients die of hyperthermia by age 3 (Herring, 2013). The mutation in *NTRK1* results in improper outgrowth of sensory axons and dendrites as well as the inability of NGF to bind to the TrkA receptor, preventing nociceptive responses (Indo et al., 1996). These patients suffer from a variety of comorbidities associated with the inability to feel pain or temperature, and have a high rate of unintentional self-mutilation.

NGFR mutations resulting in overexpression of LNGFR/p75NTR play a role in multiple diseases including Huntington's, Amyotrophic Lateral Sclerosis (ALS), and Alzheimer's (Dreyfus, 1989). In all of these diseases, LNGFR/p75NTR overexpression results in loss of neurons or functional features of neurons. For instance, in Huntington's disease LNGFR/p75NTR expression is increased in the hippocampus of diseased patients. Mouse models in which LNGFR/p75NTR is overexpressed produce a Huntington's-like disease (Brito et al., 2014). Knockdown of LNGFR/p75NTR in a mouse model of ALS results in delayed disease progression (Turner et al., 2003). It is conceivable that reducing LNGFR/p75NTR may also help slow the progression of Alzheimer's, as increased levels of LNGFR/p75NTR have been found in human patients with the disease (Costantini, Scrable, & Puglielli, 2006). Treatments that reduce the expression of LNGFR/p75NTR in these diseases are currently being investigated.

3.2 TrkB, TrkC, BDNF, and NT-3/4

TrkB shares a highly conserved homology with TrkA in both its extracellular (51%) and catalytic domains (88%) (Klein et al., 1991). However, it does not interact with NGF, but is instead a high-affinity receptor for BDNF. TrkB has also been shown to interact with NT-3 and NT-4 (Proenca et al., 2016; Rydén & Ibáñez, 1996). *Ntrk2* is expressed throughout the central and peripheral nervous system in a multitude of tissues. Mice that have homozygous mutations in the *Ntrk2* gene do not respond to typical feeding stimulus and die by postnatal day 1. According to the authors of those studies, "the significant loss of motor neurons in the facial nucleus (up to 70%) observed in the *Ntrk2* knockout mice is likely to disable their mastication muscles and therefore cause their apparent inability to suckle" (Klein et al., 1993). Despite this and the highly similar homology to TrkA and TrkC, TrkB does not have the ability to induce apoptosis. This suggests that TrkA and TrkC function as dependence receptors, requiring neurotrophins for sensory and sympathetic neuronal survival (Nikoletopoulou et al., 2010). The quote from Klein et al. is therefore more apt to suggest that the loss of motor neurons in *Ntrk2* mutants may be due to other factors.

The main ligand for TrkB is BDNF, which was initially discovered and isolated from pig brains by Barde, Edgar, and Thoenen (1982). They demonstrated that BDNF supported fiber outgrowth and survival of chick sensory neurons in culture. Since then, BDNF has been found to have a complex biology with three products of the *Bdnf* gene being produced, the pro-domain BDNF, pro-BDNF and BDNF, each having a different receptor and biological function (reviewed in detail by Pradhan, Noakes, & Bellingham, 2019). The first two products of the *Bdnf* gene activate LNGFR/p75NTR and promote apoptosis, while the third product, BDNF, binds with high affinity to TrkB and with lower affinity to LNGFR/p75NTR (Chao & Hempstead, 1995; Reichardt, 2006). The binding of BDNF to TrkB is responsible for its observed neurogenic effects. In mice, genetic deletion of *Bdnf* results in degeneration of sensory ganglia, loss of balance and coordination issues, though it does not affect neuronal numbers in the CNS (Ernfors, Lee, & Jaenisch, 1994; Rauskolb et al., 2010). Deletion of *Bdnf* is lethal postnatally, with most homozygous deletions resulting in death within a few days after birth, though a few mice live as long as 4 weeks. These animals display significant reductions in cranial and spinal sensory neurons and ataxia. However, unlike *Ntrk2* deletion, removal of *Bdnf* does not affect motor neurons (Jones, Fariñas, Backus, & Reichardt, 1994).

NT-3 was discovered by Maisonpierre et al. (1990) and became the third known neurotrophic factor after NGF and BDNF. Like the other neurotrophins, NT-3 regulates survival and differentiation of sensory neurons. NT-3 functions primarily in the peripheral nervous system, and more specifically in populations of auditory, proprioceptive, and nodose sensory and sympathetic neurons (Chalazonitis, 1996). The TrkC receptor is specifically activated by NT-3, however, NT-3 has also been shown to activate TrkB and TrkA *in vitro*, albeit with lower affinity, making it unique among neurotrophins (Ip et al., 1993; Squinto et al., 1991). Work by several groups has shown that NT-3 signaling through TrkA, and not TrkC, is responsible for promoting survival of sympathetic neurons, and *Ntrk3* (TrkC) deletion in mice does not produce defects in sympathetic neuron survival. However, an NT-3 (*Ntf3*) mutation that prevents activation of TrkA results in a loss of sympathetic neurons (Belliveau et al., 1997; Fagan et al., 1996; Tessarollo et al., 1997). In addition, neurons from *Ntrk3* knockout mice cultured with NT-3 survive *in vitro* (Davies, Minichiello, & Klein, 1995). Nonetheless, NT-3 is required for organismal survival as *Ntf3* deletion results in 80% postnatal death within 48 hours. Roughly 10% of homozygous mutants die at birth, while the remaining 10% live for up to 16 days. Interestingly, the surviving animals are able to nurse successfully and do respond to pain unlike *Ntrk1* and *Ntrk2* mutants, indicating functioning nociceptive innervation. Shortly after birth, severe neurological defects become evident, with abnormal movements and postures as well as reductions in body weight by up to 50% compared to littermates (Tessarollo, Vogel, Palko, Reid, & Parada, 1994).

TrkC is the primary high affinity receptor for NT-3. As with the other Trk receptors, TrkC is involved in neuronal survival, proliferation and differentiation. TrkC has two isoforms, catalytic and non-catalytic. Both isoforms of TrkC are expressed in NCCs (Tessarollo et al., 1993). Activation of the catalytic isoform of TrkC in these cells results in proliferation and differentiation. However, the non-catalytic TrkC isoform does not impact proliferation, but strongly drives differentiation, though requires LNGFR/p75NTR as a co-receptor in this role (Hapner, Boeshore, Large, & Lefcort, 1998). Mice in which all isoforms of *Ntrk3* are deleted survive up to 3 weeks after birth, though most die in the first 7 days (Klein et al., 1994). These mice display neuronal cell loss in multiple peripheral regions, for example the number of DRG neurons are reduced by up to 33% in *Ntrk3* mutants (Tessarollo et al., 1997). These mice also display severe neurological defects and cardiac architecture abnormalities similar to, but not as severe as, *Ntf3* deletion.

The last addition to the neurotrophin family, NT-4 (previously referred to as NT-4/5 or NT-5; reviewed in detail by Ibáñez, 1996), is the most evolutionarily divergent member (Berkemeier et al., 1991; Ibáñez, Ebendal, & Persson, 1991; Ip et al., 1992), and has the lowest, but most ubiquitous expression of the neurotrophins (Berkemeier et al., 1991; Timmusk, Belluardo, Metsis, & Persson, 1993). However, deletion of NT-4 (*Ntf4*) in mice does not result in early postnatal death, unlike the other neurotrophins (Conover et al., 1995; Liu, Ernfors, Wu, & Jaenisch, 1995). NT-4 binds with high-affinity to TrkB and LNGFR/p75NTR and is most similar to BDNF, though its interaction with the TrkB receptor differs from BDNF binding (Ibáñez, 1995; Ip et al., 1993; Klein, Lamballe, Bryant, & Barbacid, 1992). NT-4/*Ntf4* has different expression patterns and functions than BDNF/*Bdnf* during embryonic development (Conover et al., 1995; Liu et al., 1995), but both are involved, along with their cognate receptor, in the development of taste (Huang & Krimm, 2010). *Ntf4* deletion in mice results in a loss of sensory neurons in the geniculate ganglia that control taste system development while leaving the motor neurons of the facial nucleus untouched. The data from the *Ntf4* knockout mice demonstrate that *Ntf4* is not required for survival of sympathetic or motor neurons, but is required for the survival of a subset of sensory neurons (Liu et al., 1995). In another study of *Ntf4* deletion, the mutant mice did not display obvious neurological defects and had normal lifespans. In animals in which *Ntf4* and *Bdnf* are simultaneously deleted, motor neuron numbers are not impacted, which is converse to *Ntrk2* mutants, suggesting neither factor is required for motor neuron development (Conover et al., 1995).

3.3 Neuronal guidance cues

Guidance factors are secreted proteins that direct innervation throughout development by interacting with receptors on the growth cone of growing axons (Fig. 2). Initial growth cone responses were first observed by Ramon y Cajal (1892) (reviewed by Raper & Mason, 2010), in which he hypothesized that chemotactic guidance cues must be responsible for their movements. His intuition was proven correct many years later as the main protein families that serve as neuronal guidance factors were discovered: netrins, semaphorins, slits, ephrins, and one more recent addition, the cell adhesion molecules or CAMs (Tessier-Lavigne & Goodman, 1996). Each family has specific receptors and functions. Interactions of guidance factors with their receptors produce a chemotropic response, which generates either a chemoattractive or chemorepulsive cue, inducing the axonal growth cone to move toward or away from the signal (Fig. 2). These guidance factors function in a complex and coordinated manner, with the receptive cell receiving and processing many different cues at once to eventually reach its intended target.

Guidance cues fall into one of three general categories: trophic cues, modulatory cues, and adhesive cues. Trophic cues influence growth cone motility and neuronal proliferation and survival. Neurotrophins can function in this category. In addition to neurotrophins' role in growth and survival, there is evidence to suggest that they may also have some function as guidance factors (Connolly, Seeley, & Greene, 1985; Gundersen & Barrett, 1979; Letourneau, 1978). In studies by Tucker, Meyer, and Barde (2001), sensory axons in the limb buds of E10.5 mouse embryos were directed to ectopically embedded beads that were coated with any of the four neurotrophins. However, neurotrophins operate more in a short-range guidance capacity, with their primary functions falling into the previously described growth and survival roles (Genç, Özdinler, Mendoza, & Erzurumlu, 2004). Most of the well-known guidance factors (netrins, semaphorins, slits, and ephrins) fall into the trophic cues category. Adhesive cues regulate the binding of axons to the substrate, and mainly consist of the CAMs and extracellular matrix proteins. Examples of these are laminins, heparin, and integrins. Laminins promote axonal extension though interactions with the growth cone receptors. Bozyczko and Horwitz (1986) discovered that the use of a neutralizing antibody against laminin in explant cultures of neurons resulted in the disruption of neuronal adhesion to the substrate and inhibiton of process formation. Modulatory cues can influence axonal response without direct effect, for example, making an axon more or less responsive to a trophic cue. Endoglycan, a sialomucin, has been show to negatively regulate commissural axon guidance without serving as a guidance cue or receptor (Baeriswyl et al., 2021). Netrins can also function as modulatory cues, becoming a repulsive signal based upon the substrate that growth cones of axons are interacting with (Höpker, Shewan, Tessier-Lavigne, Poo, & Holt, 1999).

Netrins are a family of laminin-related extracellular proteins best known for their role in commissural axon guidance during vertebrate spinal cord development (Fazeli et al., 1997; Serafini et al., 1996, 1994). However, netrins have been shown to function in a variety of capacities including cell migration, luminal cell adhesion in mammary glands, tip development in the lung, and vascular proliferation and pathfinding (Sun, Correia, & Kennedy, 2011). The first member, uncoordinated-6 (*unc-6*), was discovered in *C. elegans* with its vertebrate homologues, netrin-1 and netrin-2, discovered in the chick embryo a few years later (Hedgecock, Culotti, & Hall, 1990; Ishii, Wadsworth, Stern, Culotti, & Hedgecock, 1992; Serafini et al., 1994). Since then, five mammalian family members have been found, the secreted netrins-1,3, and 4, and two membrane bound netrins G1 and G2. These netrins interact with six receptors UNC5A-D, neogenin, and deleted in colorectal carcinoma (DCC) (reviewed by Sun et al., 2011).

The large and varied semaphorin family has over 20 members (Reviewed in detail by Alto & Terman, 2017). It consists of membrane-bound and secreted molecules that vary in structure, with a 500 amino acid sema domain in common. Sema3a and Sema1a were the first identified semaphorin proteins. Like netrins, semaphorins were originally discovered for their role in axon guidance but have since been found to have functions in almost every tissue of the body throughout life, from embryonic development to adult organ function (Kolodkin, Matthes, & Goodman, 1993; Yazdani & Terman, 2006). Semaphorin signaling is mediated mainly through the plexin and neuropilin families of receptors. Semaphorins have been well-characterized for their roles in the nervous, vascular, and immune systems, but have more recently been investigated for their functions in bone remodeling, lung branching morphogenesis, kidney disease, the reproductive system, and liver fibrosis (De Minicis et al., 2013; Giacobini, 2015; Kagoshima & Ito, 2001; Li et al., 2017; Xia & Worzfeld, 2016; Yagai, Miyajima, & Tanaka, 2014).

Slits were discovered by Rothberg, Hartley, Walther, and Artavanis-Tsakonas (1988) while investigating the EGF-like portion of notch in *Drosophila*. The Slit family in mammals is comprised of Slit1, 2, and 3. Expression of Slit1 is dominant in the nervous system, while Slit2 is predominantly expressed in mesodermal lineage tissues with some neural expression, and Slit3 is expressed neurally as well as in mesenchymal limb buds (Yuan et al., 1999). Slit2 has also been shown to play critical roles in kidney development (Piper, Georgas, Yamada, & Little, 2000). In nervous system development, Slit proteins critically function as repulsive cues and deleting Slits in *Drosophila* results in fusion of the nerve cord at the midline (Rothberg et al., 1988; Rothberg, Jacobs, Goodman, & Artavanis-Tsakonas, 1990). The Slit proteins interact specifically with three receptors, roundabout (Robo) 1, 2 and 3 (Brose et al., 1999; Yuan et al., 1999). In the peripheral nervous system of the adult mouse, all three Slits, as well as Robo1 and 2, are expressed in sensory and motor neurons. Robo1 is also expressed in blood vessels associated with peripheral nerves (Carr, Parkinson, & Dun, 2017).

The last family of guidance factors are ephrins. Unlike the other guidance factor families, ephrins and their receptors, Ephs, are membrane bound and require cell-to-cell contact for activation and interaction (Pasquale, 2005). Ephrins act mainly as repulsive cues in the developing nervous system (Wilkinson, 2001), but serve many other roles as well. The vertebrate ephrin family consists of 8 members in 2 sub-classes, EphrinA1-5 and EphrinB1-3, and 14 Eph receptors, EphA1-8, EphA10, EphB1-4, and EphB6. The A ligands typically bind A receptors and B ligands bind B receptors, although some ligands can engage both types of receptors (Darling & Lamb, 2019; Kania & Klein, 2016).

4. Organ innervation and developmental roles

Peripheral innervation of tissues during development occurs concomitantly with organogenesis. As nerves grow, associate with, and form synapses with target cells they can modulate cellular physiology via the release of neurotransmitters (Fig. 3). Recently, studies in various organs have begun to uncover novel roles for nerves during organogenesis. The cellular responses to nerve-derived signals can impact development and resulting physiology, highlighting their significance beyond modulating mature tissue function (Fig. 3).

4.1 Lung

In mice, lung development begins around E9.5. The respiratory tract continues to develop after birth, reaching maturity around 3-4 weeks postnatally (Burri, 2006). The lungs arise from the ventral foregut endoderm, eventually

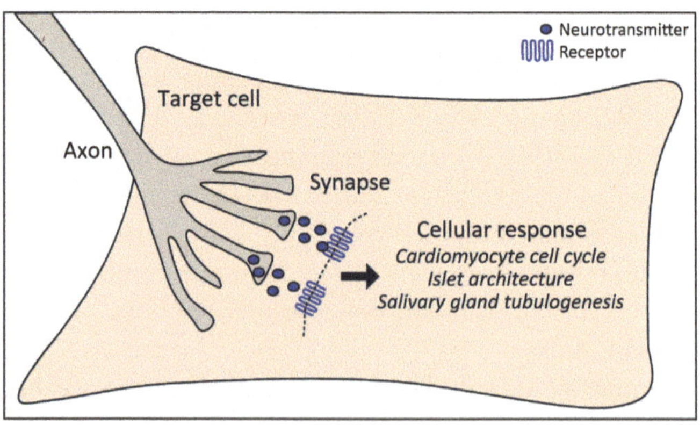

Fig. 3 Axon terminals form synapses with target cells during organogenesis. Neurotransmitters released from the neurons bind their respective receptors on the target cell to induce a cellular response. Examples of developmental processes modulated by these responses include the timing of cardiomyocyte cell cycle withdrawal, the regulation of pancreatic islet architecture, and modulation of salivary gland tubulogenesis.

forming the two primary lung buds (Cardoso & Lü, 2006). The developing lung buds will then undergo several rounds of branching while they invade the surrounding mesenchyme (Aven & Ai, 2013).

During development, the lung is innervated by intrinsic neurons as well as extrinsic parasympathetic, sympathetic, and sensory nerves. The intrinsic nerves of the lung are derived from vagal neural crest cells, which cluster and differentiate into the ganglia of the respiratory tract (Langsdorf, Radzikinas, Kroten, Jain, & Ai, 2011). These intrinsic nerves form the basis for parasympathetic innervation into the trachea and main bronchi. However, intrinsic neurons do not contribute significantly to lung innervation due to their short axon lengths, which limits their functional ability to producing localized signals in the airway (Aven & Ai, 2013; Weigand & Myers, 2010). The constraint to localized signaling by intrinsic neurons is further evident by their absence in secondary and tertiary bronchi as well as in the distal lung (Langsdorf et al., 2011).

With intrinsic neurons having a limited role, the lung is primarily innervated by extrinsic neurons. The extrinsic sympathetic nerves stem from the middle and inferior cervical region of the thoracic ganglia (Kummer, 2011). Development of the pulmonary layer of airway smooth muscle (ASM) is closely linked to extrinsic innervation. In order for proper extrinsic innervation of the ASM to occur, the appropriate gradient of the neurotrophic factors BDNF and NT-4 must be present. The timing of BDNF secretion and thus, extrinsic innervation, is coordinated by the lung mesenchyme, which expresses *Bdnf* prior to differentiation into ASM (Radzikinas et al., 2011). The lungs are also innervated by a small number of extrinsic sensory nerves which function to relay information from the lungs to the CNS. Additionally, they initiate local responses to peripheral stimuli through the release of the neuropeptides substance P, neurokinin A, neurokinin B, and calcitonin gene related peptide (CGRP) (Allen, Wharton, Polak, & Haworth, 1989; Kummer, 2011).

Interactions between the nervous system of the airway and immune cells play an important role in physiology. The lung is under constant bombardment from allergens and immune threats from outside the body. To assist in the response to these threats the airway walls contain a specialized group of epithelial cells that have both neuronal and endocrine characteristics. These cells, known as pulmonary neuroendocrine cells (PNECs), can secrete various neuropeptides and neurotransmitters to direct appropriate immune responses (Branchfield et al., 2016; Liu et al., 2020; Xu, Yu, & Sun, 2020; Sui et al., 2018). Recently, it has been demonstrated that sympathetic neurons can regulate the innate immune responses of the lung. Norepinephrine, a sympathetic neurotransmitter, and its associated receptors can elicit an immune response to treatments such as the bacterial membrane protein lipopolysaccharide (LPS). Therefore, tight regulation of this pathway is necessary to ensure a proper response upon stimulation (Liu et al., 2020)

Genetic ablation of sympathetic neurons, which is induced by deletion of the TrkA (*Ntrk1*) receptor in TH-positive neurons, leads to the development of lungs with normal size and no detectable differences in gross appearance (Liu et al., 2020). This suggests that sympathetic innervation is not strictly required for proper lung development, although further histological characterization would help clarify if more subtle phenotypes are present. However, laser-mediated ablation of intrinsic parasympathetic neurons leads to defects in lung branching. In these studies, the lungs were removed at E12.5, neurons ablated, and the tissues cultured *ex vivo* to assess effects on their growth. After ablation and culture for up to 46 h, lung lobes in which neurons were ablated displayed reduced branching. This effect was determined to be independent of cholinergic signaling, although the exact mechanism was not determined (Bower et al., 2014). Genetic ablation of the parasympathetic neurons would add additional support for the *in vivo* requirement of these nerves in lung morphogenesis.

4.2 Gastrointestinal tract

The gastrointestinal (GI) tract is one of the longest and most complex systems in the body. Eventually running throughout the head, neck and torso, the primitive GI is of endodermal origin and consists of 3 regions, termed the foregut, midgut, and hindgut (Sadler & Langman, 2012; Uesaka, Young, Pachnis, & Enomoto, 2016). As development progresses, the esophagus, stomach and duodenum eventually form from the foregut, while the small intestine and cecum arise from the midgut, with the colon developing from the midgut and hindgut (Geesman, Gesiotto, Lalani, & Tejani, 2021; Rao & Gershon, 2018).

Innervation of the GI tract begins around E9.5 with the invasion of vagal neural crest cells (VNCCs) (Young et al., 1998). These VNCCs will differentiate into enteric neurons and glia which will make up the myenteric plexus and become the ENS. The ENS consists of intrinsic enteric neurons and extrinsic sensory and sympathetic nerves. All autonomic functions of the bowels are regulated by the ENS (Duan et al., 2021). During GI organogenesis, the primitive gut undergoes several rounds of looping which will bring the midgut and hindgut into closer proximity around E11.5. During this process a population of VNCCs will migrate from the midgut to the hindgut joining with existing enteric neural crest cells (ENCCs) to form the ENS of the proximal hindgut around E13.5 (Kang, Fung, & Vanden Berghe, 2021; Nishiyama et al., 2012). A population of ENCCs originating from the myenteric plexus will form the submucosal plexus. The ENS is believed to be electrically active as early as E11.5, however activity has only been observed at E18.5 in the duodenum and at P10 in the colon thus far (Roberts et al., 2010; Roberts, Murphy, Young, & Bornstein, 2007).

The extrinsic sensory afferent and sympathetic efferent nerves of the GI tract originate from the thoracic and lumbar splanchnic nerves. In turn, these derive from two major pathways: (1) the spinal afferent pathway from the mesenteric nerves and (2) the sacral spinal pathway from the distal colon and pelvic nerves (Costa, Brookes, & Zagorodnyuk, 2004). However, it is still unknown how spinal sensory innervation of the GI tract occurs. In contrast, the sympathetic efferent nerves innervate the GI tract by following the developing arterial network beginning around E13.5 in mice. The sympathetic nerves will reach the mesentery and enter the GI wall by E15.5 (Glebova & Ginty, 2005; Hatch & Mukouyama, 2015). These nerves track along the mesenteric arteries that express the neurotrophic factor artemin (*Artn*), which signals through Ret/GFRa3 (Baloh et al., 1998; Baloh, Enomoto, Johnson, & Milbrandt, 2000). Not surprisingly, deletion of *Artn* results in defects in sympathetic innervation of the GI tract (Brunet et al., 2014; Honma et al., 2002).

In addition to artemin there is a wide variety of factors that influence and direct GI innervation during organogenesis. These include the neurotrophic factors NGF, BDNF, and NT-3. NGF is expressed by the epithelial layer of the GI tract and is a key regulator of extrinsic innervation (Kawaja et al., 2011). BDNF, NT-3 and their respective receptors, TrkB and TrkC, play an important role in the development and maintenance of the vagal sensory nerves. *Bdnf* is expressed by the mesenchyme in the fetal GI tract and GI muscle, while smooth muscle cells and vascular smooth muscle in the GI tract express *Ntf3*. TrkB is found on vagal afferent nerves and TrkC is expressed by vagal sensory nerves. Knockout of either *Ntf3* or *Ntrk3* results in a 50% decrease of nodose neurons in the GI tract (Fox, Biddinger, Baquet, Jones, & McAdams, 2013; Genç et al., 2004; Tessarollo et al., 1997). Similarly, deletion of *Bdnf* results in a significant loss of vagal sensory nerves (Murphy & Fox, 2010). However, conditional knockout of *Bdnf* in muscle leads to an increase in vagal afferent innervation in the small intestine (Biddinger & Fox, 2014).

Along with neurotrophic factors, the guidance molecules netrin-1 and Slit2 and their respective receptors, DCC and Robo1/2, also play a role in the development of the vagal nerves. The vagal nerves express the netrin receptor, DCC, which functions to attract these nerves toward the netrin-1 secreting mucosal epithelium and outer mesenchyme. DCC is also expressed by the enteric nerves, and *Ret* null mice that lack enteric neurons do not achieve vagal innervation of the stomach (Ratcliffe et al., 2006, 2011). Slit2 directs vagal innervation in the stomach and intestine through the repulsion of neurites from nodose ganglia, which express the slit receptors Robo1/2 (Goldberg et al., 2013).

Apart from neurotrophic and guidance factors, the T-Box transcription factor, *Tbx1,* also influences vagal nerve development and innervation in the GI tract. Calmont, Thapar, Scambler, and Burns (2011) and Calmont et al. (2009) found that *Tbx1* is essential for the formation of vagal ganglia and vagal nerve projection in the GI tract. *Tbx1* deletion results in hypoplastic innervation of the esophagus and a complete absence of vagal nerves in the stomach.

Genetic ablation of either sympathetic or sensory neurons revealed a reliance on sensory nerves for proper GI tract innervation by efferent neurons. Parasympathetic and sympathetic axons projected erratically in the absence of sensory neurons. Therefore, sensory afferent axons serve as scaffolds that can attract and mediate guidance for other efferent neurons in the GI tract (Niu et al., 2020). While GI innervation was analyzed in these genetic ablation experiments, the effect on GI development was not investigated. Whether nerves play an active role in driving proper development of the GI tract remains to be determined.

4.3 Pancreas

Distinct segments of the foregut endoderm give rise to the pancreas around E8.5 in mice. Pancreatic progenitor cells expressing *Pdx1* then begin to aggregate and differentiate into the exocrine, endocrine and ductal lineages of the pancreas (Jørgensen et al., 2007). The developing pancreas is innervated by vagal nerves of the autonomic nervous system, afferent nerves originating from the T7-L2 dorsal root ganglion (DRG), and adrenergic nerves from the celiac and superior mesenteric ganglia (Fasanella, Christianson, Chanthaphavong, & Davis, 2008; Love, Yi, & Smith, 2007; Travagli & Anselmi, 2016; Woods & Porte, 1974).

Sympathetic innervation in the pancreas occurs around E14.5 with the invading nerves surrounding the NGF-secreting pancreatic islet of Langerhans by postnatal day 6. Glebova and Ginty (2004) showed that pancreatic innervation is incomplete in animals in which *Ngf* has been deleted. Borden, Houtz, Leach, and Kuruvilla (2013) demonstrated that innervation was necessary for the development and maturation of the islets of Langerhans. Without the extrinsic inductive cues from sympathetic nerves the islets do not aggregate, and thus do not develop their appropriate architecture. Beta cell migratory behaviors require proper noradrenergic signaling. For example, in animals that have undergone sympathectomies pancreatic beta cells have poor cell-to-cell contact due to uneven expression of E-cadherin. Perturbations in pancreatic sympathetic innervation results in increased risk for developing glucose intolerance and diabetes (Borden et al., 2013).

4.4 Heart

Innervation of the heart begins around E9.5 with the migration of two populations of NCCs (Jiang, Rowitch, Soriano, McMahon, & Sucov, 2000). Trunk NCCs migrate to the dorsal aorta and differentiate into sympathetic and sensory neurons, then aggregate and extend axonal projections along the vasculature into the developing heart (Loring & Erickson, 1987). Conversely, cells originally thought to derive from cardiac NCCs, which were suggested to produce the parasympathetic nerves of the heart, do not aggregate but instead migrate along the vagus nerve prior to sympathetic innervation (Hildreth, Anderson, & Henderson, 2009). As previously mentioned, lineage tracing studies have more recently shown that parasympathetic ganglia and neurons develop from Schwann cell precursors (Dyachuk et al., 2014). Regardless, several in-depth reviews have covered NCC migration to the heart (Kirby & Hutson, 2010) and cardiac innervation during development in greater detail (Hasan, 2013; Hildreth et al., 2009; Kimura, Ieda, & Fukuda, 2012).

The interplay between the neurotrophins NGF, NT-3, and the TrkA receptor are necessary for proper sympathetic innervation (Fagan et al., 1996; Francis et al., 1999; Kuruvilla et al., 2004). High expression of

NT-3 from the blood vessels initially guides sympathetic nerves toward the heart. As the nerves reach the heart, NT-3 levels decrease and an increasing gradient of NGF takes over, followed by a second wave of NGF released by arteries deeper in the heart, which directs sympathetic nerves toward their final targets (Cristofaro et al., 2010; Francis et al., 1999; Nam et al., 2013). NGF levels within the heart directly relate to the degree of sympathetic innervation, with detectable amounts of cardiac NGF occurring around E12 when sympathetic nerves begin invasion of the heart, then increasing multiple fold to reach adult levels of NGF by E14. The highest expression of *Ngf* mRNA occurs in the atrium, which correlates with sympathetic nerve density (Heumann, Korsching, Scott, & Thoenen, 1984; Korsching & Thoenen, 1983, 1988).

In studies by Kuruvilla et al. (2004), deletion of *Ntf3* resulted in shorter axonal projections extending toward the heart from the stellate ganglion. As a result, little to no sympathetic innervation of the heart was present at E15.5 in mice. In addition to its role in cardiac sympathetic innervation, NT-3 is key in heart development, functioning in a non-neuronal manner as evident by the significant cardiovascular abnormalities which likely result in the early lethality observed in *Ntf3* knockout animals (Donovan, Hahn, Tessarollo, & Hempstead, 1996).

The GDNF family members neurturin and artemin function in parasympathetic and sympathetic cardiac innervation, respectively. Artemin is expressed by arteries and vascular smooth muscle, while its receptors are expressed by sympathetic neurons (Damon, teRiele, & Marko, 2007). Artemin may facilitate sympathetic nerve growth to the heart as GDNF, the founding family member, can enhance sympathetic projections toward cardiomyocytes in rat more potently that either NGF or BDNF (Miwa et al., 2010). Interestingly, artemin shares slightly greater similarity to neurturin than it does to GDNF. Neurturin conversely impacts the development of cardiac ganglion neurons (that are derived from cardiac NCCs) and thus, parasympathetic innervation. Neurturin signals through the GFRa2/Ret receptor complex which is expressed by cardiac parasympathetic neurons during development (Hiltunen, Laurikainen, Airaksinen, & Saarma, 2000). Hiltunen et al. (2000) found that deletion of either receptor component resulted in loss of cardiac ganglia and reduction in cholinergic innervation by up to 60% in some regions of the heart. In the hearts of neurturin (*Nrtn*) knockout mice, cholinergic neuron numbers were reduced by 65% and those that remained were 15% smaller (Mabe & Hoover, 2009).

Studies by Ieda et al. (2007) have found that the neuronal guidance cue *Sema3a* is highly expressed in the mouse heart between E12 and E15. Sema3a typically functions as a chemorepellent during neuronal patterning, and the authors of those studies discovered that overexpression of Sema3a in the developing heart results in reduced sympathetic innervation. Conversely, deletion of *Sema3a* results in aberrant patterning of cardiac sympathetic neurons, with roughly 80% mortality at one week after birth (Ieda et al., 2007).

Chemical sympathectomy with 6-hydroxydopamine (6-OHDA), a neurotoxin that destroys dopaminergic and noradrenergic neurons, was administered postnatally to newborn rats to assess the effects of sympathetic innervation on cardiac development. In rodents, heart development continues postnatally as cardiomyocytes switch from a proliferative state to hypertrophy. Ablation of cardiac sympathetic innervation resulted in smaller hearts. It was determined that cardiac maturation is delayed by sympathetic innervation, and β-adrenergic signaling mediates the proliferative to hypertrophic switch (Kreipke & Birren, 2015). Blocking signaling through the β-adrenergic receptor decreases the number of binucleated cardiomyocytes and promotes their proliferation (Liu et al., 2019). Together, these results highlight the significance of sympathetic innervation in modulating the cardiomyocyte cell cycle during development.

4.5 Salivary gland

In salivary gland development, parasympathetic innervation parallels development of the organ (Coughlin, 1975). Ductal tubulogenesis of the salivary gland occurs around E13.5, and is innervated solely by parasympathetic nerves and ganglia. Direct communication from the parasympathetic nerves and the epithelium plays a key role in ductal tubulogenesis. Vasoactive intestinal peptide (VIP) released from the nerves activates the cAMP and CFTR pathways, increasing ductal elongation and lumen formation, indicating that parasympathetic ganglia regulate multiple processes in the developing salivary gland (Nedvetsky et al., 2014). In the same study, the authors blocked the neurotrophic factor neurturin, which is secreted by the end buds of the submandibular gland (SMG). They thenobserved disrupted lumen formation, decreased epithelial branching, reduced nerve outgrowth and deficient innervation of the epithelium, demonstrating a role for neuronal signaling in tubulogenesis.

In prior studies, Knox et al. (2010) showed that parasympathetic innervation was necessary to maintain the epithelial progenitors in an undifferentiated state in *ex vivo* SMG cultures. Removal of the parasympathetic ganglia before *ex vivo* SMG development resulted in reductions of SMG epithelium and significant reductions in end bud number. The authors discovered through lineage tracing that the end buds of the salivary gland contain keratin 5 (K5) positive cells that function as SMG progenitor cells. The removal of the parasympathetic ganglion reduced the expression of K5 positive cells along with the number of end buds in the developing SMG. In addition, the authors investigated K5 expression in cells from the ventral prostate of postnatal mice after treatment with inhibitors of parasympathetic acetylcholine signaling and similarly found reduced K5 expression in epithelial cells of the prostate.

A more recent study by Teshima, Tucker, and Lourenço (2019) has shown that a group of cells within the parasympathetic submandibular ganglion (PSG) express the sympathetic nerve marker tyrosine hydroxylase (TH). These TH-positive neurons formed a network within the salivary gland in culture, and treatment with the dopamine receptor antagonist, 6-OHDA, resulted in the loss of TH-positive neurons in the PSG and inhibition of SMG branching. These data suggest that sympathetic nerves may also contribute to branching morphogenesis in the salivary gland.

4.6 Urogenital system

Formed from the intermediate mesoderm, the urogenital system (UGS) consists of the organs of the urinary and reproductive systems. These systems share a common embryonic origin and there is a large body of work regarding the organogenesis of these systems (see Hynes & Fraher, 2004; Little, 2011; Liaw et al., 2018; McMahon, 2016; Rehman & Ahmed, 2021). The autonomic innervation of the UGS is responsible for a number of tasks that are critical to vertebrate homeostasis, including bladder control and excretion of urine, regulation of blood pressure and sexual arousal (DiBona, 2000; Osborn, Tyshynsky, & Vulchanova, 2021; Wiese, Deal, Ireland, Cantrell, & Southard-Smith, 2017). Despite the importance of UGS innervation, its formation and function during embryogenesis has not been well studied, though recent work has begun to elucidate the process of UGS innervation.

As previously discussed in this chapter, sacral NCCs move to the hindgut around E11.5 and form pelvic ganglia (Burns & Le Douarin, 1998; Wang et al., 2011). A population of SNCCs then continue into the hindgut to become part of the ENS. The remaining SNCCs migrate from the pelvic plexus into the urogenital sinus, forming aggregates between E12.5 and E14.5, with differentiation beginning at E15.5 (Anderson et al., 2006; Wiese et al., 2012). Wiese et al. (2017) demonstrated that SNCCs migrate from the pelvic ganglia to the bladder in waves starting around E13.5, after vascularization and smooth muscle differentiation. Deal et al. (2021) uncovered a critical role for Pax3 in SNCCs, demonstrating that NCC-specific deletion of *Pax3* resulted in delayed migration of NCCs, reduced bladder wall innervation, and decreased bladder function. Using retrograde tracing, Ritter and Southard-Smith (2017) found that the serotonin receptor, 5-HT3A, was expressed in a large portion of DRG (sensory) neurons that constitute the bulk of bladder innervation.

To date only one study has looked at gonadal innervation and its role during organogenesis. McKey, Bunce, Batchvarov, Ornitz, and Capel (2019) discovered that axonal projections derived from NCCs penetrate the dorsal mesonephros of both sexes around E15.5, but then differentially innervate the ovaries and testes. These projections enter the ovaries at E16.5 and form a dense axonal network within the ovarian medulla. In contrast, the testes only develop superficial innervation of the epididymis, tunica, and vas deferens. Interestingly, in embryos in which *Fgf9* has been deleted, resulting in male to female sex reversal after sex determination, ovaries in XY animals are innervated, suggesting that ovarian organogenesis supports innervation. The authors further showed that the testes express repulsive guidance cues, preventing axonal and NCC invasion, thus resulting in sexually dimorphic innervation of the gonads (McKey et al., 2019).

Similar to the rest of the UGS, renal innervation during development has been poorly characterized. Recent work from McKey et al. (2019) demonstrated that NCC-derived sympathetic ganglia extend axonal projections into the kidney and adrenal gland by E15.5. It has been shown that parasympathetic innervation in other branching organs (lungs and salivary glands) play key roles during development, suggesting that innervation may impact renal development as well. Early studies in fetal and newborn sheep demonstrated the renal nerves regulate hemodynamics, fluid and electrolyte homeostasis, and renin secretion during development (Robillard, Guillery, Segar, Merrill, & Jose, 1993). Chemical sympathectomy in newborn rats with 6-OHDA leads to smaller kidneys with decreased DNA content and functional abnormalities (Slotkin, Levant, Orband-Miller, Queen, & Stasheff, 1988). It was also suggested through experiments with murine kidney explants that nerves could induce tubulogenesis (Sariola, Ekblom, & Henke-Fahle, 1989). Despite such promising preliminary studies in mice and other organisms, the *in vivo* relevance of innervation to kidney development remains inconclusive.

4.7 Skeletal system

The mammalian skeletal system is innervated by both sympathetic and sensory nerves that are primarily responsible for transmitting nociceptive signals and influencing skeletal metabolism (Li et al., 2017; Olsen, Reginato, & Wang, 2000; Togari, Arai, & Kondo, 2005). However, a role for innervation in mediating skeletal development was recently uncovered. Tomlinson et al. (2016) found that during formation of the skeletal system, the perichondral surfaces of the bone release NGF, which promotes the survival of invading neurons (Castañeda-Corral et al., 2011). In the Tomlinson study they further investigated the impact of skeletal NGF during development through the inactivation of TrkA signaling in sensory nerves, and observed a significant reduction in femur innervation. This reduction in femur innervation produced delays in vascularization of ossification sites and a reduction in the size of the *Osx*-expressing osteoprogenitor pool during primary ossification. These findings highlight the significance of skeletal innervation for proper development.

5. Discussion

The PNS plays significant roles in modulating mammalian physiology and maintaining homeostasis. A diverse subset of neurons innervate target organs during development. While defects in proper innervation can contribute to a number of diseases and pathophysiology, their roles in development are only beginning to be uncovered. With the generation of modern genetic tools, as well as our advancement in the knowledge of organ development, researchers can begin addressing whether nerves modulate processes such as proliferation, maturation, and other cellular behaviors during development. RNA-sequencing analyses of both neurons and tissue-specific cells they interact with during development will help uncover the mechanisms of cellular crosstalk. Such knowledge will also be informative for regenerative studies and synthetic organ generation, where the presence of nerves or nerve-derived signals may influence outcomes. More in-depth studies in all tissues are necessary to ascertain the complete repertoire of peripheral nerve function during organogenesis.

References

Allen, K. M., Wharton, J., Polak, J. M., & Haworth, S. G. (1989). A study of nerves containing peptides in the pulmonary vasculature of healthy infants and children and of those with pulmonary hypertension. *British Heart Journal*, *62*, 353–360. https://doi.org/10.1136/hrt.62.5.353.

Alto, L. T., & Terman, J. R. (2017). Semaphorins and their signaling mechanisms. *Methods in Molecular Biology*, *1493*, 1–25. https://doi.org/10.1007/978-1-4939-6448-2_1.

Anderson, R. B., Newgreen, D. F., & Young, H. M. (2006). Neural crest and the development of the enteric nervous system. *589*. *Advances in Experimental Medicine and Biology* (pp. 181–196). Springer.

Anderson, D. J., & Axel, R. (1986). A bipotential neuroendocrine precursor whose choice of cell fate is determined by NGF and glucocorticoids. *Cell*, *47*, 1079–1090. https://doi.org/10.1016/0092-8674(86)90823-8.

Anderson, D. J., Carnahan, J. F., Michelsohn, A., & Patterson, P. H. (1991). Antibody markers identify a common progenitor to sympathetic neurons and chromaffin cells in vivo and reveal the timing of commitment to neuronal differentiation in the sympathoadrenal lineage. *The Journal of Neuroscience*, *11*, 3507–3519. https://doi.org/10.1523/JNEUROSCI.11-11-03507.1991.

Anderson, R. B., Stewart, A. L., & Young, H. M. (2006). Phenotypes of neural-crest-derived cells in vagal and sacral pathways. *Cell and Tissue Research*, *323*, 11–25. https://doi.org/10.1007/s00441-005-0047-6.

Aven, L., & Ai, X. (2013). Mechanisms of respiratory innervation during embryonic development. *Organogenesis*, *9*, 194–198. https://doi.org/10.4161/org.24842.

Baeriswyl, T., Dumoulin, A., Schaettin, M., Tsapara, G., Niederkofler, V., Helbling, D., et al. (2021). Endoglycan plays a role in axon guidance by modulating cell adhesion. *eLife*, *10*, e64767. https://doi.org/10.7554/eLife.64767.

Baloh, R. H., Tansey, M. G., Lampe, P. A., Fahrner, T. J., Enomoto, H., Simburger, K. S., et al. (1998). Artemin, a novel member of the GDNF ligand family, supports peripheral and central neurons and signals through the GFRα3–RET receptor complex. *Neuron*, *21*, 1291–1302. https://doi.org/10.1016/S0896-6273(00)80649-2.

Baloh, R. H., Enomoto, H., Johnson, E. M., & Milbrandt, J. (2000). The GDNF family ligands and receptors—Implications for neural development. *Current Opinion in Neurobiology*, *10*, 103–110. https://doi.org/10.1016/S0959-4388(99)00048-3.

Barde, Y. A., Edgar, D., & Thoenen, H. (1982). Purification of a new neurotrophic factor from mammalian brain. *The EMBO Journal*, *1*, 549–553. https://doi.org/10.1002/j.1460-2075.1982.tb01207.x.

Belliveau, D. J., Krivko, I., Kohn, J., Lachance, C., Pozniak, C., Rusakov, D., et al. (1997). NGF and neurotrophin-3 both activate TrkA on sympathetic neurons but differentially regulate survival and neuritogenesis. *Journal of Cell Biology*, *136*, 375–388. https://doi.org/10.1083/jcb.136.2.375.

Berkemeier, L. R., Winslow, J. W., Kaplan, D. R., Nikolics, K., Goeddel, D. V., & Rosenthal, A. (1991). Neurotrophin-5: A novel neurotrophic factor that activates trk and trkB. *Neuron*, *7*, 857–866. https://doi.org/10.1016/0896-6273(91)90287-A.

Bibel, M., & Barde, Y. A. (2000). Neurotrophins: Key regulators of cell fate and cell shape in the vertebrate nervous system. *Genes & Development*, *14*, 2919–2937. https://doi.org/10.1101/gad.841400.

Biddinger, J. E., & Fox, E. A. (2014). Reduced intestinal brain-derived neurotrophic factor increases vagal sensory innervation of the intestine and enhances satiation. *The Journal of Neuroscience*, *34*, 10379–10393. https://doi.org/10.1523/JNEUROSCI.1042-14.2014.

Borden, P., Houtz, J., Leach, S. D., & Kuruvilla, R. (2013). Sympathetic innervation during development is necessary for pancreatic islet architecture and functional maturation. *Cell Reports*, *4*, 287–301. https://doi.org/10.1016/j.celrep.2013.06.019.

Bower, D. V., Lee, H.-K., Lansford, R., Zinn, K., Warburton, D., Fraser, S. E., et al. (2014). Airway branching has conserved needs for local parasympathetic innervation but not neurotransmission. *BMC Biology*, *12*, 92. https://doi.org/10.1186/s12915-014-0092-2.

Bozyczko, D., & Horwitz, A. F. (1986). The participation of a putative cell surface receptor for laminin and fibronectin in peripheral neurite extension. *The Journal of Neuroscience*, *6*, 1241–1251. https://doi.org/10.1523/JNEUROSCI.06-05-01241.1986.

Bradshaw, R. A., Blundell, T. L., Lapatto, R., McDonald, N. Q., & Murray-Rust, J. (1993). Nerve growth factor revisited. *Trends in Biochemical Sciences*, *18*, 48–52. https://doi.org/10.1016/0968-0004(93)90052-O.

Branchfield, K., Nantie, L., Verheyden, J. M., Sui, P., Wienhold, M. D., & Sun, X. (2016). Pulmonary neuroendocrine cells function as airway sensors to control lung immune response. *Science*, *351*, 707–710. https://doi.org/10.1126/science.aad7969.

Brito, V., Giralt, A., Enriquez-Barreto, L., Puigdellívol, M., Suelves, N., Zamora-Moratalla, A., et al. (2014). Neurotrophin receptor p75(NTR) mediates Huntington's disease-associated synaptic and memory dysfunction. *The Journal of Clinical Investigation*, *124*, 4411–4428. https://doi.org/10.1172/JCI74809.

Brose, K., Bland, K. S., Wang, K. H., Arnott, D., Henzel, W., Goodman, C. S., et al. (1999). Slit proteins bind Robo receptors and have an evolutionarily conserved role in repulsive axon guidance. *Cell*, *96*, 795–806. https://doi.org/10.1016/S0092-8674(00)80590-5.

Brunet, I., Gordon, E., Han, J., Cristofaro, B., Broqueres-You, D., Liu, C., et al. (2014). Netrin-1 controls sympathetic arterial innervation. *The Journal of Clinical Investigation*, *124*, 3230–3240. https://doi.org/10.1172/JCI75181.

Burns, A. J., & Le Douarin, N. M. (1998). The sacral neural crest contributes neurons and glia to the post-umbilical gut: Spatiotemporal analysis of the development of the enteric nervous system. *Development*, *125*, 4335–4347. https://doi.org/10.1242/dev.125.21.4335.

Burri, P. H. (2006). Structural aspects of postnatal lung development—Alveolar formation and growth. *Neonatology*, *89*, 313–322. https://doi.org/10.1159/000092868.

Butler Tjaden, N. E., & Trainor, P. A. (2013). The developmental etiology and pathogenesis of Hirschsprung disease. *Translational Research*, *162*, 1–15. https://doi.org/10.1016/j.trsl.2013.03.001.

Calmont, A., Ivins, S., Van Bueren, K. L., Papangeli, I., Kyriakopoulou, V., Andrews, W. D., et al. (2009). Tbx1 controls cardiac neural crest cell migration during arch artery development by regulating Gbx2 expression in the pharyngeal ectoderm. *Development*, *136*, 3173–3183. https://doi.org/10.1242/dev.028902.

Calmont, A., Thapar, N., Scambler, P. J., & Burns, A. J. (2011). Absence of the vagus nerve in the stomach of Tbx1 −/− mutant mice. *Neurogastroenterology and Motility*, *23*, 125–130. https://doi.org/10.1111/j.1365-2982.2010.01615.x.

Cardoso, W. V., & Lü, J. (2006). Regulation of early lung morphogenesis: Questions, facts and controversies. *Development*, *133*, 1611–1624. https://doi.org/10.1242/dev.02310.

Carr, L., Parkinson, D. B., & Dun, X. (2017). Expression patterns of Slit and Robo family members in adult mouse spinal cord and peripheral nervous system. *PLoS One*, *12*, e0172736. https://doi.org/10.1371/journal.pone.0172736.

Castañeda-Corral, G., Jimenez-Andrade, J. M., Bloom, A. P., Taylor, R. N., Mantyh, W. G., Kaczmarska, M. J., et al. (2011). The majority of myelinated and unmyelinated sensory nerve fibers that innervate bone express the tropomyosin receptor kinase A. *Neuroscience*, *178*, 196–207. https://doi.org/10.1016/j.neuroscience.2011.01.039.

Chalazonitis, A. (1996). Neurotrophin-3 as an essential signal for the developing nervous system. *Molecular Neurobiology*, *12*, 39–53. https://doi.org/10.1007/BF02740746.

Chao, M. V. (2003). Neurotrophins and their receptors: A convergence point for many signalling pathways. *Nature Reviews Neuroscience*, *4*, 299–309. https://doi.org/10.1038/nrn1078.
Chao, M. V., & Hempstead, B. L. (1995). p75 and Trk: A two-receptor system. *Trends in Neurosciences*, *18*, 321–326. https://doi.org/10.1016/0166-2236(95)93922-K.
Chrestian, N., McMillan, H., Poulin, C., Campbell, C., & Vajsar, J. (2015). Hereditary neuropathy with liability to pressure palsies in childhood: Case series and literature update. *Neuromuscular Disorders*, *25*, 693–698. https://doi.org/10.1016/j.nmd.2015.06.463.
Connolly, J. L., Seeley, P. J., & Greene, L. A. (1985). Regulation of growth cone morphology by nerve growth factor: A comparative study by scanning electron microscopy. *Journal of Neuroscience Research*, *13*, 183–198. https://doi.org/10.1002/jnr.490130113.
Conover, J. C., Erickson, J. T., Katz, D. M., Bianchi, L. M., Poueymirou, W. T., McClain, J., et al. (1995). Neuronal deficits, not involving motor neurons, in mice lacking BDNF and/or NT4. *Nature*, *375*, 235–238. https://doi.org/10.1038/375235a0.
Costa, M., Brookes, S. H. J., & Zagorodnyuk, V. (2004). How many kinds of visceral afferents? *Gut*, *53*, ii1–ii4. https://doi.org/10.1136/gut.2003.033407.
Costantini, C., Scrable, H., & Puglielli, L. (2006). An aging pathway controls the TrkA to p75NTR receptor switch and amyloid β-peptide generation. *The EMBO Journal*, *25*, 1997–2006. https://doi.org/10.1038/sj.emboj.7601062.
Coughlin, M. D. (1975). Early development of parasympathetic nerves in the mouse submandibular gland. *Developmental Biology*, *43*, 123–139. https://doi.org/10.1016/0012-1606(75)90136-0.
Cristofaro, B., Stone, O. A., Caporali, A., Dawbarn, D., Ieronimakis, N., Reyes, M., et al. (2010). Neurotrophin-3 is a novel angiogenic factor capable of therapeutic neovascularization in a mouse model of limb ischemia. *Arteriosclerosis, Thrombosis, and Vascular Biology*, *30*, 1143–1150. https://doi.org/10.1161/ATVBAHA.109.205468.
Crowley, C., Spencer, S. D., Nishimura, M. C., Chen, K. S., Pitts-Meek, S., Armaninl, M. P., et al. (1994). Mice lacking nerve growth factor display perinatal loss of sensory and sympathetic neurons yet develop basal forebrain cholinergic neurons. *Cell*, *76*, 1001–1011. https://doi.org/10.1016/0092-8674(94)90378-6.
Damon, D. H., teRiele, J. A., & Marko, S. B. (2007). Vascular-derived artemin: A determinant of vascular sympathetic innervation? *American Journal of Physiology—Heart and Circulatory Physiology*, *293*, H266–H273. https://doi.org/10.1152/ajpheart.00859.2006.
Darling, T. K., & Lamb, T. J. (2019). Emerging roles for Eph receptors and ephrin ligands in immunity. *Frontiers in Immunology*, *10*, 1473. https://doi.org/10.3389/fimmu.2019.01473.
Davey, F., & Davies, A. M. (1998). TrkB signalling inhibits p75-mediated apoptosis induced by nerve growth factor in embryonic proprioceptive neurons. *Current Biology*, *8*, 915–918. https://doi.org/10.1016/S0960-9822(07)00371-5.
Davies, A. M. (1994). The role of neurotrophins in the developing nervous system. *Journal of Neurobiology*, *25*, 1334–1348. https://doi.org/10.1002/neu.480251103.
Davies, A., Minichiello, L., & Klein, R. (1995). Developmental changes in NT3 signalling via TrkA and TrkB in embryonic neurons. *The EMBO Journal*. https://doi.org/10.1002/j.1460-2075.1995.tb00127.x.
De Minicis, S., Rychlicki, C., Agostinelli, L., Saccomanno, S., Trozzi, L., Candelaresi, C., et al. (2013). Semaphorin 7A contributes to TGF-β-mediated liver fibrogenesis. *The American Journal of Pathology*, *183*, 820–830. https://doi.org/10.1016/j.ajpath.2013.05.030.
Deal, K. K., Chandrashekar, A. S., Beaman, M. M., Branch, M. C., Buehler, D. P., Conway, S. J., et al. (2021). Altered sacral neural crest development in Pax3 spina bifida mutants underlies deficits of bladder innervation and function. *Developmental Biology*, *476*, 173–188. https://doi.org/10.1016/j.ydbio.2021.03.024.

del Barrio, M. G., & Nieto, M. A. (2002). Overexpression of Snail family members highlights their ability to promote chick neural crest formation. *Development, 129*, 1583–1593. https://doi.org/10.1242/dev.129.7.1583.

Delalande, J.-M., Barlow, A. J., Thomas, A. J., Wallace, A. S., Thapar, N., Erickson, C. A., et al. (2008). The receptor tyrosine kinase RET regulates hindgut colonization by sacral neural crest cells. *Developmental Biology, 313*, 279–292. https://doi.org/10.1016/j.ydbio.2007.10.028.

DiBona, G. F. (2000). Neural control of the kidney: Functionally specific renal sympathetic nerve fibers. *American Journal of Physiology—Regulatory, Integrative and Comparative Physiology, 279*, R1517–R1524. https://doi.org/10.1152/ajpregu.2000.279.5.R1517.

Dietrich, P., & Dragatsis, I. (2016). Familial dysautonomia: Mechanisms and models. *Genetics and Molecular Biology, 39*, 497–514. https://doi.org/10.1590/1678-4685-GMB-2015-0335.

Donovan, M. J., Hahn, R., Tessarollo, L., & Hempstead, B. L. (1996). Identification of an essential nonneuronal function of neurotrophin 3 in mammalian cardiac development. *Nature Genetics, 14*, 210–213. https://doi.org/10.1038/ng1096-210.

Dreyfus, C. F. (1989). Effects of nerve growth factor on cholinergic brain neurons. *Trends in Pharmacological Sciences, 10*, 145–149. https://doi.org/10.1016/0165-6147(89)90166-1.

Duan, H., Cai, X., Luan, Y., Yang, S., Yang, J., Dong, H., & Shao, L. (2021). Regulation of the autonomic nervous system on intestine. *Frontiers in Physiology, 12*, 700129. https://doi.org/10.3389/fphys.2021.700129.

Dyachuk, V., Furlan, A., Khatibi Shahidi, M., Giovenco, M., Kaukua, N., Konstantinidou, C., et al. (2014). Parasympathetic neurons originate from nerve-associated peripheral glial progenitors. *Science, 345*, 82–87. https://doi.org/10.1126/science.1253281.

Emanuele, E., Politi, P., Bianchi, M., Minoretti, P., Bertona, M., & Geroldi, D. (2006). Raised plasma nerve growth factor levels associated with early-stage romantic love. *Psychoneuroendocrinology, 31*, 288–294. https://doi.org/10.1016/j.psyneuen.2005.09.002.

Ernfors, P., Lee, K.-F., & Jaenisch, R. (1994). Mice lacking brain-derived neurotrophic factor develop with sensory deficits. *Nature, 368*, 147–150. https://doi.org/10.1038/368147a0.

Espinosa-Medina, I., Outin, E., Picard, C. A., Chettouh, Z., Dymecki, S., Consalez, G. G., et al. (2014). Parasympathetic ganglia derive from Schwann cell precursors. *Science, 345*, 87–90. https://doi.org/10.1126/science.1253286.

Fagan, A. M., Zhang, H., Landis, S., Smeyne, R. J., Silos-Santiago, I., & Barbacid, M. (1996). TrkA, but not TrkC, receptors are essential for survival of sympathetic neurons in vivo. *The Journal of Neuroscience, 16*, 6208–6218. https://doi.org/10.1523/JNEUROSCI.16-19-06208.1996.

Fasanella, K. E., Christianson, J. A., Chanthaphavong, R. S., & Davis, B. M. (2008). Distribution and neurochemical identification of pancreatic afferents in the mouse. *Journal of Comparative Neurology, 509*, 42–52. https://doi.org/10.1002/cne.21736.

Fazeli, A., Dickinson, S. L., Hermiston, M. L., Tighe, R. V., Steen, R. G., Small, C. G., et al. (1997). Phenotype of mice lacking functional Deleted in colorectal cancer (Dcc) gene. *Nature, 386*, 796–804. https://doi.org/10.1038/386796a0.

Fox, E. A., Biddinger, J. E., Baquet, Z. C., Jones, K. R., & McAdams, J. (2013). Loss of neurotrophin-3 from smooth muscle disrupts vagal gastrointestinal afferent signaling and satiation. *American Journal of Physiology—Regulatory, Integrative and Comparative Physiology, 305*, R1307–R1322. https://doi.org/10.1152/ajpregu.00337.2013.

Francis, N., Farinas, I., Brennan, C., Rivas-Plata, K., Backus, C., Reichardt, L., et al. (1999). NT-3, like NGF, Is Required for Survival of Sympathetic Neurons, but Not Their Precursors. *Developmental Biology, 210*, 411–427. https://doi.org/10.1006/dbio.1999.9269.

Freeman, R. S., Burch, R. L., Crowder, R. J., Lomb, D. J., Schoell, M. C., Straub, J. A., et al. (2004). NGF deprivation-induced gene expression: After ten years, where do we stand? In *Progress in brain research, NGF and related molecules in health and disease* (pp. 111–126). Elsevier. https://doi.org/10.1016/S0079-6123(03)46008-1.

Geesman, G., Gesiotto, Q. J., Lalani, Z., & Tejani, N. (2021). Anatomy of the gastrointestinal system. In D. Narayan, S. E. Kapadia, G. Kodumudi, & N. Vadivelu (Eds.), *Surgical and perioperative management of patients with anatomic anomalies* (pp. 145–177). Cham: Springer International Publishing. https://doi.org/10.1007/978-3-030-55660-0_8.

Genç, B., Özdinler, P. H., Mendoza, A. E., & Erzurumlu, R. S. (2004). A chemoattractant role for NT-3 in proprioceptive axon guidance. *PLoS Biology*, *2*, e403. https://doi.org/10.1371/journal.pbio.0020403.

George, L., Chaverra, M., Todd, V., Lansford, R., & Lefcort, F. (2007). Nociceptive sensory neurons derive from contralaterally migrating, fate-restricted neural crest cells. *Nature Neuroscience*, *10*, 1287–1293. https://doi.org/10.1038/nn1962.

Gershon, M. D. (1999). The enteric nervous system: A second brain. *Hospital Practice*, *34*, 31–52. https://doi.org/10.3810/hp.1999.07.153.

Giacobini, P. (2015). Shaping the reproductive system: Role of semaphorins in gonadotropin-releasing hormone development and function. *NEN*, *102*, 200–215. https://doi.org/10.1159/000431021.

Glebova, N. O., & Ginty, D. D. (2004). Heterogeneous requirement of NGF for sympathetic target innervation *in vivo*. *The Journal of Neuroscience*, *24*, 743. https://doi.org/10.1523/JNEUROSCI.4523-03.2004.

Glebova, N. O., & Ginty, D. D. (2005). Growth and survival signals controlling sympathetic nervous system development. *Annual Review of Neuroscience*, *28*, 191–222. https://doi.org/10.1146/annurev.neuro.28.061604.135659.

Goldberg, D., Borojevic, R., Anderson, M., Chen, J. J., Gershon, M. D., & Ratcliffe, E. M. (2013). Slit/Robo-mediated chemorepulsion of vagal sensory axons in the fetal gut. *Developmental Dynamics*, *242*, 9–15. https://doi.org/10.1002/dvdy.23898.

Gundersen, R. W., & Barrett, J. N. (1979). Neuronal chemotaxis: Chick dorsal-root axons turn toward high concentrations of nerve growth factor. *Science*, *206*, 1079–1080. https://doi.org/10.1126/science.493992.

Hapner, S. J., Boeshore, K. L., Large, T. H., & Lefcort, F. (1998). Neural differentiation promoted by truncated trk C receptors in collaboration with p75NTR. *Developmental Biology*, *201*, 90–100. https://doi.org/10.1006/dbio.1998.8970.

Hasan, W. (2013). Autonomic cardiac innervation. *Organogenesis*, *9*, 176–193. https://doi.org/10.4161/org.24892.

Hatch, J., & Mukouyama, Y. (2015). Spatiotemporal mapping of vascularization and innervation in the fetal murine intestine. *Developmental Dynamics*, *244*, 56–68. https://doi.org/10.1002/dvdy.24178.

Hedgecock, E. M., Culotti, J. G., & Hall, D. H. (1990). The unc-5, unc-6, and unc-40 genes guide circumferential migrations of pioneer axons and mesodermal cells on the epidermis in C. elegans. *Neuron*, *4*, 61–85. https://doi.org/10.1016/0896-6273(90)90444-K.

Herring, J. A. (2013). *Tachdjian's pediatric orthopaedics: From the Texas Scottish Rite Hospital for Children* (pp. 285–319). Elsevier Health Sciences. https://www.elsevier.com/books/tachdjians-pediatric-orthopaedics-from-the-texas-scottish-rite-hospital-for-children/herring/978-1-4377-1549-1 (accessed 8.16.21).

Heumann, R., Korsching, S., Scott, J., & Thoenen, H. (1984). Relationship between levels of nerve growth factor (NGF) and its messenger RNA in sympathetic ganglia and peripheral target tissues. *The EMBO Journal*, *3*, 3183–3189. https://doi.org/10.1002/j.1460-2075.1984.tb02277.x.

Hildreth, V., Anderson, R. H., & Henderson, D. J. (2009). Autonomic innervation of the developing heart: Origins and function. *Clinical Anatomy*, *22*, 36–46. https://doi.org/10.1002/ca.20695.

Hiltunen, J. O., Laurikainen, A., Airaksinen, M. S., & Saarma, M. (2000). GDNF family receptors in the embryonic and postnatal rat heart and reduced cholinergic innervation in mice hearts lacking Ret or GFRα2. *Developmental Dynamics*, *219*, 28–39. https://doi.org/10.1002/1097-0177(2000)9999:9999<::AID-DVDY1031>3.0.CO;2-P.

Hjerling-Leffler, J., Marmigère, F., Heglind, M., Cederberg, A., Koltzenburg, M., Enerbäck, S., et al. (2005). The boundary cap: A source of neural crest stem cells that generate multiple sensory neuron subtypes. *Development*, *132*, 2623–2632. https://doi.org/10.1242/dev.01852.

Honma, Y., Araki, T., Gianino, S., Bruce, A., Heuckeroth, R. O., Johnson, E. M., et al. (2002). Artemin is a vascular-derived neurotropic factor for developing sympathetic neurons. *Neuron*, *35*, 267–282. https://doi.org/10.1016/S0896-6273(02)00774-2.

Höpker, V. H., Shewan, D., Tessier-Lavigne, M., Poo, M., & Holt, C. (1999). Growth-cone attraction to netrin-1 is converted to repulsion by laminin-1. *Nature*, *401*, 69–73. https://doi.org/10.1038/43441.

Horstadius, S. O. (1950). *The neural crest: Its properties and derivatives in the light of experimental research*. London: Oxford University Press. https://doi.org/10.1001/archneurpsyc.1951.02320060135017.

Huang, T., & Krimm, R. F. (2010). Developmental expression of Bdnf, Ntf4/5, and TrkB in the mouse peripheral taste system. *Developmental Dynamics*, *239*, 2637–2646. https://doi.org/10.1002/dvdy.22412.

Huang, E. J., & Reichardt, L. F. (2001). Neurotrophins: Roles in neuronal development and function. *Annual Review of Neuroscience*, *24*, 677–736. https://doi.org/10.1146/annurev.neuro.24.1.677.

Huang, E. J., & Reichardt, L. F. (2003). Trk receptors: Roles in neuronal signal transduction. *Annual Review of Biochemistry*, *72*, 609–642. https://doi.org/10.1146/annurev.biochem.72.121801.161629.

Huang, X., & Saint-Jeannet, J.-P. (2004). Induction of the neural crest and the opportunities of life on the edge. *Developmental Biology*, *275*, 1–11. https://doi.org/10.1016/j.ydbio.2004.07.033.

Huber, K. (2006). The sympathoadrenal cell lineage: Specification, diversification, and new perspectives. *Developmental Biology*, *298*, 335–343. https://doi.org/10.1016/j.ydbio.2006.07.010.

Hynes, P. J., & Fraher, J. P. (2004). The development of the male genitourinary system. I. The origin of the urorectal septum and the formation of the perineum. *British Journal of Plastic Surgery*, *57*, 27–36. https://doi.org/10.1016/j.bjps.2003.08.019.

Ibáñez, C. F. (1995). Neurotrophic factors: From structure-function studies to designing effective therapeutics. *Trends in Biotechnology*, *13*, 217–227. https://doi.org/10.1016/S0167-7799(00)88949-0.

Ibáñez, C. F. (1996). Neurotrophin-4: The odd one out in the neurotrophin family. *Neurochemical Research*, *21*, 787–793. https://doi.org/10.1007/BF02532301.

Ibáñez, C. F., Ebendal, T., & Persson, H. (1991). Chimeric molecules with multiple neurotrophic activities reveal structural elements determining the specificities of NGF and BDNF. *The EMBO Journal*, *10*, 2105–2110. https://doi.org/10.1002/j.1460-2075.1991.tb07743.x.

Ieda, M., Kanazawa, H., Kimura, K., Hattori, F., Ieda, Y., Taniguchi, M., & Fukuda, K. (2007). Sema3a maintains normal heart rhythm through sympathetic innervation patterning. *Nature Medicine*, *13*, 604–612. https://doi.org/10.1038/nm1570.

Indo, Y. (1993). NTRK1 congenital insensitivity to pain with anhidrosis. In M. P. Adam, H. H. Ardinger, R. A. Pagon, S. E. Wallace, L. J. Bean, G. Mirzaa, & A. Amemiya (Eds.), *GeneReviews®*. Seattle, WA: University of Washington, Seattle. https://www.ncbi.nlm.nih.gov/books/NBK1769/.

Indo, Y., Tsuruta, M., Hayashida, Y., Karim, M. A., Ohta, K., Kawano, T., et al. (1996). Mutations in the TRKA/NGF receptor gene in patients with congenital insensitivity to pain with anhidrosis. *Nature Genetics*, *13*, 485–488. https://doi.org/10.1038/ng0896-485.

Ip, N. Y., Ibáñez, C. F., Nye, S. H., McClain, J., Jones, P. F., Gies, D. R., et al. (1992). Mammalian neurotrophin-4: Structure, chromosomal localization, tissue distribution, and receptor specificity. *Proceedings of the National Academy of Sciences of the United States of America*, *89*, 3060–3064. https://doi.org/10.1073/pnas.89.7.3060.

Ip, N. Y., Stitt, T. N., Tapley, P., Klein, R., Glass, D. J., Fandl, J., et al. (1993). Similarities and differences in the way neurotrophins interact with the Trk receptors in neuronal and nonneuronal cells. *Neuron*, *10*, 137–149. https://doi.org/10.1016/0896-6273(93)90306-C.

Ishii, N., Wadsworth, W. G., Stern, B. D., Culotti, J. G., & Hedgecock, E. M. (1992). UNC-6, a laminin-related protein, guides cell and pioneer axon migrations in C. elegans. *Neuron*, *9*, 873–881. https://doi.org/10.1016/0896-6273(92)90240-E.

Jacobson, A. G. (1991). Experimental analyses of the shaping of the neural plate and Tube1. *American Zoologist*, *31*, 628–643. https://doi.org/10.1093/icb/31.4.628.

Jiang, X., Rowitch, D. H., Soriano, P., McMahon, A. P., & Sucov, H. M. (2000). Fate of the mammalian cardiac neural crest. *Development*, *127*, 1607–1616. https://doi.org/10.1242/dev.127.8.1607.

Jones, K. R., Fariñas, I., Backus, C., & Reichardt, L. F. (1994). Targeted disruption of the BDNF gene perturbs brain and sensory neuron development but not motor neuron development. *Cell*, *76*, 989–999. https://doi.org/10.1016/0092-8674(94)90377-8.

Jørgensen, M. C., Ahnfelt-Rønne, J., Hald, J., Madsen, O. D., Serup, P., & Hecksher-Sørensen, J. (2007). An illustrated review of early pancreas development in the mouse. *Endocrine Reviews*, *28*, 685–705. https://doi.org/10.1210/er.2007-0016.

Kagoshima, M., & Ito, T. (2001). Diverse gene expression and function of semaphorins in developing lung: Positive and negative regulatory roles of semaphorins in lung branching morphogenesis. *Genes to Cells*, *6*, 559–571. https://doi.org/10.1046/j.1365-2443.2001.00441.x.

Kang, Y.-N., Fung, C., & Vanden Berghe, P. (2021). Gut innervation and enteric nervous system development: A spatial, temporal and molecular tour de force. *Development*, *148*. https://doi.org/10.1242/dev.182543.

Kania, A., & Klein, R. (2016). Mechanisms of ephrin–Eph signalling in development, physiology and disease. *Nature Reviews Molecular Cell Biology*, *17*, 240–256. https://doi.org/10.1038/nrm.2015.16.

Kaplan, D. R., & Miller, F. D. (2000). Neurotrophin signal transduction in the nervous system. *Current Opinion in Neurobiology*, *10*, 381–391. https://doi.org/10.1016/S0959-4388(00)00092-1.

Kapur, R. P., Yost, C., & Palmiter, R. D. (1992). A transgenic model for studying development of the enteric nervous system in normal and aganglionic mice. *Development*, *116*, 167–175. https://doi.org/10.1242/dev.116.1.167.

Kawaja, M. D., Smithson, L. J., Elliott, J., Trinh, G., Crotty, A.-M., Michalski, B., et al. (2011). Nerve growth factor promoter activity revealed in mice expressing enhanced green fluorescent protein. *Journal of Comparative Neurology*, *519*, 2522–2545. https://doi.org/10.1002/cne.22629.

Kimura, K., Ieda, M., & Fukuda, K. (2012). Development, maturation, and transdifferentiation of cardiac sympathetic nerves. *Circulation Research*, *110*, 325–336. https://doi.org/10.1161/CIRCRESAHA.111.257253.

Kirby, M. L., & Hutson, M. R. (2010). Factors controlling cardiac neural crest cell migration. *Cell Adhesion & Migration*, *4*, 609–621. https://doi.org/10.4161/cam.4.4.13489.

Klein, R., Nanduri, V., Jing, S., Lamballe, F., Tapley, P., Bryant, S., et al. (1991). The trkB tyrosine protein kinase is a receptor for brain-derived neurotrophic factor and neurotrophin-3. *Cell*, *66*, 395–403. https://doi.org/10.1016/0092-8674(91)90628-C.

Klein, R., Smeyne, R. J., Wurst, W., Long, L. K., Auerbach, B. A., Joyner, A. L., et al. (1993). Targeted disruption of the trkB neurotrophin receptor gene results in nervous system lesions and neonatal death. *Cell*, *75*, 113–122. https://doi.org/10.1016/S0092-8674(05)80088-1.

Klein, R., Silos-Santiago, I., Smeyne, R. J., Lira, S. A., Brambilla, R., Bryant, S., et al. (1994). Disruption of the neurotrophin-3 receptor gene trkC eliminates la muscle afferents and results in abnormal movements. *Nature*, *368*, 249–251. https://doi.org/10.1038/368249a0.

Klein, R., Lamballe, F., Bryant, S., & Barbacid, M. (1992). The trkB tyrosine protein kinase is a receptor for neurotrophin-4. *Neuron*, *8*, 947–956. https://doi.org/10.1016/0896-6273(92)90209-V.

Knox, S. M., Lombaert, I. M. A., Reed, X., Vitale-Cross, L., Gutkind, J. S., & Hoffman, M. P. (2010). Parasympathetic innervation maintains epithelial progenitor cells during salivary organogenesis. *Science*, *329*, 1645–1647. https://doi.org/10.1126/science.1192046.

Kolodkin, A. L., Matthes, D. J., & Goodman, C. S. (1993). The semaphorin genes encode a family of transmembrane and secreted growth cone guidance molecules. *Cell*, *75*, 1389–1399. https://doi.org/10.1016/0092-8674(93)90625-Z.

Korsching, S., & Thoenen, H. (1983). Nerve growth factor in sympathetic ganglia and corresponding target organs of the rat: Correlation with density of sympathetic innervation. *Proceedings of the National Academy of Sciences of the United States of America*, *80*, 3513–3516. https://doi.org/10.1073/pnas.80.11.3513.

Korsching, S., & Thoenen, H. (1988). Developmental changes of nerve growth factor levels in sympathetic ganglia and their target organs. *Developmental Biology*, *126*, 40–46. https://doi.org/10.1016/0012-1606(88)90236-9.

Kramer, I., Sigrist, M., de Nooij, J. C., Taniuchi, I., Jessell, T. M., & Arber, S. (2006). A role for Runx transcription factor signaling in dorsal root ganglion sensory neuron diversification. *Neuron*, *49*, 379–393. https://doi.org/10.1016/j.neuron.2006.01.008.

Kreipke, R. E., & Birren, S. J. (2015). Innervating sympathetic neurons regulate heart size and the timing of cardiomyocyte cell cycle withdrawal. *The Journal of Physiology*, *593*, 5057–5073. https://doi.org/10.1113/JP270917.

Kummer, W. (2011). Pulmonary vascular innervation and its role in responses to hypoxia. *Proceedings of the American Thoracic Society*, *8*, 471–476. https://doi.org/10.1513/pats.201101-013MW.

Kuruvilla, R., Zweifel, L. S., Glebova, N. O., Lonze, B. E., Valdez, G., Ye, H., et al. (2004). A neurotrophin signaling cascade coordinates sympathetic neuron development through differential control of TrkA trafficking and retrograde signaling. *Cell*, *118*, 243–255. https://doi.org/10.1016/j.cell.2004.06.021.

LaBonne, C., & Bronner-Fraser, M. (1999). Molecular mechanisms of neural crest formation. *Annual Review of Cell and Developmental Biology*, *15*, 81–112. https://doi.org/10.1146/annurev.cellbio.15.1.81.

Lambiase, A., Centofanti, M., Micera, A., Manni, G. L., Mattei, E., De Gregorio, A., et al. (1997). Nerve growth factor (NGF) reduces and NGF antibody exacerbates retinal damage induced in rabbit by experimental ocular hypertension. *Graefe's Archive for Clinical and Experimental Ophthalmology*, *235*, 780–785. https://doi.org/10.1007/BF02332863.

Langsdorf, A., Radzikinas, K., Kroten, A., Jain, S., & Ai, X. (2011). Neural crest cell origin and signals for intrinsic neurogenesis in the mammalian respiratory tract. *American Journal of Respiratory Cell and Molecular Biology*, *44*, 293–301. https://doi.org/10.1165/rcmb.2009-0462OC. 20139349.

Le Douarin, N., & Kalcheim, C. (1999). The neural crest. *Developmental and cell biology series* (2nd ed.). Cambridge University Press, Cambridge. https://doi.org/10.1017/CBO9780511897948.

Lee, H.-Y., Kléber, M., Hari, L., Brault, V., Suter, U., Taketo, M. M., et al. (2004). Instructive role of Wnt/ß-catenin in sensory fate specification in neural crest stem cells. *Science, 303*, 1020–1023. https://doi.org/10.1126/science.1091611.

Lee, R., Kermani, P., Teng, K. K., & Hempstead, B. L. (2001). Regulation of cell survival by secreted proneurotrophins. *Science, 294*, 1945–1948. https://doi.org/10.1126/science.1065057.

Lee, F. S., Kim, A. H., Khursigara, G., & Chao, M. V. (2001). The uniqueness of being a neurotrophin receptor. *Current Opinion in Neurobiology, 11*, 281–286. https://doi.org/10.1016/S0959-4388(00)00209-9.

Letourneau, P. C. (1978). Chemotactic response of nerve fiber elongation to nerve growth factor. *Developmental Biology, 66*, 183–196. https://doi.org/10.1016/0012-1606(78)90283-X.

Levi-Montalcini, R., & Booker, B. (1960). Destruction of the sympathetic ganglia in mammals by an antiserum to a nerve-growth protein. *Proceedings of the National Academy of Sciences. United States of America, 46*, 384. https://doi.org/10.1073/pnas.46.3.384.

Levi-Montalcini, R., & Cohen, S. (1956). In vitro and in vivo effects of a nerve growth-stimulating agent isolated from snake venom. *Proceedings of the National Academy of Sciences, 42*, 695–699. https://doi.org/10.1073/pnas.42.9.695.

Li, Z., Hao, J., Duan, X., Wu, N., Zhou, Z., Yang, F., et al. (2017). The role of semaphorin 3A in bone remodeling. *Frontiers in Cellular Neuroscience, 11*, 40. https://doi.org/10.3389/fncel.2017.00040.

Liaw, A., Cunha, G. R., Shen, J., Cao, M., Liu, G., Sinclair, A., et al. (2018). Development of the human bladder and ureterovesical junction. *Differentiation, 103*, 66–73. https://doi.org/10.1016/j.diff.2018.08.004.

Little, M.H. (2011). Renal organogenesis, Organogenesis, 7:4, 229-241, DOI: 10.4161/org.7.4.18057.

Liu, H., Zhang, C.-H., Ammanamanchi, N., Suresh, S., Lewarchik, C., Rao, K., et al. (2019). Control of cytokinesis by β-adrenergic receptors indicates an approach for regulating cardiomyocyte endowment. *Science Translational Medicine, 11*, eaaw6419. https://doi.org/10.1126/scitranslmed.aaw6419.

Liu, T., Yang, L., Han, X., Ding, X., Li, J., & Yang, J. (2020). Local sympathetic innervations modulate the lung innate immune responses. *Science Advances, 6*, eaay1497. https://doi.org/10.1126/sciadv.aay1497.

Liu, X., Ernfors, P., Wu, H., & Jaenisch, R. (1995). Sensory but not motor neuron deficits in mice lacking NT4 and BDNF. *Nature, 375*, 238–241. https://doi.org/10.1038/375238a0.

Loring, J. F., & Erickson, C. A. (1987). Neural crest cell migratory pathways in the trunk of the chick embryo. *Developmental Biology, 121*, 220–236. https://doi.org/10.1016/0012-1606(87)90154-0.

Love, J. A., Yi, E., & Smith, T. G. (2007). Autonomic pathways regulating pancreatic exocrine secretion. *Autonomic Neuroscience: Basic & Clinical, 133*, 19–34. https://doi.org/10.1016/j.autneu.2006.10.001.

Ma, Q., Fode, C., Guillemot, F., & Anderson, D. J. (1999). Neurogenin1 and neurogenin2 control two distinct waves of neurogenesis in developing dorsal root ganglia. *Genes & Development, 13*, 1717–1728. https://doi.org/10.1101/gad.13.13.1717.

Mabe, A. M., & Hoover, D. B. (2009). Structural and functional cardiac cholinergic deficits in adult neurturin knockout mice. *Cardiovascular Research, 82*, 93–99. https://doi.org/10.1093/cvr/cvp029.

Maisonpierre, P. C., Belluscio, L., Squinto, S., Ip, N. Y., Furth, M. E., Lindsay, R. M., et al. (1990). Neurotrophin-3: A neurotrophic factor related to NGF and BDNF. *Science, 247*, 1446–1451. https://doi.org/10.1126/science.2321006.

Martin-Zanca, D., Hughes, S. H., & Barbacid, M. (1986). A human oncogene formed by the fusion of truncated tropomyosin and protein tyrosine kinase sequences. *Nature, 319*, 743–748. https://doi.org/10.1038/319743a0.

McKey, J., Bunce, C., Batchvarov, I. S., Ornitz, D. M., & Capel, B. (2019). Neural crest-derived neurons invade the ovary but not the testis during mouse gonad development. *Proceedings of the National Academy of Sciences of the United States of America, 116*, 5570–5575. https://doi.org/10.1073/pnas.1814930116.

McMahon, A. P. (2016). Development of the mammalian kidney. *117. Current Topics in Developmental Biology* (pp. 31–64). Elsevier Inc.

Meulemans, D., & Bronner-Fraser, M. (2004). Gene-regulatory interactions in neural crest evolution and development. *Developmental Cell*, 7, 291–299. https://doi.org/10.1016/j.devcel.2004.08.007.

Miwa, K., Lee, J.-K., Takagishi, Y., Opthof, T., Fu, X., & Kodama, I. (2010). Glial cell line-derived neurotrophic factor (GDNF) enhances sympathetic neurite growth in rat hearts at early developmental stages. *Biomedical Research, 31*, 353–361. https://doi.org/10.2220/biomedres.31.353.

Murphy, M. C., & Fox, E. A. (2010). Mice deficient in brain-derived neurotrophic factor have altered development of gastric vagal sensory innervation. *Journal of Comparative Neurology, 518*, 2934–2951. https://doi.org/10.1002/cne.22372.

Nagy, N., & Goldstein, A. M. (2017). Enteric nervous system development: A crest cell's journey from neural tube to colon. *Seminars in Cell & Developmental Biology, 66*, 94–106. https://doi.org/10.1016/j.semcdb.2017.01.006.

Nam, J., Onitsuka, I., Hatch, J., Uchida, Y., Ray, S., Huang, S., & Mukouyama, Y. (2013). Coronary veins determine the pattern of sympathetic innervation in the developing heart. *Development, 140*, 1475–1485. https://doi.org/10.1242/dev.087601.

Nedvetsky, P. I., Emmerson, E., Finley, J. K., Ettinger, A., Cruz-Pacheco, N., Prochazka, J., et al. (2014). Parasympathetic innervation regulates tubulogenesis in the developing salivary gland. *Developmental Cell, 30*, 449–462. https://doi.org/10.1016/j.devcel.2014.06.012.

Niederreither, K., Vermot, J., Roux, I. L., Schuhbaur, B., Chambon, P., & Dollé, P. (2003). The regional pattern of retinoic acid synthesis by RALDH2 is essential for the development of posterior pharyngeal arches and the enteric nervous system. *Development, 130*, 2525–2534. https://doi.org/10.1242/dev.00463.

Nikoletopoulou, V., Lickert, H., Frade, J. M., Rencurel, C., Giallonardo, P., Zhang, L., et al. (2010). Neurotrophin receptors TrkA and TrkC cause neuronal death whereas TrkB does not. *Nature, 467*, 59–63. https://doi.org/10.1038/nature09336.

Nishiyama, C., Uesaka, T., Manabe, T., Yonekura, Y., Nagasawa, T., Newgreen, D. F., et al. (2012). Trans-mesenteric neural crest cells are the principal source of the colonic enteric nervous system. *Nature Neuroscience, 15*, 1211–1218. https://doi.org/10.1038/nn.3184.

Niu, X., Liu, L., Wang, T., Chuan, X., Yu, Q., Du, M., et al. (2020). Mapping of extrinsic innervation of the gastrointestinal tract in the mouse embryo. *The Journal of Neuroscience*. https://doi.org/10.1523/JNEUROSCI.0309-20.2020. JN-RM-0309-20.

Olsen, B. R., Reginato, A. M., & Wang, W. (2000). Bone development. *Annual Review of Cell and Developmental Biology, 16*, 191–220. https://doi.org/10.1146/annurev.cellbio.16.1.191.

Oppenheim, R. W., Prevette, D., Qin-Wei, Y., Collins, F., & Mac Donald, J. (1991). Control of embryonic motoneuron survival in vivo by ciliary neurotrophic factor. *Science, 251*, 1616–1618. https://doi.org/10.1126/science.2011743.

O'Donnell, M., Chance, R. K., & Bashaw, G. J. (2009). Axon growth and guidance: Receptor regulation and signal transduction. *Annual Review of Neuroscience, 32,* 383–412. https://doi.org/10.1146/annurev.neuro.051508.135614.

Osborn, J. W., Tyshynsky, R., & Vulchanova, L. (2021). Function of renal nerves in kidney physiology and pathophysiology. *Annual Review of Physiology, 83,* 429–450. https://doi.org/10.1146/annurev-physiol-031620-091656.

Pasquale, E. B. (2005). Eph receptor signalling casts a wide net on cell behaviour. *Nature Reviews. Molecular Cell Biology, 6,* 462–475. https://doi.org/10.1038/nrm1662.

Patapoutian, A., & Reichardt, L. F. (2001). Trk receptors: Mediators of neurotrophin action. *Current Opinion in Neurobiology, 11,* 272–280. https://doi.org/10.1016/S0959-4388(00)00208-7.

Patel, T. D., Jackman, A., Rice, F. L., Kucera, J., & Snider, W. D. (2000). Development of sensory neurons in the absence of NGF/TrkA signaling in vivo. *Neuron, 25,* 345–357. https://doi.org/10.1016/S0896-6273(00)80899-5.

Pavan, W. J., & Raible, D. W. (2012). Specification of neural crest into sensory neuron and melanocyte lineages. *Developmental Biology, 366,* 55–63. https://doi.org/10.1016/j.ydbio.2012.02.038.

Pierucci, D., Cicconi, S., Bonini, P., Ferrelli, F., Pastore, D., Matteucci, C., et al. (2001). NGF-withdrawal induces apoptosis in pancreatic beta cells in vitro. *Diabetologia, 44,* 1281–1295. https://doi.org/10.1007/s001250100650.

Piña-Garza, J. E. (2013). Chapter 9—Sensory and autonomic disturbances. In J. E. Piña-Garza (Ed.), *Fenichel's clinical pediatric neurology* (7th ed., pp. 207–214). London: W.B. Saunders. https://doi.org/10.1016/B978-1-4557-2376-8.00009-8.

Piper, M., Georgas, K., Yamada, T., & Little, M. (2000). Expression of the vertebrate Slit Gene family and their putative receptors, the Robo genes, in the developing murine kidney. *Mechanisms of Development, 94,* 213–217. https://doi.org/10.1016/S0925-4773(00)00313-0.

Pradhan, J., Noakes, P. G., & Bellingham, M. C. (2019). The role of altered BDNF/TrkB signaling in amyotrophic lateral sclerosis. *Frontiers in Cellular Neuroscience.* https://doi.org/10.3389/fncel.2019.00368.

Proenca, C. C., Song, M., & Lee, F. S. (2016). Differential effects of BDNF and neurotrophin 4 (NT4) on endocytic sorting of TrkB receptors. *Journal of Neurochemistry, 138,* 397–406. https://doi.org/10.1111/jnc.13676.

Radzikinas, K., Aven, L., Jiang, Z., Tran, T., Paez-Cortez, J., Boppidi, K., et al. (2011). A Shh/miR-206/BDNF cascade coordinates innervation and formation of airway smooth muscle. *The Journal of Neuroscience, 31,* 15407–15415. https://doi.org/10.1523/JNEUROSCI.2745-11.2011.

Ramon y Cajal, S. (1892). La Cellule. *La retine des vertebres, 9,* 121–247.

Rao, M., & Gershon, M. D. (2018). Enteric nervous system development: What could possibly go wrong? *Nature Reviews Neuroscience, 19,* 552–565. https://doi.org/10.1038/s41583-018-0041-0.

Raper, J., & Mason, C. (2010). Cellular strategies of axonal pathfinding. *Cold Spring Harbor Perspectives in Biology, 2,* a001933. https://doi.org/10.1101/cshperspect.a001933.

Ratcliffe, E. M., Setru, S. U., Chen, J. J., Li, Z. S., D'Autréaux, F., & Gershon, M. D. (2006). Netrin/DCC-mediated attraction of vagal sensory axons to the fetal mouse gut. *Journal of Comparative Neurology, 498,* 567–580. https://doi.org/10.1002/cne.21027.

Ratcliffe, E. M., Fan, L., Mohammed, T. J., Anderson, M., Chalazonitis, A., & Gershon, M. D. (2011). Enteric neurons synthesize netrins and are essential for the development of the vagal sensory innervation of the fetal gut. *Developmental Neurobiology, 71,* 362–373. https://doi.org/10.1002/dneu.20869.

Ratto, M. H., Leduc, Y. A., Valderrama, X. P., van Straaten, K. E., Delbaere, L. T. J., Pierson, R. A., et al. (2012). The nerve of ovulation-inducing factor in semen.

Proceedings of the National Academy of Sciences of the United States of America, 109, 15042–15047. https://doi.org/10.1073/pnas.1206273109.

Rauskolb, S., Zagrebelsky, M., Dreznjak, A., Deogracias, R., Matsumoto, T., Wiese, S., et al. (2010). Global deprivation of brain-derived neurotrophic factor in the CNS reveals an area-specific requirement for dendritic growth. *The Journal of Neuroscience, 30*, 1739–1749. https://doi.org/10.1523/JNEUROSCI.5100-09.2010.

Rehman, S., & Ahmed, D. (2021). Embryology, kidney, bladder, and ureter (Updated 2021 Aug 11). In *StatPearls [Internet]*. Treasure Island (FL): StatPearls Publishing. https://www.ncbi.nlm.nih.gov/books/NBK547747/.

Reichardt, L. F. (2006). Neurotrophin-regulated signalling pathways. *Philosophical Transactions of the Royal Society, B: Biological Sciences, 361*, 1545–1564. https://doi.org/10.1098/rstb.2006.1894.

Reissmann, E., Ernsberger, U., Francis-West, P. H., Rueger, D., Brickell, P. M., & Rohrer, H. (1996). Involvement of bone morphogenetic protein-4 and bone morphogenetic protein-7 in the differentiation of the adrenergic phenotype in developing sympathetic neurons. *Development, 122*, 2079–2088. https://doi.org/10.1242/dev.122.7.2079.

Ritter, K. E., & Southard-Smith, E. M. (2017). Dynamic expression of serotonin receptor 5-HT3A in developing sensory innervation of the lower urinary tract. *Frontiers in Neuroscience, 10*, 592. https://doi.org/10.3389/fnins.2016.00592.

Roberts, R. R., Ellis, M., Gwynne, R. M., Bergner, A. J., Lewis, M. D., Beckett, E. A., et al. (2010). The first intestinal motility patterns in fetal mice are not mediated by neurons or interstitial cells of Cajal. *The Journal of Physiology, 588*, 1153–1169. https://doi.org/10.1113/jphysiol.2009.185421.

Roberts, R. R., Murphy, J. F., Young, H. M., & Bornstein, J. C. (2007). Development of colonic motility in the neonatal mouse-studies using spatiotemporal maps. *American Journal of Physiology - Gastrointestinal and Liver Physiology, 292*, G930–G938. https://doi.org/10.1152/ajpgi.00444.2006.

Robillard, J. E., Guillery, E. N., Segar, J. L., Merrill, D. C., & Jose, P. A. (1993). Influence of renal nerves on renal function during development. *Pediatric Nephrology, 7*, 667–671. https://doi.org/10.1007/BF00852576.

Rothberg, J. M., Hartley, D. A., Walther, Z., & Artavanis-Tsakonas, S. (1988). slit: An EGF-homologous locus of D. melanogaster involved in the development of the embryonic central nervous system. *Cell, 55*, 1047–1059. https://doi.org/10.1016/0092-8674(88)90249-8.

Rothberg, J. M., Jacobs, J. R., Goodman, C. S., & Artavanis-Tsakonas, S. (1990). Slit: An extracellular protein necessary for development of midline glia and commissural axon pathways contains both EGF and LRR domains. *Genes & Development, 4*, 2169–2187. https://doi.org/10.1101/gad.4.12a.2169.

Rydén, M., & Ibáñez, C. F. (1996). Binding of neurotrophin-3 to p75LNGFR, TrkA, and TrkB mediated by a single functional epitope distinct from that recognized by TrkC (*). *Journal of Biological Chemistry, 271*, 5623–5627. https://doi.org/10.1074/jbc.271.10.5623.

Sadler, T. W., & Langman, J. (2012). In T. W. Sadler (Ed.), *Langman's medical embryology* (12th ed.). Philadelphia: Wolters Kluwer Health/Lippincott Williams & Wilkins, ISBN:978-1-4511-1342-6. Print.

Sanes, D. H., Reh, T. A., & Harris, W. A. (2011). Wiring up the Brain: Axon Navigation. *Development of the nervous system*. Elsevier.

Saporta, M. A., & Shy, M. E. (2013). Inherited peripheral neuropathies. *Neurologic Clinics, 31*, 597–619. https://doi.org/10.1016/j.ncl.2013.01.009.

Sariola, H., Ekblom, P., & Henke-Fahle, S. (1989). Embryonic neurons as in vitro inducers of differentiation of nephrogenic mesenchyme. *Developmental Biology, 132*, 271–281. https://doi.org/10.1016/0012-1606(89)90224-8.

Schneider, C., Wicht, H., Enderich, J., Wegner, M., & Rohrer, H. (1999). Bone morphogenetic proteins are required in vivo for the generation of sympathetic neurons. *Neuron*, *24*, 861–870. https://doi.org/10.1016/S0896-6273(00)81033-8.

Scott-Solomon, E., Boehm, E., & Kuruvilla, R. (2021). The sympathetic nervous system in development and disease. *Nature Reviews Neuroscience*, *22*, 685–702. https://doi.org/10.1038/s41583-021-00523-y.

Seidl, K., & Unsicker, K. (1989a). Survival and neuritic growth of sympathoadrenal (chromaffin) precursor cells in vitro. *International Journal of Developmental Neuroscience*, *7*, 465–473. https://doi.org/10.1016/0736-5748(89)90006-3.

Seidl, K., & Unsicker, K. (1989b). The determination of the adrenal medullary cell fate during embryogenesis. *Developmental Biology*, *136*, 481–490. https://doi.org/10.1016/0012-1606(89)90273-x.

Serafini, T., Kennedy, T. E., Gaiko, M. J., Mirzayan, C., Jessell, T. M., & Tessier-Lavigne, M. (1994). The netrins define a family of axon outgrowth-promoting proteins homologous to C. elegans UNC-6. *Cell*, *78*, 409–424. https://doi.org/10.1016/0092-8674(94)90420-0.

Serafini, T., Colamarino, S. A., Leonardo, E. D., Wang, H., Beddington, R., Skarnes, W. C., et al. (1996). Netrin-1 is required for commissural axon guidance in the developing vertebrate nervous system. *Cell*, *87*, 1001–1014. https://doi.org/10.1016/s0092-8674(00)81795-x.

Silos-Santiago, I., Molliver, D. C., Ozaki, S., Smeyne, R. J., Fagan, A. M., Barbacid, M., et al. (1995). Non-TrkA-expressing small DRG neurons are lost in TrkA deficient mice. *The Journal of Neuroscience*, *15*, 5929–5942. https://doi.org/10.1523/JNEUROSCI.15-09-05929.1995.

Simkin, J. E., Zhang, D., Rollo, B. N., & Newgreen, D. F. (2013). Retinoic acid upregulates Ret and induces chain migration and population expansion in vagal neural crest cells to colonise the embryonic gut. *PLoS One*, *8*, e64077. https://doi.org/10.1371/journal.pone.0064077.

Slotkin, T. A., Levant, B., Orband-Miller, L., Queen, K. L., & Stasheff, S. (1988). Do sympathetic neurons coordinate cellular development in the heart and kidney? Effects of neonatal central and peripheral catecholaminergic lesions on cardiac and renal nucleic acids and proteins. *The Journal of Pharmacology and Experimental Therapeutics*, *244*, 166. PMID: 2447274.

Smeyne, R. J., Klein, R., Schnapp, A., Long, L. K., Bryant, S., Lewin, A., et al. (1994). Severe sensory and sympathetic neuropathies in mice carrying a disrupted Trk/NGF receptor gene. *Nature*, *368*, 246–249. https://doi.org/10.1038/368246a0.

Squinto, S. P., Stitt, T. N., Aldrich, T. H., Davis, S., Blanco, S. M., RadzieJewski, C., et al. (1991). trkB encodes a functional receptor for brain-derived neurotrophic factor and neurotrophin-3 but not nerve growth factor. *Cell*, *65*, 885–893. https://doi.org/10.1016/0092-8674(91)90395-F.

Sui, P., Wiesner, D. L., Xu, J., Zhang, Y., Lee, J., Van Dyken, S., et al. (2018). Pulmonary neuroendocrine cells amplify allergic asthma responses. *Science*, *360*. https://doi.org/10.1126/science.aan8546.

Sun, K. L. W., Correia, J. P., & Kennedy, T. E. (2011). Netrins: Versatile extracellular cues with diverse functions. *Development*, *138*, 2153–2169. https://doi.org/10.1242/dev.044529.

Takahashi, Y., Sipp, D., & Enomoto, H. (2013). Tissue interactions in neural crest cell development and disease. *Science*, *341*, 860–863. https://doi.org/10.1126/science.1230717.

Teshima, T. H. N., Tucker, A. S., & Lourenço, S. V. (2019). Dual sympathetic input into developing salivary glands. *Journal of Dental Research*, *98*, 1122–1130. https://doi.org/10.1177/0022034519865222.

Tessarollo, L., Tsoulfas, P., Martin-Zanca, D., Gilbert, D. J., Jenkins, N. A., Copeland, N. G., et al. (1993). trkC, a receptor for neurotrophin-3, is widely expressed in the developing nervous system and in non-neuronal tissues. *Development*, *118*, 463–475. https://doi.org/10.1242/dev.118.2.463.

Tessarollo, L., Tsoulfas, P., Donovan, M. J., Palko, M. E., Blair-Flynn, J., Hempstead, B. L., et al. (1997). Targeted deletion of all isoforms of the trkC gene suggests the use of alternate receptors by its ligand neurotrophin-3 in neuronal development and implicates trkC in normal cardiogenesis. *Proceedings of the National Academy of Sciences of the United States of America, 94*, 14776–14781. https://doi.org/10.1073/pnas.94.26.14776.

Tessarollo, L., Vogel, K. S., Palko, M. E., Reid, S. W., & Parada, L. F. (1994). Targeted mutation in the neurotrophin-3 gene results in loss of muscle sensory neurons. *Proceedings of the National Academy of Sciences of the United States of America, 91*, 11844–11848. https://doi.org/10.1073/pnas.91.25.11844.

Tessier-Lavigne, M., & Goodman, C. S. (1996). The molecular biology of axon guidance. *Science, 274*, 1123–1133. https://doi.org/10.1126/science.274.5290.1123.

Timmusk, T., Belluardo, N., Metsis, M., & Persson, H. (1993). Widespread and developmentally regulated expression of neurotrophin-4 mRNA in rat brain and peripheral tissues. *European Journal of Neuroscience, 5*, 605–613. https://doi.org/10.1111/j.1460-9568.1993.tb00526.x.

Togari, A., Arai, M., & Kondo, A. (2005). The role of the sympathetic nervous system in controlling bone metabolism. *Expert Opinion on Therapeutic Targets, 9*, 931–940. https://doi.org/10.1517/14728222.9.5.931.

Tomlinson, R. E., Li, Z., Zhang, Q., Goh, B. C., Li, Z., Thorek, D. L. J., et al. (2016). NGF-TrkA signaling by sensory nerves coordinates the vascularization and ossification of developing endochondral bone. *Cell Reports, 16*, 2723–2735. https://doi.org/10.1016/j.celrep.2016.08.002.

Travagli, R. A., & Anselmi, L. (2016). Vagal neurocircuitry and its influence on gastric motility. *Nature Reviews Gastroenterology & Hepatology, 13*, 389–401. https://doi.org/10.1038/nrgastro.2016.76.

Tucker, K. L., Meyer, M., & Barde, Y.-A. (2001). Neurotrophins are required for nerve growth during development. *Nature Neuroscience, 4*, 29–37. https://doi.org/10.1038/82868.

Turner, B. J., Cheah, I. K., Macfarlane, K. J., Lopes, E. C., Petratos, S., Langford, S. J., et al. (2003). Antisense peptide nucleic acid-mediated knockdown of the p75 neurotrophin receptor delays motor neuron disease in mutant SOD1 transgenic mice. *Journal of Neurochemistry, 87*, 752–763. https://doi.org/10.1046/j.1471-4159.2003.02053.x.

Uesaka, T., Young, H. M., Pachnis, V., & Enomoto, H. (2016). Development of the intrinsic and extrinsic innervation of the gut. *Developmental Biology, 417*, 158–167. https://doi.org/10.1016/j.ydbio.2016.04.016.

Wang, X., Chan, A. K. K., Sham, M. H., Burns, A. J., & Chan, W. Y. (2011). Analysis of the sacral neural crest cell contribution to the hindgut enteric nervous system in the mouse embryo. *Gastroenterology, 141*, 992–1002.e6. https://doi.org/10.1053/j.gastro.2011.06.002.

Weigand, L. A., & Myers, A. C. (2010). Synaptic and membrane properties of parasympathetic ganglionic neurons innervating mouse trachea and bronchi. *American Journal of Physiology—Lung Cellular and Molecular Physiology, 298*, L593–L599. https://doi.org/10.1152/ajplung.00386.2009.

Wiese, C., Ireland, S., Fleming, N., Yu, J., Valerius, M., Georgas, K., et al. (2012). A genome-wide screen to identify transcription factors expressed in pelvic ganglia of the lower urinary tract. *Frontiers in Neuroscience, 6*, 130. https://doi.org/10.3389/fnins.2012.00130.

Wiese, C. B., Deal, K. K., Ireland, S. J., Cantrell, V. A., & Southard-Smith, E. M. (2017). Migration pathways of sacral neural crest during development of lower urogenital tract innervation. *Developmental Biology, 429*, 356–369. https://doi.org/10.1016/j.ydbio.2017.04.011.

Wilkinson, D. G. (2001). Multiple roles of eph receptors and ephrins in neural development. *Nature Reviews. Neuroscience*, *2*, 155–164. https://doi.org/10.1038/35058515.
Woods, S. C., & Porte, D. (1974). Neural control of the endocrine pancreas. *Physiological Reviews*, *54*, 596–619. https://doi.org/10.1152/physrev.1974.54.3.596.
Xia, J., & Worzfeld, T. (2016). Semaphorins and plexins in kidney disease. *Nefrología*, *132*, 93–100. https://doi.org/10.1159/000443645.
Xu, J., Yu, H., & Sun, X. (2020). Less is more: Rare pulmonary neuroendocrine cells function as critical sensors in lung. *Developmental Cell*, *55*, 123–132. https://doi.org/10.1016/j.devcel.2020.09.024.
Yagai, T., Miyajima, A., & Tanaka, M. (2014). Semaphorin 3E secreted by damaged hepatocytes regulates the sinusoidal regeneration and liver fibrosis during liver regeneration. *The American Journal of Pathology*, *184*, 2250–2259. https://doi.org/10.1016/j.ajpath.2014.04.018.
Yazdani, U., & Terman, J. R. (2006). The semaphorins. *Genome Biology*, *7*, 211. https://doi.org/10.1186/gb-2006-7-3-211.
Young, H. M., Hearn, C. J., Ciampoli, D., Southwell, B. R., Brunet, J.-F., & Newgreen, D. F. (1998). A single rostrocaudal colonization of the rodent intestine by enteric neuron precursors is revealed by the expression of Phox 2b, Ret, and p75 and by explants grown under the kidney capsule or in organ culture. *Developmental Biology*, *202*, 67–84. https://doi.org/10.1006/dbio.1998.8987.
Young, H. M., Bergner, A. J., Anderson, R. B., Enomoto, H., Milbrandt, J., Newgreen, D. F., et al. (2004). Dynamics of neural crest-derived cell migration in the embryonic mouse gut. *Developmental Biology*, *270*, 455–473. https://doi.org/10.1016/j.ydbio.2004.03.015.
Yuan, W., Zhou, L., Chen, J., Wu, J. Y., Rao, Y., & Ornitz, D. M. (1999). The mouse SLIT family: Secreted ligands for ROBO expressed in patterns that suggest a role in morphogenesis and axon guidance. *Developmental Biology*, *212*, 290–306. https://doi.org/10.1006/dbio.1999.9371.

CHAPTER NINE

Growth control of the kidney

Leif Oxburgh*
The Rogosin Institute, New York, NY, United States
*Corresponding author: e-mail address: leo9022@nyp.org

Contents

Abstract

The functional mass of kidney tissue in an adult is an important determinant of human health. Kidney formation during development is an essential determinant of the final nephron endowment of the adult organ, and no evidence has been reported that mice or humans are able to generate new nephrons after the developmental period. Mechanisms controlling organ growth after development are essential to establish the final adult organ size. The potential for organ growth is maintained in adult life and the size of one kidney may be significantly increased by loss of the contralateral kidney. The mouse has provided a model system for investigators to critically explore genetic, cell biological, and hormonal control of developmental and juvenile kidney growth. This article reviews three basic aspects of kidney size regulation: (1) Mechanisms that control nephron formation and how these are altered by the cessation of nephrogenesis at the end of the developmental period. (2) Applicability of the general model for growth hormone-insulin like growth factor control to kidney growth both pre- and postnatally. (3) Commonalities between mechanisms of juvenile kidney growth and the compensatory growth that is stimulated in adult life by reduction of kidney mass. Understanding the mechanisms that determine set-points for cell numbers and size in the kidney may inform ongoing efforts to generate kidney tissue from stem cells.

Current Topics in Developmental Biology, Volume 148
ISSN 0070-2153
https://doi.org/10.1016/bs.ctdb.2021.12.007

Abbreviations

AP1	activator protein 1
BMP	bone morphogenetic protein
CD	collecting duct
CM	cap mesenchyme
FGF	fibroblast growth factor
GBM	glomerular basement membrane
GH	growth hormone
IGF	insulin-like growth factor
IUGR	intrauterine growth restriction
MAPK	mitogen activated protein kinase
MM	metanephric mesenchyme
mTOR	mammalian target of rapamycin
NPC	nephron progenitor cell
PI3K	phosphatidylinositol 3-kinase
SMAD	SMA- and MAD-like transcription factor
TSC	tuberous sclerosis complex
UB	ureteric bud
Wnt	wingless/integrated

1. Introduction

The mouse kidney has provided an invaluable model for many aspects of kidney development. Genetic technologies have allowed us to establish a framework for mammalian kidney development based on studies of gene inactivation and overexpression in mice. Its applicability to human kidney development is being actively interrogated using single cell transcriptomics, novel imaging technologies, and the investigation of human gene variants that affect kidney development. Likewise, mouse genetics has been an essential tool in understanding adult homeostasis/injury/repair and physiological function. Many comparative studies highlight the congruence between mouse and human kidney biology, but species-specific biological processes have also been identified that inform the limits of the mouse model system. Examples include differences in speed of nephron formation and maturation (Lindström et al., 2018), the layout of nephrons in the adult organ (Saxén, 1987), and differential gene usage in core developmental pathways (O'Brien et al., 2016).

Cell and organ culture have been essential for mechanistic studies of kidney biology (Saxén, 1987), and the advent of methods to generate kidney organoids *in vitro* from pluripotent stem cells has provided an important new tool for the study of human kidney tissue (Morizane et al., 2015; Takasato et al., 2015). Combined with state-of-the-art genetic technologies that enable gene knockout and knockdown as well as subtle modifications of gene and protein structure (Schwank et al., 2013), this culture-based model system has become a powerful tool to directly investigate human kidney biology. Organoids reproduce a complex repertoire of kidney cell types that can be used to model important aspects of human kidney biology, such as cilia function (Forbes et al., 2018). However, one limitation to their use is that they most accurately model human fetal kidney tissue and may not reproduce mature physiological functions required to study important aspects of the human kidney (Wu et al., 2018). Tissue growth and differentiation are thus major challenges in refinement of kidney organoid technology to reproduce adult human kidney biology. Research into growth regulation of the kidney has a long history (Gladstone, 1924; Smith & Moise, 1927; Walter & Addis, 1939); mouse and rat modeling using hormone treatments, surgical techniques and more recently genetic technologies have provided important insights into how organ growth is regulated, and this article aims to review core pathways that these studies have uncovered.

2. Kidney anatomy

The adult kidney is an array of nephrons, each with a vascular connection at its proximal pole and a connection to the collecting duct (CD) system at its distal pole (Fig. 1A,B). CDs collect the filtrate that has been conditioned by the tubule epithelium of the nephron and act as the conduit to the pelvis from which urine passes to the bladder. Individual nephrons serve to filter blood and concentrate urine and by arraying single nephrons in cassettes around CDs and blood vessels, a countercurrent system is formed that coordinates water and electrolyte resorption by nephrons, blood vessels, and CDs.

Approximately 15% of the cardiac output reaches the kidney through the renal artery and is distributed through a network of arterial branches to the glomeruli (Fig. 1C). Afferent arterioles carry oxygenated blood into the glomerular capillary network, and blood passes out through the efferent arteriole into the post-glomerular capillary network that surrounds the nephron, providing oxygen and nutrients for the highly metabolically active tubule epithelium. The glomerulus contains a fenestrated capillary surrounded by podocytes that produce a thick extracellular membrane known

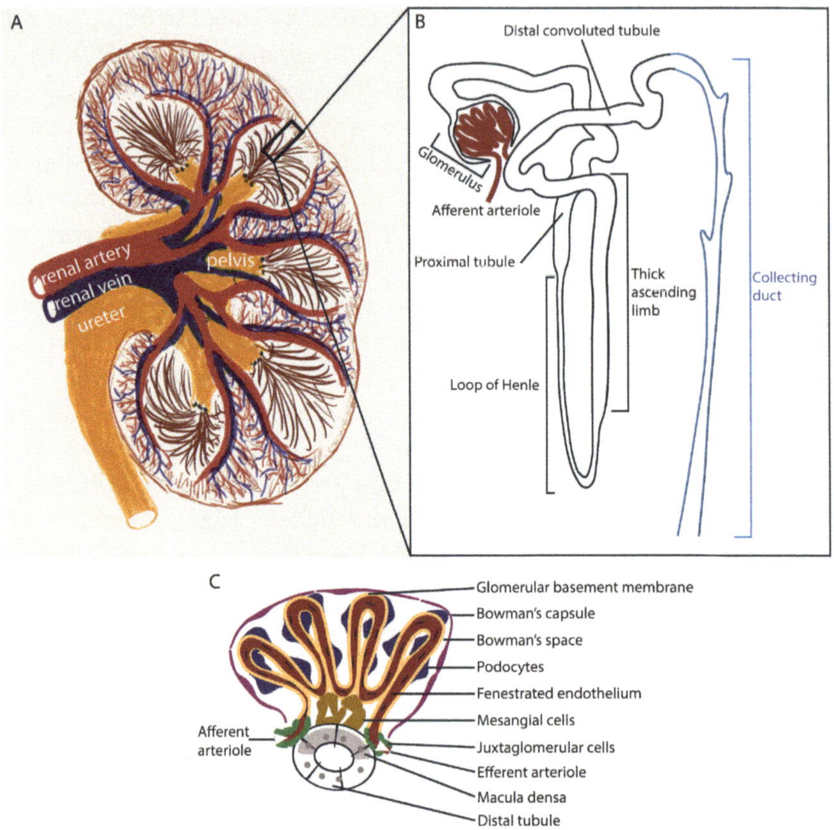

Fig. 1 Kidney anatomy. (A) Schematic of a cross-section of a human kidney. Arterial blood enters through the renal artery, passes through the glomerulus into the vascular bed of the tissue and out through the renal vein. Filtrate that passes through the basement membrane of the glomerulus is conditioned through the nephron and collecting duct (box) and collected in the pelvis from where it is transported to the bladder *via* the ureter. (B) Schematic of nephron and collecting duct showing microanatomical structures described in the text. (C) Schematic of the glomerulus.

as the glomerular basement membrane (GBM) together with the endothelium. Acellular filtrate passes across the GBM into Bowman's space which funnels the filtrate into the nephron tubule. Glomerular capillary size can be regulated by the contraction of mesangial cells located around the glomerular vessels. Contractility is regulated by neural input and local sensory input from the macula densa, a specialized region of distal nephron epithelium that produces vasoconstrictive factors in response to osmolality changes of the filtrate (Grassi, Mark, & Esler, 2015). The macula densa also contacts the juxtaglomerular apparatus at the vascular pole of the glomerulus and controls renin output from these cells in conjunction with blood pressure changes in the arteriole that feeds the glomerulus. Renin is a potent regulator of systemic blood pressure. The epithelial tubule of the nephron is divided into functionally discrete segments: proximal tubule, loop of Henle, thick ascending limb of the loop of Henle, distal convoluted tubule, and connecting tubule. The proximal tubule resorbs vital molecules such as glucose, protein and amino acids that escape through the glomerular filter, in addition to balancing electrolytes and resorbing water. The loop of Henle and distal portions of the tubular nephron are focused on water resorption and electrolyte balance. After passing through the nephron, filtrate enters the CD, where water is further resorbed, and electrolytes and pH are balanced.

While the microanatomy of the kidney is very similar between the human and the mouse, species differences are pronounced in higher-order architecture, presumably as a body-size adaptation. Both species have a cortical zone defined by the presence of glomeruli, an outer medullary zone, and an inner medullary zone or papilla where CDs converge and empty into the renal pelvis. However, while the mouse has a single papilla, the human has several.

3. Kidney development

There are three stages of kidney development in the mouse and human: pronephric, mesonephric and metanephric. The latter leads to formation of the permanent kidney that is retained into adult life. Formation of the metanephros is initiated when the epithelial ureteric bud (UB) migrates from the cloaca into the field of intermediate mesoderm known as the metanephric mesenchyme (MM). This occurs on approximately the 11th day of gestation in the mouse and on the 26th day of gestation in the human (Otis & Brent, 1954). Following this key event, the mouse kidney enters an 11-day long period of metanephric organogenesis in which all nephrons are formed. In the human this organogenetic period occurs between approximately day 29 of gestation and 36–40 weeks (term) (Ryan et al., 2018).

The homeotic code specifies the MM by regulating expression of the growth factor *Gdnf* (Wellik, Hawkes, & Capecchi, 2002). GDNF from MM acts as a chemotactic factor for the UB, which expresses the cognate c-Ret and GFR1a receptors (Pichel et al., 1996; Sainio et al., 1997; Schuchardt, D'Agati, Larsson-Blomberg, Costantini, & Pachnis, 1994; Vega, Worby, Lechner, Dixon, & Dressler, 1996). After this contact the

UB promotes condensation of MM into a cap mesenchyme (CM) containing nephron progenitor cells (NPCs) and undergoes a program of repetitive branching. The resulting epithelial tree is referred to as the collecting duct system. Tips of the CD tree are located at the periphery of the kidney with the nephrogenic zone, where they induce formation of new nephrons from CM in a Wnt9b-dependent inductive process (Carroll, Park, Hayashi, Majumdar, & McMahon, 2005). Repetitive addition of nephrons at the periphery of the kidney ensures radial growth of the organ. CM remains associated with CD tips throughout branching, and a nephron forms through a process in which a subset of these cells condenses, undergoes epithelialization and fuses to the CD to ensure a connection between the differentiating nephron and the collecting system. In the human, these principles were described by careful microdissection and reconstruction of fetal kidneys at a variety of developmental stages (Osathanondh & Potter, 1963). Because of its small size, the mouse kidney offers many more possibilities for imaging than the human kidney, and various state-of-the-art technologies have been applied to understanding the dynamic changes in its morphology that occur over the developmental period. Culture of the rodent kidney represented an important technological advance because it enabled time-lapse imaging, so that the process of mesoderm condensation around the UB tip and subsequent CD branching could be studied in real time (Saxen & Wartiovaara, 1966). An important observation from these studies was that the CM is carried forward at the tip of each CD and redistributed during each branching event so that a subset of undifferentiated NPCs is available for nephron formation at each branch tip. Tracking of cell movements within the CM of cultured kidneys shows that individual NPCs are motile, suggesting a form of swarming behavior around the CD tip that allows the CM structure to rearrange dynamically in response to changes in CD morphology (Combes, Lefevre, Wilson, Hamilton, & Little, 2016). The whole organ culture technique was adapted specifically for the study of nephron formation by separation of the MM and UB immediately following the first contact between the two and prior to any branching (Grobstein, 1953). Experimental separation of UB and MM led to the essential finding that the CD epithelium and nephron epithelium derive from separate cellular origins. Reporter gene marking of distinct structures and tomography coupled with computer modeling have revealed the dynamics of key morphogenetic processes in the mouse kidney at very high spatiotemporal resolution (Short et al., 2014).

4. Cessation of nephrogenesis

The formation of new nephrons ceases on approximately the second day after birth in the mouse in a final wave of NPC differentiation (Hartman et al. 2007), and the mouse model has allowed the field to understand genetic and cell signaling events that set the limit for nephron number in an individual. Although nephrogenesis ends abruptly there is a gradual change in the morphology of the nephrogenic zone preceding cessation, with a reduction in size of both CD tips and CMs over the course of several days and subtle changes in the orientation of nephron differentiation within the nephrogenic zone (Rumballe et al., 2011; Short et al., 2014). A possible connection between the gradual thinning of the CM and cessation of nephrogenesis has been proposed; theoretical modeling predicts that once CM size falls below a certain threshold the final wave of differentiation and cessation will ensue (Zubkov et al., 2015). The alternate possibility that nephron differentiation at CD tips prevents their further branching, thus removing tips from the nephron-inducing pool has been tested but is not supported by experimental analyses (Short et al., 2018).

Changes in activities of key signaling pathways responsible for CM maintenance have been associated with the terminal alterations in the nephrogenic zone associated with cessation of nephron formation. BMP/SMAD signaling increases within the CM over the last 3 days of kidney development (Brown, Muthukrishnan, & Oxburgh, 2015). The balance between MAPK and SMAD signaling downstream of BMP7 plays an important role in balancing renewal and differentiation of NPCs; increased SMAD signaling is predicted to increase the sensitivity of these cells to WNT/β-catenin mediated differentiation, thus depleting the CM more rapidly. Heterochronic engraftment studies in which postnatal CM cells were engrafted into CMs of E12.5 kidneys revealed that the older cells are more prone to exit the CM and differentiate than the E12.5 cells (Chen et al., 2015). The aging CM thus appears fundamentally altered in its capacity for renewal, and engraftment studies with mutant kidneys indicate that reduced FGF expression may be associated with this age-related change. FGF9 and 20 are essential for the maintenance of the CM in its undifferentiated state, and a reduction in FGF20 is predicted to tip the balance in favor of differentiation and more rapid exit of cells from the CM. A coordinated shift of BMP signaling from activation of MAPK to SMAD with a reduction in FGF signaling amplitude is predicted to reduce proliferation of CM cells, since BMP/MAPK and FGF signaling control proliferation of these cells by coordinately regulating the abundance of the AP1 transcription factor which promotes transcription of cell cycle genes (Fig. 2) (Muthukrishnan, Yang, Friesel, & Oxburgh, 2015). Global genome landscape analysis of CM cells isolated from mouse kidneys early and late in the organogenetic period demonstrates an age-related reduction in accessibility of predicted enhancers for the CM self-renewal transcription factor SIX2 and an age-related increase in the accessibility of predicted enhancers for the BATF and BACH2 transcription factors, which compete with AP1 for binding to target genes (Hilliard et al., 2019). These multiple strands of evidence point to a model in which cell signaling and genome accessibility change as the CM cells age, coordinately reducing AP1 activation of cell cycle genes.

Morphometric studies based on measurement of proliferation within the CM support a model in which two distinct subsets of cells coexist—one rapidly dividing and the other dividing more slowly (Short et al., 2014). The proportion of slowly cycling cells increases in an age-dependent manner concomitantly with thinning of the CM at each CD tip. All of this data suggests a model in which CM cells early in development receive strong BMP/MAPK and FGF signals promoting AP1 activation of the cell cycle.

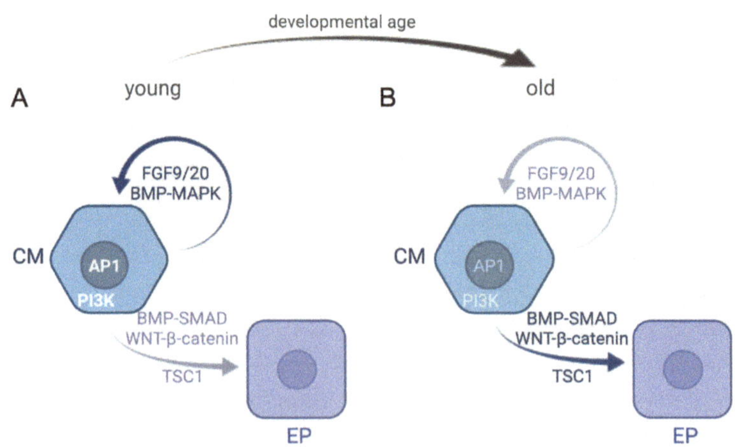

Fig. 2 Model for cessation of nephrogenesis. (A) Schematic of events in the developmentally young kidney in which cap mesenchyme (CM) differentiation to committed nephron epithelial precursors (EP) is balanced resulting in progenitor renewal and differentiation. (B) Events in the kidney immediately prior to cessation of nephrogenesis in which differentiation to committed nephron epithelial precursors exceeds renewal of cap mesenchyme resulting in terminal depletion of progenitors.

As development proceeds, the proportion of these rapidly cycling cells is reduced due to diminished signaling amplitude and competition for AP1 binding sites, and the slowly cycling cells that are more prone to differentiation are recruited out of the CM (Fig. 2).

Given the roles for BMP and FGF signaling in cessation of nephrogenesis, it is interesting to ask if there is also a role for Wnt, which functions as the inducer of nephron formation. Two fundamentally different Wnt response programs have been identified in cells of the CM; a low-amplitude program that is required for renewal of these cells in their undifferentiated state, and a high-amplitude program that promotes their exit into the nephron differentiation program (Karner et al., 2011). The essential source of Wnt ligand in this process is the CD tip that expresses Wnt9b (Carroll et al., 2005) and it is possible that as the CM thins more of the cells come into direct contact with the CD tip, exposing them to higher-amplitude signaling and accelerating their differentiation. Additionally, CM cells late in organogenesis display a higher sensitivity to Wnt/β-catenin induced differentiation than CM cells isolated from kidneys early in organogenesis. Fuel usage by cells of the CM changes over developmental time, and young CM cells utilize glycolysis to a greater extent than old CM cells (Liu et al., 2017). Cell and organ culture studies have revealed that the susceptibility to pro-differentiation effects of Wnt/β-catenin signaling is inversely proportional to glycolytic activity in these cells, and these investigations indicate that the age-associated shift from glycolysis to oxidative phosphorylation in CM cells potentiates their Wnt/β-catenin-mediated recruitment into the differentiation program and out of the CM.

Why would fuel source utilization in CM cells change over the organogenetic period? Inhibition of the PI3K pathway reduces glycolysis in CM cells (Liu et al., 2017), and one model supported by this data is that PI3K signaling is reduced in CM cells as they age, suppressing their glycolytic fuel usage. The cause of the reduction in PI3K activation in CM as these cells age has not been determined. Studies have shown that PI3K is essential for the proliferative effect of FGF on CM cells (Brown et al., 2011) and that PI3K is activated by both BMP/MAPK and FGF signaling (Lindström, Carragher, & Hohenstein, 2015). Thus, the reduction of BMP/MAPK and FGF signaling seen in the aging CM may reduce PI3K amplitude and diminish glycolysis. Extrapolating from other cell biological systems, it would be anticipated that reduced PI3K activity would diminish mTOR activation because PI3K-AKT inhibits the tuberous sclerosis (TSC) complex, which suppresses mTOR activation. However, single cell transcriptomic data comparing CM cells at different developmental stages reveal an age-related increase in mTOR expression and age-related increases in ribosomal biogenesis pathways, which are important downstream effectors of mTOR signaling (Chen, Brunskill, et al., 2015). A series of conditional gene inactivation experiments to explore the role of mTOR and its inhibitory TSC1 protein in the CM revealed an essential role for mTOR in nephrogenesis, but no specific role in cessation (Volovelsky et al., 2018). Interestingly, heterozygous inactivation of *Tsc1* delayed cessation of nephrogenesis with a concomitant increase in final nephron number (Fig. 2). An intuitive hypothesis is that relieving TSC1 suppression of mTOR activity by eliminating one gene copy increases mTOR activity causing this effect, but compound genetic inactivation of both *Tsc1* and *Mtor* in CM shows that the effect of *Tsc1* is independent of *Mtor*. While the genetic analysis shows that *Tsc1* is operating through a *Mtor*-independent pathway, it also shows that the delay of cessation caused by *Tsc1* heterozygosity is dependent on Raptor, which encodes an mTOR scaffolding protein. mTOR-independent downstream activities of TSC have been proposed in other cell- and organ systems, such as regulation of the PAK2 kinase (Alves et al., 2015). However, the pathway(s) downstream of TSC1 responsible for controlling the timing of cessation remains unknown.

5. Nephron endowment

Nephron endowment is an important determinant of human health (Luyckx & Brenner, 2010), and nephron number varies up to 13-fold between humans. A variety of causes for this enormous variation have been proposed, and conditions predisposing to low birth weight and preterm birth have a strong influence (Luyckx & Brenner, 2020). Determinants of nephron endowment in the preterm and term newborn human are extremely complex, and an action plan to promote nephron health through understanding these underlying causes has been formulated (Luyckx et al., 2017). Mouse studies have shown that modulating BMP/Smad, TSC, and GDNF expression can alter the timing of cessation and the ultimate nephron number, providing insights into candidate pathways that may be therapeutically tractable (Brown et al., 2015; Li et al., 2021; Volovelsky et al., 2018).

In addition to providing the genetic and cell biological tools to understand the molecular wiring of the nephron formation program, the mouse has been used to study common pathologies that alter nephron endowment, such as gestational diabetes which affects approximately 7% of pregnancies in the USA and Canada (Casagrande, Linder, & Cowie, 2018; Feig et al., 2014). A mouse model of uncontrolled gestational hyperglycemia following streptozotocin-induced pancreatic injury revealed a reduction in nephron endowment in offspring of approximately 40%, with increased postnatal cell death in nephrons (Tran et al., 2008). The Akita genetic model of type 1 diabetes revealed a more modest reduction in nephron endowment of approximately 20%, with perturbations in Wnt and Notch signaling and reduced differentiation of CM to nephrons but no evidence of increased cell death (Cerqueira et al., 2019). While both studies support an inhibitory effect of maternal hyperglycemia on nephron formation, the proposed mechanisms differ. One possible explanation is that the chemically induced model of diabetes results in intrauterine growth restriction (IUGR), while the genetic model does not. IUGR may itself be associated with low nephron endowment, and the magnitude of nephron reduction and extensive cell death in nephrons seen in the streptozotocin model may reflect the compounded effects of gestational hyperglycemia and IUGR on nephron abundance.

6. Insulin like growth factors in prenatal growth control

During organogenesis, new nephrons are being induced while previous generations of nephrons, CDs and vasculature are growing. At any time during this period nephrons of a variety of sizes and stages of differentiation can be found in the same kidney, with the youngest nephrons located at the periphery of the organ. After cessation of nephrogenesis when nephron endowment is initially determined there is no evidence of any natural mechanism to reactivate nephron formation in either the mouse or the human. DNA quantification and radioisotope incorporation studies to determine DNA replication indicate that the kidney grows through vigorous cell proliferation during organogenesis through to approximately the 5th postnatal week in the rat (Enesco & LeBlond, 1962; Winick & Noble, 1965). The birds-eye view of kidney growth extrapolated from these cellular proliferation studies provides a framework with which to understand kidney growth. The ratio of protein content to nuclear content is unchanged during the period of organogenesis through to approximately the 14th postnatal day (P14), while nuclear content increases rapidly, indicating that the organ grows through cell proliferation during this first phase of its growth. After P14 the kidney enters a phase during which nuclear content continues to rise but the ratio of protein to nuclear content in the organ also rises, suggesting growth due to an increase in both number and size of cells, and perhaps also increases in extracellular protein content. Around P34, nuclear content stabilizes abruptly but the ratio of protein content to nuclear content continues to rise, indicating that the kidney has entered a third phase of growth dependent on increases in cell size and/or non-cellular components. Extrapolating this data to mouse, it is anticipated that the cell proliferative phase ends and the phase of simultaneous cell proliferation and increased protein content per cell begins 2 days earlier compared with rat. The timing of transition to the third phase is unclear. Based on landmark events such as weaning, eye-opening and gonadal hormone production, it is anticipated that timing of postnatal growth is only 2 days earlier in the mouse compared with the rat during the first 3 postnatal weeks, but that the subsequent juvenile maturation period in the rat which extends beyond P34 is almost double that of the mouse (Bell, 2018).

Insulin-like growth factors (IGFs) play central roles in organismal growth, both during organogenesis and in the postnatal period. IGF1 and IGF2 are structurally related to the insulin pro-peptide and signal through insulin like growth factor receptor 1 (IGF1R) at the cell surface (Fig. 3). IGF1R is similar to the insulin receptor both at the level of primary sequence and structural topology, and has tyrosine kinase activity (Ullrich et al., 1986). Receptor activation by the ligand can engage the MAPK/ERK and the PI3K signaling pathways (reviewed in Taguchi & White, 2008). In concert these pathways control cell cycle, synthesis of new cellular components, and cellular fuel source usage. IGF2R is not related to the insulin receptor,

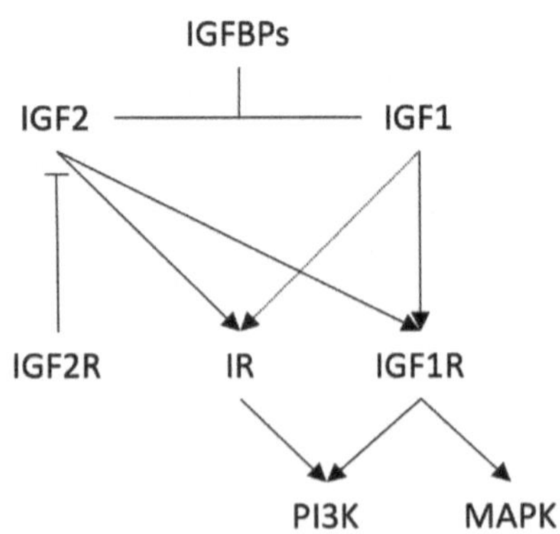

Fig. 3 IGF signaling schematic.

lacks a tyrosine kinase domain and can act as a high-affinity sink specific for IGF2, trapping circulating molecules and targeting them for degradation (Brown et al., 2008). IGF1R is thus the primary initiator of signaling in the pathway, and IGF2R is proposed to counteract IGF signaling. While IGF1 and IGF2 both bind IGF1R, IGF1 displays many-fold higher affinity (Germain-Lee, Janicot, Lammers, Ullrich, & Casella, 1992). The signaling relationship between IGFs and insulin may be complex because insulin binds IGF1R and the IGFs bind the insulin receptor (Hintz, Clemmons, Underwood, & Van Wyk, 1972; Marshall, Underwood, Voina, Foushee, & Van Wyk, 1974). In both cases the cognate ligand has significantly higher activity, but biological effects of these cross-pathway signaling interactions have been documented (Morcavallo et al., 2011; Versteyhe et al., 2013). Of these interactions, the IGF2-insulin receptor interaction is best understood, and compound genetic analyses indicate that IGF2 exerts its growth-promoting activities in the prenatal period through both IGF1R and the insulin receptor (Louvi, Accili, & Efstratiadis, 1997).

Genetic inactivation of *Igf1* or *Igf2* causes proportionate prenatal growth reduction and mutant embryos are approximately 60% the size of wild types (DeChiara, Efstratiadis, & Robertson, 1990; DeChiara, Robertson, & Efstratiadis, 1991; Liu, Baker, Perkins, Robertson, & Efstratiadis, 1993), demonstrating that both are essential for prenatal growth. Control of placental growth is a confounding factor in the analysis of embryonic growth, as small placental size is likely to secondarily depress embryo growth. Genetic inactivation of *Igf1* does not affect placental development, and is predicted to control growth by signaling directly to embryonic tissues (Baker, Liu, Robertson, & Efstratiadis, 1993). Similarly, *Igf1r* inactivation does not affect placental size. However, inactivation of *Igf2* causes reduction in placental size concomitantly with reductions in embryonic organ size, suggesting that there may be a contribution of placental deficiency to the growth impairment seen in *Igf2* mutant embryos. Genetic inactivation of a placenta-specific transcript of *Igf2* demonstrates its essential role in placenta development and shows severe secondary IUGR of the embryo (Constância et al., 2002). The genetics of *Igf1* and *Igf2* differ significantly, in that *Igf2* is a parentally imprinted gene, with the active copy being paternally inherited. Thus, an *Igf2* heterozygote with a paternally inherited mutant allele is functionally null, except for expression in choroid plexus and leptomeninges where both maternal and paternal copies are active (DeChiara et al., 1991). Dysregulation of *Igf2* imprinting is seen in Beckwith Wiedemann overgrowth syndrome, in which kidney enlargement is common and which also predisposes to pediatric kidney cancer (Maher & Reik, 2000; Mussa et al., 2012). Mouse modeling has been essential to understanding origins of pathological features seen in this syndrome, and show that the kidney is disproportionally enlarged in animals with transgenically elevated *Igf2* (Sun, Dean, Kelsey, Allen, & Reik, 1997).

Igf1 and *Igf2* are expressed in the kidney and thus it is expected that they signal in an autocrine/paracrine manner. *In situ* hybridization studies show that expression of *Igf1* is ubiquitous, and that *Igf2* is close to ubiquitous but lacks expression in CD and the glomerular mesangium (Lindenbergh-Kortleve et al., 1997). More recent transcriptome analyses of mouse embryonic kidneys are ambiguous, but single cell transcriptomics on human fetal kidney reveals an expression pattern for *Igf2* that is supportive of the *in situ* hybridization results (https://www.rebuildingakidney.org/id/17-E9KE) (Harding et al., 2011). Interactions with the IGF-binding proteins allow the IGFs to circulate in a stabilized form with extended half-life (Zapf, Hauri, Waldvogel, & Froesch, 1986) and analyses of serum from rat embryos at E17 reveals circulation of both IGF1 and IGF2 (Daughaday, Parker, Borowsky, Trivedi, & Kapadia, 1982), suggesting they may act both as circulating hormones and paracrine factors during organogenesis. Histological inspection of kidneys from newborn *Igf1* null mice reveal the expected size reduction, but it appears that nephron and CD structures are represented in proportions similar to wild type controls, and no developmental perturbations were identified in this analysis, indicating a proportionate size reduction with regard to kidney microanatomy (Rogers, Powell-Braxton, & Hammerman, 1999).

While the proportional dwarfing reported in *Igf* knockout studies suggests a hormonal role for IGFs in coordinating prenatal organ growth, the relative contributions of hormonal *versus* paracrine IGF signaling to prenatal growth control are not clear. Given the potential ubiquity of IGF1 and IGF2 ligands, the phenotype of inactivating their common receptor IGF1R yields the most straight-forward measure of their importance in prenatal kidney growth. Global *Igf1r* inactivation results in detectable growth difference from embryonic day 13 in the mouse, only 2 days after the formation of the metanephric kidney (Baker et al., 1993), and the size differential between mutant and wild type increases throughout the period of kidney organogenesis. Studies of *Igf1r* expression in the mouse kidney using *in situ* hybridization reveal close to ubiquitous expression (Lindenbergh-Kortleve et al., 1997; Wada et al., 1993). Transcriptional microarray and RNA-seq experiments show widespread expression largely confirming conclusions from *in situ* hybridization. Antisense oligonucleotide treatment of *in vitro* cultured E13 kidneys impairs growth profoundly compared to sense oligonucleotide controls, and both nephron formation and CD branching are reduced (Wada et al., 1993). Although these findings were not morphometrically quantified, the histological analyses clearly show reductions in both CD branching and nephron induction in the absence of overt cytotoxicity. This analysis was conducted at a very early timepoint in metanephric development and points to a role for IGF in promoting the nephrogenic program, but does not answer if IGF signaling controls the growth of differentiated nephrons. A mouse strain with loxP-flanked *Igf1r* has been used to study the role of IGF signaling in growth of several organ systems, but no reports on its use to understand the role of IGF signaling in developing kidney were found for this review.

The general model for IGF signaling in which MAPK/ERK and PI3K pathways stimulate cellular growth in target cells by promoting concerted synthesis of cellular components and proliferation is assumed to operate in the kidney. The widespread expression of *Igfr* and the proportionate dwarfing seen in *Igf1r* and *Igf1* mutants suggests that a growth-promoting influence of IGF signaling is generalized across most or all components of the organ during development. One caveat is that morphometric studies confirming the proportionate reduction of microanatomical structures in the developing kidney has not been reported.

Although the genetic and biochemical evidence for IGF control of kidney growth during development is convincing, it is a challenge to integrate global growth control by IGF signaling with the specific signaling events that control formation and differentiation of the constituent components of the kidney. Receptor tyrosine kinase pathways initiated by FGFR and Ret that control differentiation of specific microanatomical components are known to activate intracellular signaling cascades like those activated by IGF. CM cells isolated from developing kidneys are responsive to IGF1 and IGF2, and their long-term propagation in culture requires continual IGF treatment, supporting a generally mitogenic role of IGFs in this cell compartment (Brown et al., 2015). However, genetic inactivation studies have identified FGF signaling as essential for CM proliferation *in vivo* based on the profound loss of CM seen following inactivating of *Fgf9* and *Fgf20* (Barak et al., 2012; Brown et al., 2011). Inactivation of the ERK signaling pathway in these cells suggests that FGFs regulate MAPK–ERK signaling in these cells *in vivo* (Ihermann-Hella et al., 2018). If FGF and IGF were redundantly activating survival and proliferation pathways in these cells, a severe phenotype of

FGF9/20 inactivation would not be anticipated. A similar point can be made for the CD whose growth and branching is governed by GDNF-Ret, FGF-FGFR and ERK signaling (Ihermann-Hella et al., 2018; Michos et al., 2010; Shakya, Watanabe, & Costantini, 2005). While these reports focus on the requirement for MAPK–ERK signaling in CM and CD, the IGF system also acts through PI3K, which *in vitro* studies have shown is required for CM proliferation and CD branching (Kim & Dressler, 2007; Lindström et al., 2015, 2018; Tang, Cai, Tsai, Wang, & Dressler, 2002). It is possible that IGF-PI3K may act to augment or modulate cellular responses that are initiated by FGFR/Ret-MAPK–ERK signaling. The relationship between these developmental and growth pathways is unlikely to be a question of simple redundancy, and understanding how they interact to coordinately control nephrogenesis and tissue growth will require more detailed mechanistic dissection.

7. Growth hormone and insulin like growth factor in postnatal growth control

Following birth, growth hormone (GH) becomes a primary driver of kidney growth. GH is produced by the pituitary, and analyses of congenitally hypopituitary dwarf mice (Winick, 1968) and GH receptor mutants (Zhou et al., 1997) revealed that growth differences *versus* wild type become apparent at postnatal day 10, indicating that there is no role for GH in controlling organ growth during the embryonic or immediate postnatal period. Studies involving genetic ablation of hormone producing cells (Behringer, Mathews, Palmiter, & Brinster, 1988) and *Gh* gene mutation (Meyer et al., 2004) have revealed that GH is required to establish and maintain IGF1 expression (Behringer et al., 1988; Lupu, Terwilliger, Lee, Segre, & Efstratiadis, 2001; Schwander, Hauri, Zapf, & Froesch, 1983). IGF2 is not thought to play a role in postnatal growth regulation, since *Igf2* null mice display an identical postnatal growth curve to wild type littermates after birth (DeChiara et al., 1990). *Igf1r* is expressed at some level in most cell types of the kidney suggesting the potential for IGF1-IGF1R activation of growth-promoting signaling cascades (Behringer et al., 1988; Lupu et al., 2001; Schwander et al., 1983). Cell type specific inactivation of *Igf1r* in the proximal tubule using the gamma-glutamyltransferase 1—cre (γGT-cre) deleter showed only a modest size reduction of the adult kidney (Li et al., 2016). γGT-cre is expressed in the differentiated proximal tubule and would be expected to primarily reflect the postnatal role of IGF signaling.

The explanation for the modest phenotype may be technical rather than biological, as expression analysis for IGF1R in mutant kidneys revealed a reduction rather than an ablation. Activation of the PI3K signaling pathway by deletion of the gene encoding the antagonist PTEN with γGT-cre potently induces hypertrophy in the proximal tubule by 5 weeks of age, supporting an essential role for PI3K signaling in postnatal growth (Chen et al., 2015). In addition to their roles in kidney growth, IGFs and GH influence aspects of renal physiology including sodium, phosphate and calcium resorption and tubule gluconeogenesis, which may complicate the interpretation of their roles in growth control of the adult organ (reviewed in Kamenický, Mazziotti, Lombès, Giustina, & Chanson, 2014).

It has been proposed that GH controls organ growth by promoting liver production of IGF1, which in turn maintains circulating levels of this protein allowing remote hormonal effects (Fig. 4) (Schwander et al., 1983). However, liver-specific inactivation of *Igf1* does not perturb body size or bone growth, raising questions about this model (Yakar et al., 1999). A confounding issue in these genetic experiments is that IGF1 negatively regulates GH secretion, and subsequent studies attempting to maintain equivalent GH levels in mice with normal *versus* reduced IGF1 expression from the liver suggest that circulating liver-derived IGF1 is in fact an important determinant of organ growth (Nordstrom, Tran, Sos, Wagner, & Weiss, 2011). A consensus on the relative importance of hormonal *versus* paracrine activation of IGF by GH has not been reached, and there is support for local growth-regulatory action of the GH-IGF axis in the kidney (Fig. 4). *Igf1* is expressed locally in the adult kidney, and GH injection into hypophysectomized mice increases

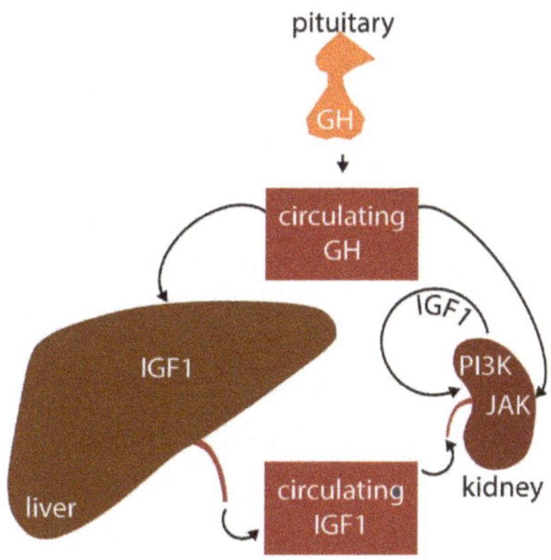

Fig. 4 The growth hormone axis in postnatal growth. GH controls circulating IGF1 largely from the liver, and there is also support for local IGF1 expression in the kidney.

the kidney's *Igf1* expression (Lowe, Roberts, Lasky, & LeRoith, 1987). Whole organ expression studies show that GH receptor (*Ghr*) is locally expressed in the kidney and that its expression increases from birth to adult (Mathews, Enberg, & Norstedt, 1989), supporting the idea of a local GHR-IGF1 axis. *In situ* hybridization studies of adult rat kidneys aiming to determine where *Ghr* and *Igf1* co-localize show strong *Ghr* expression in proximal tubule but *Igf1* expression is limited to the thick ascending limb of the loop of Henle, where *Ghr* expression is modest (Chin, Zhou, & Bondy, 1992). Single cell analysis of adult mouse kidneys reveals *Ghr* expression throughout the proximal tubule, while *Igf1* is expressed in the vasculature and cells of Bowman's capsule (Ransick et al., 2019). Distal tubule and vascular expression of *Igf1* have previously been identified by *in situ* hybridization (Lindenbergh-Kortleve et al., 1997). While the expression data lend some support to the concept of a local GHR-IGF1 axis, this would be localized to a relatively small subset of the nephron epithelium. One technical caveat is that most of these inferences are based on expression analyses of adult kidneys. Pathway components may be dynamically regulated in the period between birth and adult, and expression data at intermediate stages would be required to resolve models for GH-GHR-IGF1-IGF1R growth control of the kidney. The strong expression of *Ghr* in cells of the proximal tubule suggests that GH may act directly in an IGF1-independent manner on these cells. The latter is supported by gene inactivation studies; while *Igf1* knockout results in reduction of kidney size proportionate to the reduction in overall organismal size (Liu et al., 1993), *Gh* knockout results in disproportionate size reduction of the kidneys compared to overall size (Zhou et al., 1997). GHR is a cytokine receptor and activates JAK–STAT (Argetsinger et al., 1993) and ERK (Hodge et al., 1998) pathways. Reports assessing *in vivo* how this IGF-independent signaling may affect growth in strongly *Ghr* expressing kidney structures such as the proximal tubule were not found for this review.

8. Reactivation of kidney growth in the adult

Reduction of kidney mass through unilateral nephrectomy causes a compensatory increase in the size of the remaining kidney. Following unilateral nephrectomy in the rodent, the rate of filtration through glomeruli falls acutely by approximately 50%. Despite the increase in cardiac output to the remaining kidney, studies in rats show that the glomerular filtration does not start to increase until after approximately 3 days and returns to almost 90% of the pre-nephrectomy level by 3 weeks (Malt, 1969).

The magnitude of compensation varies between species, and the mass increase associated with functional compensation in the mouse generally exceeds 150% (Malt & Lemaitre, 1968). This is a reversible response, as transplantation of an enlarged kidney into a unilaterally nephrectomized rat restores the size of both the transplanted and the recipient kidneys to approximately their original sizes (Silber & Malvin, 1974).

As predicted based on the reversibility of organ growth in the adult, this response is based primarily on increased cell size rather than cell number, although both are reported to increase. There is no evidence for formation of new nephrons, and existing glomeruli and proximal tubules are significantly enlarged (Arataki, 1926), suggesting predominantly a hypertrophic growth response. However, radiolabeling studies in the mouse reveal a short spike of proliferation peaking 48 h after nephrectomy, primarily in the proximal tubule epithelium (Johnson & Vera Roman, 1966; Rollason, 1949). Based on these studies, proliferation is predicted to account for approximately 7% of the size increase of the remaining kidney, making a relatively small contribution to overall size increase to the remaining organ.

Circulating GH increases severalfold following unilateral nephrectomy (Haramati, Lumpkin, & Mulroney, 1994), and pharmacological antagonism of GHR shows that the compensatory growth response is GH dependent (Flyvbjerg et al., 1999). Thus, GH signaling appears to be an essential component of the compensatory growth response. Circulating and local IGF1 in the remaining kidney also increase (Stiles, Sosenko, D'Ercole, & Smith, 1985), suggesting that the GH-IGF1 axis may operate similarly in compensatory and somatic growth. Administration of rapamycin to unilaterally nephrectomized mice reduces compensatory hypertrophy of the remaining kidney, implicating mTOR signaling through the mTORC1 complex (Chen, Chen, Neilson, & Harris, 2005). mTORC1 controls cell size by promoting activity of the translation initiation factor eIF4E and by promoting protein synthesis by increasing phosphorylation of the 40S ribosomal protein S6 (Fingar, Salama, Tsou, Harlow, & Blenis, 2002). Inactivation of the S6 kinase S6K1 largely abrogates the compensatory growth of the remaining kidney after unilateral nephrectomy, identifying one potent mechanism by which mTOR controls compensatory kidney growth (Chen, Chen, Thomas, Kozma, & Harris, 2009). In support of the connection between PI3K and mTORC1, proximal tubule-specific inactivation of the class III PI3K pathway by knockout of *Pik3c3* prevents S6K1 activity and inhibits hypertrophic growth of proximal tubule cells in the remaining kidney following unilateral nephrectomy (Liu et al., 2020). While no reports were found for this review that genetically identify a specific growth factor upstream of PI3K-mTOR signaling in the compensatory response, one possible candidate is IGF signaling through IGF1R in the remaining kidney. The compensatory growth response is most pronounced in the proximal tubule (Hayslett, Kashgarian, & Epstein, 1968), where the ribosome content and vesicle load increase hours after nephrectomy (Anderson, 1967). This, together with the availability of efficient cell type specific cre drivers, has focused attention on defining the signaling circuit for hypertrophic growth of the proximal tubule epithelium.

9. Perspectives

Mechanisms governing kidney growth have intrigued generations of scientists and investigating this process has led to important discoveries in cell signaling, physiology and the understanding of kidney disease. The timeline of kidney development in the mouse is now highly refined, and morphometric studies have explored subtle changes in tissue structure during this period. The event or signal that ends the developmental period remains unknown, but we are beginning to understand the signaling events associated with cessation of nephrogenesis, which experimental systems show strongly affects the nephron endowment or functional mass of the kidney. There is strong support for the IGF system controlling organ growth in the prenatal period, and one open question is how the IGF growth control system interfaces with pathways such as FGF that control critical developmental steps, since both IGFs and FGFs can activate the same intracellular signaling cascades. Postnatally, there is strong support that the general model for postnatal organ growth based on GH and IGF1 controls kidney growth. Many studies have sought to spatially and qualitatively determine how IGF signals in the organ, but a larger-scale systematic analysis of activation of growth control pathways superimposed on a microanatomical map of growth is now a realistic goal given available technologies. In the proliferative phase following cessation of nephrogenesis, labeling of cycling cells indicates that microanatomical structures coordinately stop proliferating in a temporal sequence that begins with the inner medulla and ends with the outer cortex (Adams & Oxburgh, 2009). These coordinated exits from the cell cycle occur during the period of GH-IGF growth control. Coordinated down-regulation of a cassette of growth-promoting genes has been defined across organ systems (Lui, Finkielstain, Barnes, & Baron, 2008), but the event that initiates the deceleration of growth in the kidney is unclear.

Interestingly, delaying postnatal growth also delays initiation of this coordinated program, indicating that it may be regulated by growth itself (Lui et al., 2010). Some tissues decelerate their growth by producing growth-suppressive factors, providing an endogenous circuit that determines the set-point for tissue mass, for example negative control of skeletal muscle growth by myostatin (McPherron, Lawler, & Lee, 1997), but no such factor has been identified for the kidney. Understanding the mechanisms that determine set-points for cell numbers and sizes in the kidney may inform ongoing efforts to generate tissue from stem cells that accurately mimics the kidney.

Acknowledgments

This work was supported by grant RC2DK125960 to LO from the National Institutes of Health/National Institute of Diabetes and Digestive and Kidney Diseases. The content is solely the responsibility of the author and does not necessarily represent the official views of the National Institutes of Health.

References

Adams, D. C., & Oxburgh, L. (2009). The long-term label retaining population of the renal papilla arises through divergent regional growth of the kidney. *American Journal of Physiology. Renal Physiology*, *297*, F809–F815.

Alves, M. M., Fuhler, G. M., Queiroz, K. C., Scholma, J., Goorden, S., Anink, J., et al. (2015). PAK2 is an effector of TSC1/2 signaling independent of mTOR and a potential therapeutic target for tuberous sclerosis complex. *Scientific Reports*, *5*, 14534.

Anderson, W. A. (1967). The fine structure of compensatory growth in the rat kidney after unilateral nephrectomy. *The American Journal of Anatomy*, *121*, 217–247.

Arataki, M. (1926). On the postnatal growth of the kidney, with special reference to the number and size of the glomeruli (albino rat). *The American Journal of Anatomy*, *36*, 399–436.

Argetsinger, L. S., Campbell, G. S., Yang, X., Witthuhn, B. A., Silvennoinen, O., Ihle, J. N., et al. (1993). Identification of JAK2 as a growth hormone receptor-associated tyrosine kinase. *Cell*, *74*, 237–244.

Baker, J., Liu, J. P., Robertson, E. J., & Efstratiadis, A. (1993). Role of insulin-like growth factors in embryonic and postnatal growth. *Cell*, *75*, 73–82.

Barak, H., Huh, S. H., Chen, S., Jeanpierre, C., Martinovic, J., Parisot, M., et al. (2012). FGF9 and FGF20 maintain the stemness of nephron progenitors in mice and man. *Developmental Cell*, *22*, 1191–1207.

Behringer, R. R., Mathews, L. S., Palmiter, R. D., & Brinster, R. L. (1988). Dwarf mice produced by genetic ablation of growth hormone-expressing cells. *Genes & Development*, *2*, 453–461.

Bell, M. R. (2018). Comparing postnatal development of gonadal hormones and associated social Behaviors in rats, mice, and humans. *Endocrinology*, *159*, 2596–2613.

Brown, A. C., Adams, D., de Caestecker, M., Yang, X., Friesel, R., & Oxburgh, L. (2011). FGF/EGF signaling regulates the renewal of early nephron progenitors during embryonic development. *Development (Cambridge, England)*, *138*, 5099–5112.

Brown, J., Delaine, C., Zaccheo, O. J., Siebold, C., Gilbert, R. J., van Boxel, G., et al. (2008). Structure and functional analysis of the IGF-II/IGF2R interaction. *The EMBO Journal*, *27*, 265–276.

Brown, A. C., Muthukrishnan, S. D., & Oxburgh, L. (2015). A synthetic niche for nephron progenitor cells. *Developmental Cell*, *34*, 229–241.

Carroll, T. J., Park, J. S., Hayashi, S., Majumdar, A., & McMahon, A. P. (2005). Wnt9b plays a central role in the regulation of mesenchymal to epithelial transitions underlying organogenesis of the mammalian urogenital system. *Developmental Cell*, *9*, 283–292.

Casagrande, S. S., Linder, B., & Cowie, C. C. (2018). Prevalence of gestational diabetes and subsequent type 2 diabetes among U.S. women. *Diabetes Research and Clinical Practice*, *141*, 200–208.

Cerqueira, D. M., Hemker, S. L., Bodnar, A. J., Ortiz, D. M., Oladipupo, F. O., Mukherjee, E., et al. (2019). In utero exposure to maternal diabetes impairs nephron progenitor differentiation. *American Journal of Physiology. Renal Physiology*, *317*, F1318–F1330.

Chen, S., Brunskill, E. W., Potter, S. S., Dexheimer, P. J., Salomonis, N., Aronow, B. J., et al. (2015). Intrinsic age-dependent changes and cell-cell contacts regulate nephron progenitor lifespan. *Developmental Cell*, *35*, 49–62.

Chen, J. K., Chen, J., Neilson, E. G., & Harris, R. C. (2005). Role of mammalian target of rapamycin signaling in compensatory renal hypertrophy. *Journal of the American Society of Nephrology*, *16*, 1384–1391.

Chen, J. K., Chen, J., Thomas, G., Kozma, S. C., & Harris, R. C. (2009). S6 kinase 1 knockout inhibits uninephrectomy- or diabetes-induced renal hypertrophy. *American Journal of Physiology. Renal Physiology*, *297*, F585–F593.

Chen, J. K., Nagai, K., Chen, J., Plieth, D., Hino, M., Xu, J., et al. (2015). Phosphatidylinositol 3-kinase signaling determines kidney size. *The Journal of Clinical Investigation*, *125*, 2429–2444.

Chin, E., Zhou, J., & Bondy, C. A. (1992). Renal growth hormone receptor gene expression: Relationship to renal insulin-like growth factor system. *Endocrinology*, *131*, 3061–3066.

Combes, A. N., Lefevre, J. G., Wilson, S., Hamilton, N. A., & Little, M. H. (2016). Cap mesenchyme cell swarming during kidney development is influenced by attraction, repulsion, and adhesion to the ureteric tip. *Developmental Biology*, *418*, 297–306.

Constância, M., Hemberger, M., Hughes, J., Dean, W., Ferguson-Smith, A., Fundele, R., et al. (2002). Placental-specific IGF-II is a major modulator of placental and fetal growth. *Nature*, *417*, 945–948.

Daughaday, W. H., Parker, K. A., Borowsky, S., Trivedi, B., & Kapadia, M. (1982). Measurement of somatomedin-related peptides in fetal, neonatal, and maternal rat serum by insulin-like growth factor (IGF) I radioimmunoassay, IGF-II radioreceptor assay (RRA), and multiplication-stimulating activity RRA after acid-ethanol extraction. *Endocrinology*, *110*, 575–581.

DeChiara, T. M., Efstratiadis, A., & Robertson, E. J. (1990). A growth-deficiency phenotype in heterozygous mice carrying an insulin-like growth factor II gene disrupted by targeting. *Nature*, *345*, 78–80.

DeChiara, T. M., Robertson, E. J., & Efstratiadis, A. (1991). Parental imprinting of the mouse insulin-like growth factor II gene. *Cell*, *64*, 849–859.

Enesco, M., & LeBlond, C. P. (1962). Increase in cell number as a factor in the growth of the organs and tissues of the young male rat. *Journal of Embryology and Experimental Morphology*, *10*, 530–562.

Feig, D. S., Hwee, J., Shah, B. R., Booth, G. L., Bierman, A. S., & Lipscombe, L. L. (2014). Trends in incidence of diabetes in pregnancy and serious perinatal outcomes: A large, population-based study in Ontario, Canada, 1996–2010. *Diabetes Care*, *37*, 1590–1596.

Fingar, D. C., Salama, S., Tsou, C., Harlow, E., & Blenis, J. (2002). Mammalian cell size is controlled by mTOR and its downstream targets S6K1 and 4EBP1/eIF4E. *Genes & Development, 16*, 1472–1487.

Flyvbjerg, A., Bennett, W. F., Rasch, R., van Neck, J. W., Groffen, C. A., Kopchick, J. J., et al. (1999). Compensatory renal growth in uninephrectomized adult mice is growth hormone dependent. *Kidney International, 56*, 2048–2054.

Forbes, T. A., Howden, S. E., Lawlor, K., Phipson, B., Maksimovic, J., Hale, L., et al. (2018). Patient-iPSC-derived kidney organoids show functional validation of a ciliopathic renal phenotype and reveal underlying pathogenetic mechanisms. *American Journal of Human Genetics, 102*, 816–831.

Germain-Lee, E. L., Janicot, M., Lammers, R., Ullrich, A., & Casella, S. J. (1992). Expression of a type I insulin-like growth factor receptor with low affinity for insulin-like growth factor II. *The Biochemical Journal, 281*(Pt. 2), 413–417.

Gladstone, R. J. (1924). A note on the post-natal growth of the kidney, thyroid gland and liver. *Journal of Anatomy, 58*, 170–177.

Grassi, G., Mark, A., & Esler, M. (2015). The sympathetic nervous system alterations in human hypertension. *Circulation Research, 116*, 976–990.

Grobstein, C. (1953). Inductive epitheliomesenchymal interaction in cultured organ rudiments of the mouse. *Science, 118*, 52–55.

Haramati, A., Lumpkin, M. D., & Mulroney, S. E. (1994). Early increase in pulsatile growth hormone release after unilateral nephrectomy in adult rats. *The American Journal of Physiology, 266*, F628–F632.

Harding, S. D., Armit, C., Armstrong, J., Brennan, J., Cheng, Y., Haggarty, B., et al. (2011). The GUDMAP database—An online resource for genitourinary research. *Development (Cambridge, England), 138*, 2845–2853.

Hayslett, J. P., Kashgarian, M., & Epstein, F. H. (1968). Functional correlates of compensatory renal hypertrophy. *The Journal of Clinical Investigation, 47*, 774–799.

Hilliard, S., Song, R., Liu, H., Chen, C. H., Li, Y., Baddoo, M., et al. (2019). Defining the dynamic chromatin landscape of mouse nephron progenitors. *Biology Open, 8*(5), bio042754.

Hintz, R. L., Clemmons, D. R., Underwood, L. E., & Van Wyk, J. J. (1972). Competitive binding of somatomedin to the insulin receptors of adipocytes, chondrocytes, and liver membranes. *Proceedings of the National Academy of Sciences of the United States of America, 69*, 2351–2353.

Hodge, C., Liao, J., Stofega, M., Guan, K., Carter-Su, C., & Schwartz, J. (1998). Growth hormone stimulates phosphorylation and activation of elk-1 and expression of c-fos, egr-1, and junB through activation of extracellular signal-regulated kinases 1 and 2. *The Journal of Biological Chemistry, 273*, 31327–31336.

Ihermann-Hella, A., Hirashima, T., Kupari, J., Kurtzeborn, K., Li, H., Kwon, H. N., et al. (2018). Dynamic MAPK/ERK activity sustains nephron progenitors through niche regulation and primes precursors for differentiation. *Stem Cell Reports, 11*, 912–928.

Johnson, H. A., & Vera Roman, J. M. (1966). Compensatory renal enlargement. Hypertrophy versus hyperplasia. *American Journal of Pathology, 49*, 1–13.

Kamenický, P., Mazziotti, G., Lombès, M., Giustina, A., & Chanson, P. (2014). Growth hormone, insulin-like growth factor-1, and the kidney: Pathophysiological and clinical implications. *Endocrine Reviews, 35*, 234–281.

Karner, C. M., Das, A., Ma, Z., Self, M., Chen, C., Lum, L., et al. (2011). Canonical Wnt9b signaling balances progenitor cell expansion and differentiation during kidney development. *Development (Cambridge, England), 138*, 1247–1257.

Kim, D., & Dressler, G. R. (2007). PTEN modulates GDNF/RET mediated chemotaxis and branching morphogenesis in the developing kidney. *Developmental Biology, 307*, 290–299.

Li, L., Byrd, M., Doh, K., Dixon, P. D., Lee, H., Tiwari, S., et al. (2016). Absence of renal enlargement in fructose-fed proximal-tubule-select insulin receptor (IR), insulin-like-growth factor receptor (IGF1R) double knockout mice. *Physiological Reports*, *4*(23), e13052.

Li, H., Kurtzeborn, K., Kupari, J., Gui, Y., Siefker, E., Lu, B., et al. (2021). Postnatal prolongation of mammalian nephrogenesis by excess fetal GDNF. *Development*, *148*.

Lindenbergh-Kortleve, D. J., Rosato, R. R., van Neck, J. W., Nauta, J., van Kleffens, M., Groffen, C., et al. (1997). Gene expression of the insulin-like growth factor system during mouse kidney development. *Molecular and Cellular Endocrinology*, *132*, 81–91.

Lindström, N. O., Carragher, N. O., & Hohenstein, P. (2015). The PI3K pathway balances self-renewal and differentiation of nephron progenitor cells through β-catenin signaling. *Stem Cell Reports*, *4*, 551–560.

Lindström, N. O., McMahon, J. A., Guo, J., Tran, T., Guo, Q., Rutledge, E., et al. (2018). Conserved and divergent features of human and mouse kidney organogenesis. *Journal of the American Society of Nephrology*, *29*, 785–805.

Liu, J. P., Baker, J., Perkins, A. S., Robertson, E. J., & Efstratiadis, A. (1993). Mice carrying null mutations of the genes encoding insulin-like growth factor I (Igf-1) and type 1 IGF receptor (Igf1r). *Cell*, *75*, 59–72.

Liu, J., Edgington-Giordano, F., Dugas, C., Abrams, A., Katakam, P., Satou, R., et al. (2017). Regulation of nephron progenitor cell self-renewal by intermediary metabolism. *Journal of the American Society of Nephrology*, *28*, 3323–3335.

Liu, T., Yuan, J., Dai, C., Xu, J., Li, S., Humphreys, B. D., et al. (2020). Cre/loxP approach-mediated downregulation of Pik3c3 inhibits the hypertrophic growth of renal proximal tubule cells. *Journal of Cellular Physiology*, *235*, 9958–9973.

Louvi, A., Accili, D., & Efstratiadis, A. (1997). Growth-promoting interaction of IGF-II with the insulin receptor during mouse embryonic development. *Developmental Biology*, *189*, 33–48.

Lowe, W. L., Roberts, C. T., Lasky, S. R., & LeRoith, D. (1987). Differential expression of alternative 5′ untranslated regions in mRNAs encoding rat insulin-like growth factor I. *Proceedings of the National Academy of Sciences of the United States of America*, *84*, 8946–8950.

Lui, J. C., Finkielstain, G. P., Barnes, K. M., & Baron, J. (2008). An imprinted gene network that controls mammalian somatic growth is down-regulated during postnatal growth deceleration in multiple organs. *American Journal of Physiology. Regulatory, Integrative and Comparative Physiology*, *295*, R189–R196.

Lui, J. C., Forcinito, P., Chang, M., Chen, W., Barnes, K. M., & Baron, J. (2010). Coordinated postnatal down-regulation of multiple growth-promoting genes: Evidence for a genetic program limiting organ growth. *The FASEB Journal*, *24*, 3083–3092.

Lupu, F., Terwilliger, J. D., Lee, K., Segre, G. V., & Efstratiadis, A. (2001). Roles of growth hormone and insulin-like growth factor 1 in mouse postnatal growth. *Developmental Biology*, *229*, 141–162.

Luyckx, V. A., & Brenner, B. M. (2010). The clinical importance of nephron mass. *Journal of the American Society of Nephrology*, *21*, 898–910.

Luyckx, V. A., & Brenner, B. M. (2020). Clinical consequences of developmental programming of low nephron number. *Anatomical Record (Hoboken)*, *303*, 2613–2631.

Luyckx, V. A., Perico, N., Somaschini, M., Manfellotto, D., Valensise, H., Cetin, I., et al. (2017). A developmental approach to the prevention of hypertension and kidney disease: A report from the low birth weight and nephron number working group. *Lancet*, *390*, 424–428.

Maher, E. R., & Reik, W. (2000). Beckwith-Wiedemann syndrome: Imprinting in clusters revisited. *The Journal of Clinical Investigation*, *105*, 247–252.

Malt, R. A. (1969). Compensatory growth of the kidney. *The New England Journal of Medicine, 280*, 1446–1459.
Malt, R. A., & Lemaitre, D. A. (1968). Accretion and turnover of RNA in the renoprival kidney. *The American Journal of Physiology, 214*, 1041–1047.
Marshall, R. N., Underwood, L. E., Voina, S. J., Foushee, D. B., & Van Wyk, J. J. (1974). Characterization of the insulin and somatomedin-C receptors in human placental cell membranes. *The Journal of Clinical Endocrinology and Metabolism, 39*, 283–292.
Mathews, L. S., Enberg, B., & Norstedt, G. (1989). Regulation of rat growth hormone receptor gene expression. *The Journal of Biological Chemistry, 264*, 9905–9910.
McPherron, A. C., Lawler, A. M., & Lee, S. J. (1997). Regulation of skeletal muscle mass in mice by a new TGF-beta superfamily member. *Nature, 387*, 83–90.
Meyer, C. W., Korthaus, D., Jagla, W., Cornali, E., Grosse, J., Fuchs, H., et al. (2004). A novel missense mutation in the mouse growth hormone gene causes semidominant dwarfism, hyperghrelinemia, and obesity. *Endocrinology, 145*, 2531–2541.
Michos, O., Cebrian, C., Hyink, D., Grieshammer, U., Williams, L., D'Agati, V., et al. (2010). Kidney development in the absence of Gdnf and Spry1 requires Fgf10. *PLoS Genetics, 6*, e1000809.
Morcavallo, A., Gaspari, M., Pandini, G., Palummo, A., Cuda, G., Larsen, M. R., et al. (2011). Research resource: New and diverse substrates for the insulin receptor isoform a revealed by quantitative proteomics after stimulation with IGF-II or insulin. *Molecular Endocrinology, 25*, 1456–1468.
Morizane, R., Lam, A. Q., Freedman, B. S., Kishi, S., Valerius, M. T., & Bonventre, J. V. (2015). Nephron organoids derived from human pluripotent stem cells model kidney development and injury. *Nature Biotechnology, 33*, 1193–1200.
Mussa, A., Peruzzi, L., Chiesa, N., De Crescenzo, A., Russo, S., Melis, D., et al. (2012). Nephrological findings and genotype-phenotype correlation in Beckwith-Wiedemann syndrome. *Pediatric Nephrology, 27*, 397–406.
Muthukrishnan, S. D., Yang, X., Friesel, R., & Oxburgh, L. (2015). Concurrent BMP7 and FGF9 signalling governs AP-1 function to promote self-renewal of nephron progenitor cells. *Nature Communications, 6*, 10027.
Nordstrom, S. M., Tran, J. L., Sos, B. C., Wagner, K. U., & Weiss, E. J. (2011). Liver-derived IGF-I contributes to GH-dependent increases in lean mass and bone mineral density in mice with comparable levels of circulating GH. *Molecular Endocrinology, 25*, 1223–1230.
O'Brien, L. L., Guo, Q., Lee, Y., Tran, T., Benazet, J. D., Whitney, P. H., et al. (2016). Differential regulation of mouse and human nephron progenitors by the six family of transcriptional regulators. *Development (Cambridge, England), 143*, 595–608.
Osathanondh, V., & Potter, E. L. (1963). Development of human kidney as shown by micro-dissection. II. Renal pelvis, calyces, and papillae. *Archives of Pathology, 76*, 277–289.
Otis, E. M., & Brent, R. (1954). Equivalent ages in mouse and human embryos. *The Anatomical Record, 120*, 33–63.
Pichel, J. G., Shen, L., Sheng, H. Z., Granholm, A. C., Drago, J., Grinberg, A., et al. (1996). Defects in enteric innervation and kidney development in mice lacking GDNF. *Nature, 382*, 73–76.
Ransick, A., Lindström, N. O., Liu, J., Zhu, Q., Guo, J. J., Alvarado, G. F., et al. (2019). Single-cell profiling reveals sex, lineage, and regional diversity in the mouse kidney. *Developmental Cell, 51*, 399–413.e397.
Rogers, S. A., Powell-Braxton, L., & Hammerman, M. R. (1999). Insulin-like growth factor I regulates renal development in rodents. *Developmental Genetics, 24*, 293–298.
Rollason, H. D. (1949). Compensatory hypertrophy of the kidney of the young rat with special emphasis on the role of cellular hyperplasia. *The Anatomical Record, 104*, 263–285.

Rumballe, B. A., Georgas, K. M., Combes, A. N., Ju, A. L., Gilbert, T., & Little, M. H. (2011). Nephron formation adopts a novel spatial topology at cessation of nephrogenesis. *Developmental Biology*, *360*, 110–122.

Ryan, D., Sutherland, M. R., Flores, T. J., Kent, A. L., Dahlstrom, J. E., Puelles, V. G., et al. (2018). Development of the human fetal kidney from mid to late gestation in male and female infants. *eBioMedicine*, *27*, 275–283.

Sainio, K., Suvanto, P., Davies, J., Wartiovaara, J., Wartiovaara, K., Saarma, M., et al. (1997). Glial-cell-line-derived neurotrophic factor is required for bud initiation from ureteric epithelium. *Development (Cambridge, England)*, *124*, 4077–4087.

Saxén, L. (1987). *Organogenesis of the kidney*.

Saxen, L., & Wartiovaara, J. (1966). Cell contact and cell adhesion during tissue organization. *International Journal of Cancer*, *1*, 271–290.

Schuchardt, A., D'Agati, V., Larsson-Blomberg, L., Costantini, F., & Pachnis, V. (1994). Defects in the kidney and enteric nervous system of mice lacking the tyrosine kinase receptor ret. *Nature*, *367*, 380–383.

Schwander, J. C., Hauri, C., Zapf, J., & Froesch, E. R. (1983). Synthesis and secretion of insulin-like growth factor and its binding protein by the perfused rat liver: Dependence on growth hormone status. *Endocrinology*, *113*, 297–305.

Schwank, G., Koo, B. K., Sasselli, V., Dekkers, J. F., Heo, I., Demircan, T., et al. (2013). Functional repair of CFTR by CRISPR/Cas9 in intestinal stem cell organoids of cystic fibrosis patients. *Cell Stem Cell*, *13*, 653–658.

Shakya, R., Watanabe, T., & Costantini, F. (2005). The role of GDNF/Ret signaling in ureteric bud cell fate and branching morphogenesis. *Developmental Cell*, *8*, 65–74.

Short, K. M., Combes, A. N., Lefevre, J., Ju, A. L., Georgas, K. M., Lamberton, T., et al. (2014). Global quantification of tissue dynamics in the developing mouse kidney. *Developmental Cell*, *29*, 188–202.

Short, K. M., Combes, A. N., Lisnyak, V., Lefevre, J. G., Jones, L. K., Little, M. H., et al. (2018). Branching morphogenesis in the developing kidney is not impacted by nephron formation or integration. *eLife*, *7*.

Silber, S., & Malvin, R. L. (1974). Compensatory and obligatory renal growth in rats. *The American Journal of Physiology*, *226*, 114–117.

Smith, A. H., & Moise, T. S. (1927). Diet and tissue growth: IV. The rate of compensatory renal enlargement after unilateral nephrectomy in the white rat. *The Journal of Experimental Medicine*, *45*, 263–276.

Stiles, A. D., Sosenko, I. R., D'Ercole, A. J., & Smith, B. T. (1985). Relation of kidney tissue somatomedin-C/insulin-like growth factor I to postnephrectomy renal growth in the rat. *Endocrinology*, *117*, 2397–2401.

Sun, F. L., Dean, W. L., Kelsey, G., Allen, N. D., & Reik, W. (1997). Transactivation of Igf2 in a mouse model of Beckwith-Wiedemann syndrome. *Nature*, *389*, 809–815.

Taguchi, A., & White, M. F. (2008). Insulin-like signaling, nutrient homeostasis, and life span. *Annual Review of Physiology*, *70*, 191–212.

Takasato, M., Er, P. X., Chiu, H. S., Maier, B., Baillie, G. J., Ferguson, C., et al. (2015). Kidney organoids from human iPS cells contain multiple lineages and model human nephrogenesis. *Nature*, *526*, 564–568.

Tang, M. J., Cai, Y., Tsai, S. J., Wang, Y. K., & Dressler, G. R. (2002). Ureteric bud outgrowth in response to RET activation is mediated by phosphatidylinositol 3-kinase. *Developmental Biology*, *243*, 128–136.

Tran, S., Chen, Y. W., Chenier, I., Chan, J. S., Quaggin, S., Hébert, M. J., et al. (2008). Maternal diabetes modulates renal morphogenesis in offspring. *Journal of the American Society of Nephrology*, *19*, 943–952.

Ullrich, A., Gray, A., Tam, A. W., Yang-Feng, T., Tsubokawa, M., Collins, C., et al. (1986). Insulin-like growth factor I receptor primary structure: Comparison with insulin receptor suggests structural determinants that define functional specificity. *The EMBO Journal, 5*, 2503–2512.

Vega, Q. C., Worby, C. A., Lechner, M. S., Dixon, J. E., & Dressler, G. R. (1996). Glial cell line-derived neurotrophic factor activates the receptor tyrosine kinase RET and promotes kidney morphogenesis. *Proceedings of the National Academy of Sciences of the United States of America, 93*, 10657–10661.

Versteyhe, S., Klaproth, B., Borup, R., Palsgaard, J., Jensen, M., Gray, S. G., et al. (2013). IGF-I, IGF-II, and insulin stimulate different gene expression responses through binding to the IGF-I receptor. *Frontiers in Endocrinology (Lausanne), 4*, 98.

Volovelsky, O., Nguyen, T., Jarmas, A. E., Combes, A. N., Wilson, S. B., Little, M. H., et al. (2018). Hamartin regulates cessation of mouse nephrogenesis independently of Mtor. *Proceedings of the National Academy of Sciences of the United States of America, 115*, 5998–6003.

Wada, J., Liu, Z. Z., Alvares, K., Kumar, A., Wallner, E., Makino, H., et al. (1993). Cloning of cDNA for the alpha subunit of mouse insulin-like growth factor I receptor and the role of the receptor in metanephric development. *Proceedings of the National Academy of Sciences of the United States of America, 90*, 10360–10364.

Walter, F., & Addis, T. (1939). Organ work and organ weight. *The Journal of Experimental Medicine, 69*, 467–483.

Wellik, D. M., Hawkes, P. J., & Capecchi, M. R. (2002). Hox11 paralogous genes are essential for metanephric kidney induction. *Genes & Development, 16*, 1423–1432.

Winick, M. (1968). Cellular response with increased feeding in pituitary dwarf mice. *The Journal of Nutrition, 94*, 121–124.

Winick, M., & Noble, A. (1965). Quantitative changes in DNA, RNA, and protein during prenatal and postnatal growth in the rat. *Developmental Biology, 12*, 451–466.

Wu, H., Uchimura, K., Donnelly, E. L., Kirita, Y., Morris, S. A., & Humphreys, B. D. (2018). Comparative analysis and refinement of human PSC-derived kidney organoid differentiation with single-cell transcriptomics. *Cell Stem Cell, 23*, 869–881.e868.

Yakar, S., Liu, J. L., Stannard, B., Butler, A., Accili, D., Sauer, B., et al. (1999). Normal growth and development in the absence of hepatic insulin-like growth factor I. *Proceedings of the National Academy of Sciences of the United States of America, 96*, 7324–7329.

Zapf, J., Hauri, C., Waldvogel, M., & Froesch, E. R. (1986). Acute metabolic effects and half-lives of intravenously administered insulin like growth factors I and II in normal and hypophysectomized rats. *The Journal of Clinical Investigation, 77*, 1768–1775.

Zhou, Y., Xu, B. C., Maheshwari, H. G., He, L., Reed, M., Lozykowski, M., et al. (1997). A mammalian model for Laron syndrome produced by targeted disruption of the mouse growth hormone receptor/binding protein gene (the Laron mouse). *Proceedings of the National Academy of Sciences of the United States of America, 94*, 13215–13220.

Zubkov, V. S., Combes, A. N., Short, K. M., Lefevre, J., Hamilton, N. A., Smyth, I. M., et al. (2015). A spatially-averaged mathematical model of kidney branching morphogenesis. *Journal of Theoretical Biology, 379*, 24–37.

FSC
www.fsc.org
MIX
Papier | Fördert
gute Waldnutzung
FSC® C083411

Kolibri 360 GmbH
Ferdinand-Jühlke-Straße 7
99095 Erfurt, Deutschland
produktsicherheit@kolibri360.de